ELECTROSTATICS

ELECTROSTATICS

Hilary D. Brewster

Oxford Book Company
Jaipur ▪ New Delhi

Published By
Oxford Book Company
Regd. Off. "Bony Residency", Gate No. 2
Opp. Tilak Public School, Vishweswariya Nagar,
Gopalpura Road, Jaipur - 302018, India
Ph: 0141 - 2761280, Telefax: 2761381
email - oxfordbookjpr@gmail.com
website - oxfordbookcompany.com

Delhi Office: 102, 1st Floor, Satyam House,
4327/3, Ansari Road, Darya Ganj,
New Delhi - 110002, India
Ph: 011-45652440

Published in India, 2025

ISBN : 9789380179001

Price : ₹ 2295/-

Typeset by Shivangi Computers, Jaipur

Printed and bound in India by Thomson Press Ltd.,
New Delhi

Preface

This book 'Electrostatics ' on the physics of electricity, magnetism, and electromagnetic fields and waves. It is written mainly with the physics student in mind, although it will also be of use to students of electrical and electronic engineering. The major aspects of the electrical field including atomic structure and basic electricity, direct and alternating current, basic circuit theory, three-phase circuits, single phase, transformers, generators, and motors.

All students of physics need to understand the basic concepts of electricity and magnetism. E&M is central to the study of physics, and central tc understanding the developments of the last two hundred years of not just science, but technology. But the core of electricity and magnetism can be difficult to understand - many of the ideas are counterintuitive and difficult to appreciate.

This as a basis, it is possible to develop all of electromagnetic theory from a single experimental postulate founded on Coulomb's law. An enriched understanding of magnetism results, and the Biot-Savart law is a consequence rather than a postulate. The Lorentz force law is seen to be a transformation of Coulomb's law occasioned by the relativistic interpretation of force.

Upon accepting the Lorentz force law as fundamental, one is able to derive Faraday's emf law and Maxwell's equations as additional consequences. This procedure provides the further satisfaction of demonstrating that the fields contained in the Lorentz force law and in Maxwell's equations are one and the same, a conclusion not possible in the conventional development of the subject.

Hilary. D. Brewster

Contents

Chapter 1

Matter and Electricity

ELECTRIC CHARGE: A PROPERTY OF MATTER

As with all other physical phenomena associated with motion, the "explanation" of electrical phenomena starts only with observations. The experiments are more difficult to replicate and harder to explain than those relevant for motion.

Still, if you compare enough observations involving *static* electrical phenomena (e.g. dragging your feet across a carpet and touching a door handle, rubbing plastic piping with cotten, etc.) you can, as Benjamin Franklin did, arrive at the following conclusion: Matter can have a physical property which we call electrical charge

Electrical charge, like mass, is an intrinsic property of matter. It shares many of the properties of mass in fact. The most important ones that we have discovered over the ages are that

- Charges create and are subject to electrical forces.
- Charges can neither be created nor destroyed, i.e. they are intrinsic to matter.

In addition, there are some things that charges do not share in common with mass:

- Charges come in two types: designated as positive and negative.
- Charge is conserved as stated above, but it can readily be transferred from one object to another as long as the object is not a fundamental particle (e.g. an electron).

That last caveat is important. It turns out that, ultimately, charge and mass do share the property of being *intrinsic* to matter, i.e. elementary particles have a definite mass and a it fixed charge. Thus, the fact that we can transfer electrical charges between objects is explained by stating that this takes place by the actual exchange of charged *particles*.

Thus, the charges involved are not created or destroyed anywhere, just transferred from one object to another by the exchange of small (essentially invisible) charged particles (which we now know are usually, but not always, electrons).

As neat as the above explanation is, it still does not explain all electrical observations you can make though.

For example, two elementary particles (e.g. an electron and a proton) exert forces on each other without the exchange of anything "charged" between them. We can also see that macroscopic objects which are "neutral" can be subject to electrical forces without making any physical connection between them and any other object (this is referred to as induced charging). Hence we have to infer that "something" permeates the space between electrically charged objects to exert a force on them. We explain this "something" as being the electric field. Before describing it's properties, let's first work backwards, or operationally, i.e. let's say what the field does and let that serve as our definition of what it is.

COULOMB'S LAW

The force between charged particles was described on the basis of sensitive experiments by Charles Augustin Coulomb in a formula known as Coulomb's Law. Assume two charges, 1 and 2, are separated by a distance r_{12}. Since charges come in two types, we can have two cases: the charges are the same "sign" (i.e. both positive or both negative) or they are opposite (i.e. one is positive and the other negative). For either of these two cases, Coulomb found that the magnitude of the force between the particles was

$$F = \frac{k|q_1||q_2|}{r_{12}^2}$$

where the magnitude of the charges are expressed in units of Coulombs (C) in the SI system of units. The proportionality constant k has an experimentally determined value of 9.0×10^9 N · m²/C². To complete the picture, we have to define the direction of the force and define what it acts on. Experimentally, we observe for our two cases (i.e. both charges have the same sign or the charges have opposite signs) that the forces acting look as follows:

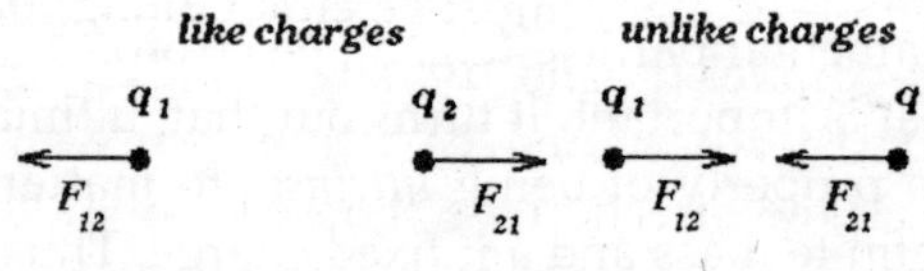

Thus, to complete our definitions of the forces produced, we go to the vector expression of Coulomb's Law.

$$F_{21} = \frac{kq_1q_2r_{12}}{r_{12}^3}$$

This is the force on charge q_2 due to charge q_1. To get the proper direction, we need to note two things.

- The quantities q_1 and q_2 now have signs which reflect the sign of the charges they represent, i.e. they can be either positive or negative.
- Note that the vector direction is given by the unit vector r_{12}/r_{12}. The vector r_{12} extends from q_1 to q_2, i.e. *from* the charge providing the force *to* the charge that the force acts on.

Let's consider an application of Coulomb's Law. Since the elementary case of a single charge is already explained, let's consider three stationary charges. We will not specify the magnitudes or even the signs of the charges to emphasize that the approach to the problem of calculating the forces is independent of these actual values.

The answer certainly depends on having this information, but the solution depends only on the mathematical nature of electrical forces as expressed through Coulomb's Law.

Consider three electric charges arranged as a triangle. Find the net force acting on charge q_3.

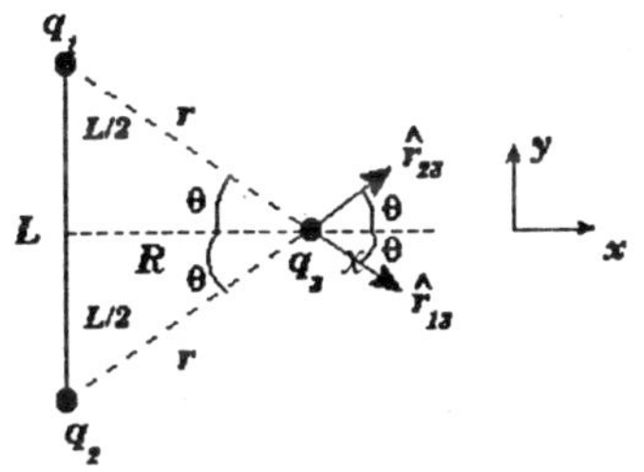

The forces exerted by two charges, q_1 and q_2 on a third charge, q_3, must lie parallel or antiparallel to the directions given by the unit vectors $\hat{r}_{13}$ and $\hat{r}_{23}$, respectively.

Charge q_1 must exert a force on q_3 which lies either along (q_1 and q_3 have the same sign) or anti-parallel (q_1 and q_3 have opposite signs) to the vector r_{13}.

The force of charge q_2 on q_3 must lie either along (q_2 and q_3 have the same sign) or anti-parallel (q_2 and q_3 have opposite signs) to the vector r_{23}. This will automatically be reflected in our answer if we adopt the convention of letting the direction we choose be from the charge causing the force *toward* the charge which experiences the force.

ELECTRIC FIELDS AND ELECTRIC FORCES

Although we could go on to describe experimental observations of forces determined from endless combinations of electrical charges with various numbers, signs, magnitudes of charges, distances, and orientations, let's take an approach that will prove enormously useful. That is to introduce a mechanism for how the force is created. Since charge and mass are analogous, let's assume that the mechanism for how mass "creates" gravitational force is copied by charge for "creating" electrical forces. We call this mechanism the electric field. We have to make the properties of electric fields consistent with the experimentally observed properties of electrical forces. A simple scheme for doing this is to make the field and the force proportional to one another, hence we define the electric field, E, as E = F/q

That is to say, the force on a charge q which is in the presence of an electric field E is F = qE. Note that the bold letters used indicate that E and F are vectors. The relationship we have used makes it easy to define the properties of E in terms of the point charges that produce them. Since we have already seen how one point charge (the charge, say on an elementary particle) can exert a force on a second, we should state that the electric field produced by a positive (negative) point charge goes radially outward (inward) in all directions from (toward) the charge.

ELECTRIC FIELDS

The field lines are a (rather gross) characterization of the actual electric field. The field extends radially, in all directions in 3-dimensional space, from the charge. If the charge is positive, the field is said to point away from the charge.

If the charge is negative, the field points iPn towards the charge. The field lines give a general idea of the direction, but it is important to remember that the actual field in 3 dimensions extends to every point in space out to infinity if there are no other charges or matter. The weird shape of the field for two or more charges comes about strictly for geometrical reasons.

The electric force, we find experimentally, is both conservative and obeys the principal of superposition (just as gravitational forces do). Therefore, electric fields must also obey these rules.

The superposition property means that electric fields from two different sources add *vectorially* to produce a *net* field. Let's look at two point charges as in lecture 1. The four possible combinations of positive or negative charges are shown.

Note that we have chosen the same position (relative to the charges) to calculate the net electric field in each of the four cases. For cases 1 and 2 (both charges the same sign), we note that the horizontal or x components of the fields tend to cancel, leaving the y components to add. This will be true for any point which is close to the line which is parallel to y and located at an x position close to the midpoint between the charges. You should note that the field lines indicate that the field is nearly vertical for points close to the midpoint line as we have described it.

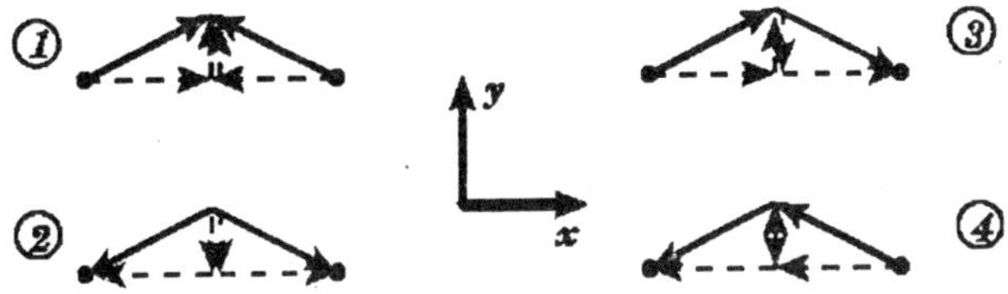

Fig. Vector Components for Electric Fields from two Charges.

For cases (unlike charges), the position we have chosen for calculating the net electric field tends to have the y components canceling and the x components adding. Hence any point near the midpoint line for unlike charges will have electric fields pointing nearly horizontally. Right along the midpoint line, the y component is exactly zero (assuming the two charges are equal magnitude).

We need to find the vector sum of the electric fields extending radially from or toward each charge (depending on whether the charge is positive or negative) at each point in space, then, for some small subset of points, find the direction of the net electric field. Thus, the shape looks complex, but it is arrived at by doing the vector sum of radial field lines from each charge, so nothing complex is going on.

To find the force on one charge due to other charges then, we note that:

- No point particle is subject to forces from electric fields produced by the charge of the particle itself
- Superposition says that the net force on a charged point particle due to the presence of other charged point particles can be derived by first finding the electric field due to the other particles, then using qE = F.

Let's try this out by find the net force on the lower left particle in the arrangement shown.

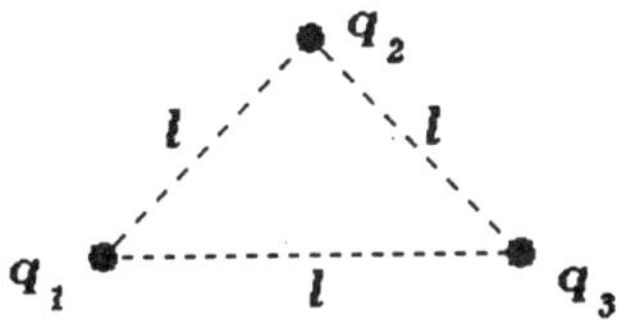

Fig. Three Charges Fixed in a Triangular Pattern.

Assume that the red particles are positively charged and the blue particle is negatively charged. Since the particles are along the vertices of an equilateral triangle, the relative angles with respect to a convenient x-y axis is easy to determine. We observe the following for the contributions to the electric field at the lower left corner:

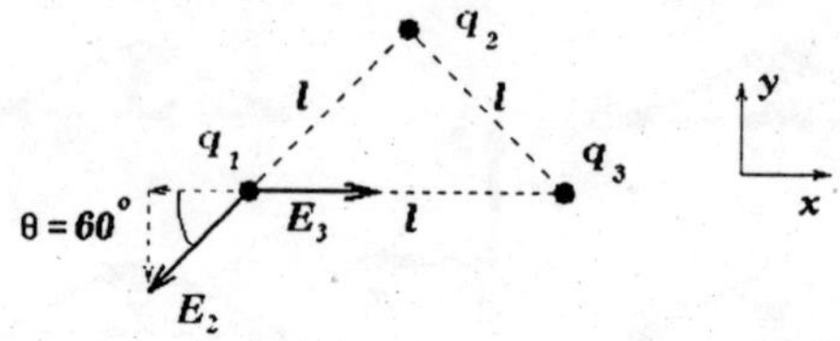

Fig. Electric Fields on Charge q_1.

Note that only the fields due to q_2 and q_3 have to be considered as the force on q_1 is determined only by *external* charges (for the case of point particles).

CONTINUOUS CHARGE DISTRIBUTIONS

Naturally, we cannot be content with considering only the fields or forces between point particles. Real objects, however, can be considered as being made up of a large number of very small pieces which *approximate* points.

If we understand how charge is distributed over an object and if the object and its charge distribution have a high degree of symmetry, then we can easily calculate the electric field due to that object, and hence the force it can produce on other charged objects, by the usual techniques of integral calculus.

As an example of the above statement, consider a charged line segment of length l and charge +Q. Here, we have not a point, but an infinitely thin line of charge.

Without any other information to go on, we assume that +Q is uniformly distributed throughout the line. If we wish to know what the electric field is at a point a distance h above the midpoint of the line, how do we calculate this?

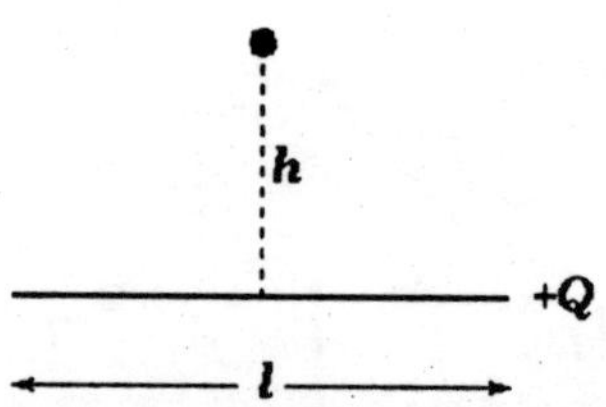

Fig. A Linearly Distributed Charge.

We approach this problem by breaking the line into small segments of infinitesimal length.

The lengths are small enough that we can assume that each little segment can be considered as a point. Before we go through any calculation though, we consider how symmetry can reduce the amount of work we need to do.

In this case, if we consider the result of two infinitesimal segments, one on either side of the midpoint line, we note that, as in the previous case of two point charges, the horizontal components of the electric fields tend to cancel in making the net field.

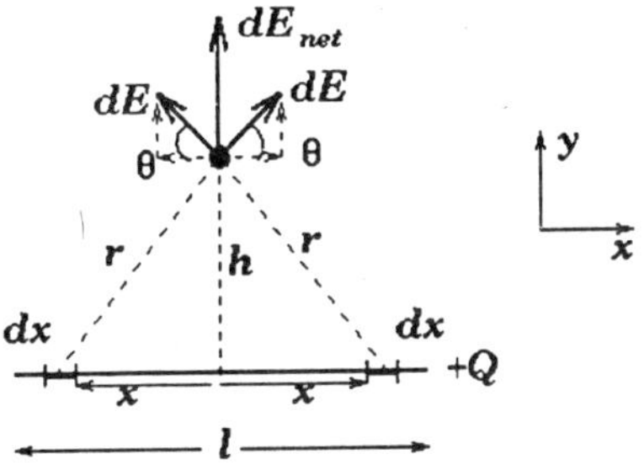

Fig. Calculating the Electric Field due to a Charged Line Segment.

Thus, we note that we only need to keep track of the vertical components for the line segments as the horizontal ones will cancel out. Furthermore, we can get away with only adding up the contributions to the net electric field from one half of the line segment and multiplying by two to take care of the contribution of the other half.

This is not much work in this case, but it's nice to be clever when you're sure you're right. In this case, the integration can proceed once we know how to write down dE.

To get that, we need to consider the charge of the infinitesimal line segment. If we assume that the charge is uniformly distributed, then the charge of each infinitesimal segment must be

$$dQ = Q\frac{dx}{1}$$

Another way of saying this is to consider the *linear charge density* as being Q/l. Then the charge of any length, dx, of the line segment is (Q/l) dx. The vertical component of the electric field due to any little length, dx, is therefore

$$dE\frac{\sin}{2} = \frac{kQdx}{lr^2}\frac{h}{r}$$

where $r^2 = (x^2 + h^2)$. Now we can set up the integration and finish the problem. For convenience, let's place the origin of our coordinate system at the midpoint of the line segment. Maple makes quick work of the integral.

Let's go back and check that we have self-consistency with our original assumption about very short line segments looking like point charges.

If we consider the above answer, we can ask the question: what does the electric field look like if l, the length of the line segment, is itself very small compared to h.

In other words, if h is big, then the point where we want to calculate the field is very far away from the segment and we expect that it will be difficult to tell the difference between the field due to a point charge and the field due to a small line segment. On the other hand, if h is not very large but l is very small, again, we assume the field looks like that due to a point charge. Thus, the formula for the net electric field does go to that of a point charge (kQ/h^2).

ELECTRIC FIELDS OF FINITE OBJECTS

We can define the electric field for any number of distributions now that we know how to apply Coulomb's Law in combination with integral calculus. One of the hallmarks of truly "learning" a subject though is that you first learn "what" neat things you can do, then you learn when not to do them.

In other words, it's much more important to understand the importance of symmetry in determining electric fields than it is to memorize the formulas for any particular charge distribution.

Symmetry reduces the amount of effort required to get from charge distributions to E fields to the point where the memorization is actually more work than deriving the answer when you need it. A couple of examples are in order:

ELECTRIC FIELD FOR A RING OF CHARGE

The ring of charge has a charge +q uniformly spread around a ring of radius R.

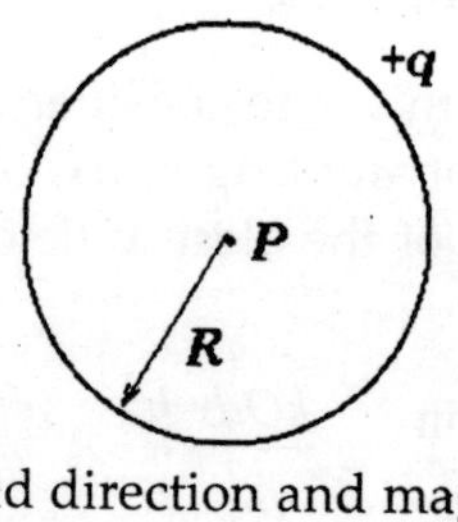

What is the electric field direction and magnitude at the centre of the ring?

In this case no calculation is necessary. If we break the ring into infinitesimal pairs, each with charge dq and arc-length ds, we see that the electric fields from the pairs cancel as long as the members of the pairs are chosen so as to be on opposite sides of the ring.

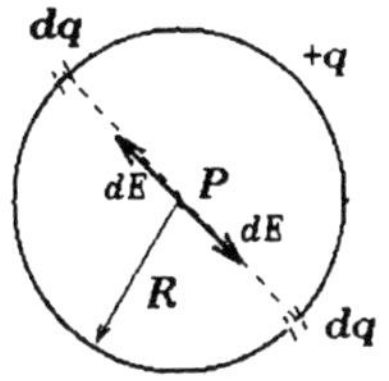

Fig. Electric Field at the Centre of a Uniform Ring of Charge +q.

That is to say, the electric field is zero since each infinitesimal dq we pick around the ring has its field canceled by another dq diametrically opposite it.

ELECTRIC FIELD FOR A DIPOLE RING CHARGE

Suppose the ring above has the top half possessing a positive charge +q and a negative charge -q on the bottom half.

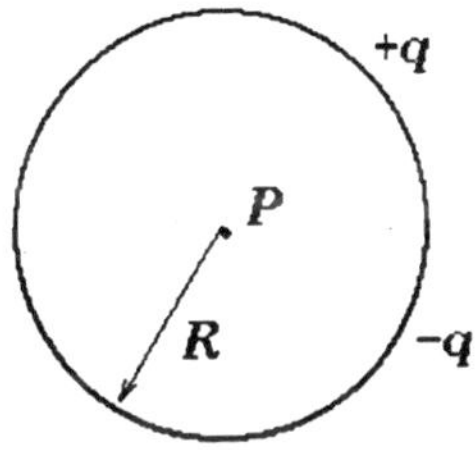

Now we need to do some work. First note that there are *two* symmetries at work here..

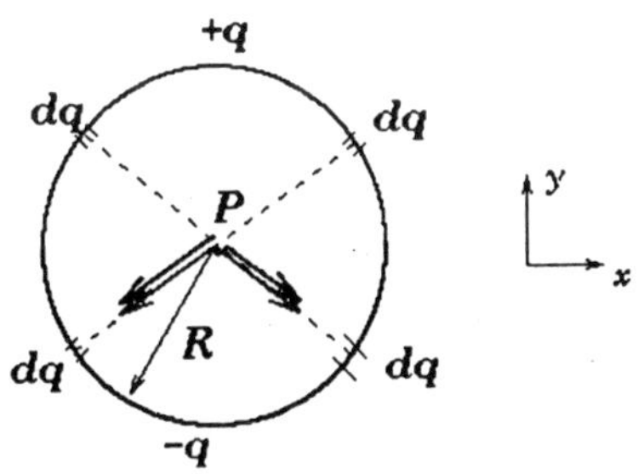

We see that for any pair of infinitesimal arc-lengths on the positive half-ring we can choose the elements so that the horizontal (i.e. x) components cancel while the vertical (i.e. y) components add. If we choose pairs of infinitesimal arc-lengths from the negative half-ring in the same way, we see that their field components also add vertically and cancel horizontally.

Thus, we can say on the basis of symmetry alone that the *net* electric field points downward at point P. With the direction known it is now just a question of calculating the magnitude of the net E field. For each infinitesimal arc-length, we have $dE = k\, dq/R^2$ as the magnitude of E field produced at the centre.

To give each dq a location, we can use the angle w.r.t. the vertical (note that we could use the angle w.r.t. the horizontal just as easily). Then, this angle defines the vertical component of dE at point P.

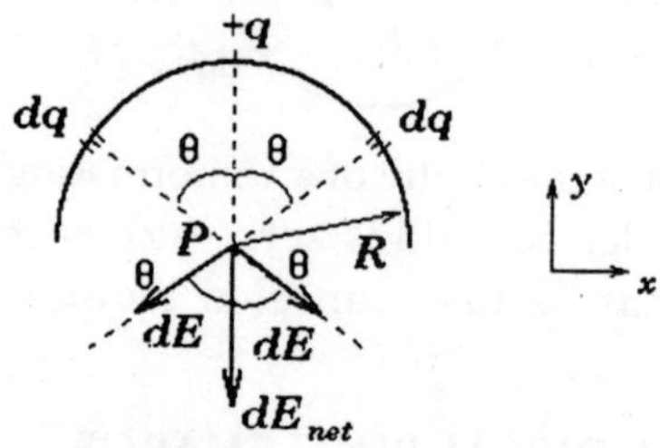

Therefore, we can find the net field contribution of the pair of infinitesimal arc-lengths as

$$dE_{net} = dE_y + dE_y = \frac{dq \cos\theta}{R^2}$$

Now we take advantage of one of our symmetries to only calculate the vertical components since the horizontal components all add. We can take advantage of the second symmetry when we are done calculating the field due to the top half-ring since the bottom must contribute the same field at point P.

Thus, we do the calculation for the top half then multiply by two to get the answer for the entire charge distribution (top half-ring positive and bottom half-ring negative).

To do the calculation for the top half-ring, we need to parameterize dq in terms of theta.

That's easy enough in that the linear charge density of the top half-ring is q/(p*R) since p*R is the circumference of a half-ring.

The arc-length of dq is R*dq. So, now we are set to do our integration to find the net field due to the top half-ring.

Note that, since we are doing pairs of dq's, we only need to integrate from 0 to λ/2 (i.e. a quarter-circle). Also note that we will use λ for the linear charge density.

ELECTRIC FIELD FOR A RECTANGULAR SHEET OF CHARGE

Find the direction and magnitude of the electric field for a sheet of charge –Q uniformly spread over an infinitely thin rectangular sheet of length *l* and width w, at a point P located a distance h above the midpoint of the sheet.

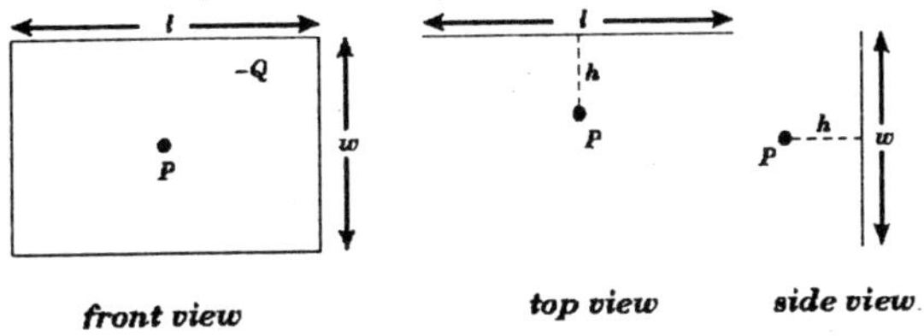

An infinitesimally thin sheet of charge –dQ with length l and width w. The electric field is evaluated at point P a distance h above the midpoint of the sheet.

ELECTRIC FIELD FOR A RECTANGULAR SHEET OF CHARGE

We could do this problem by breaking the sheet of charge into tiny rectangles of length dx and width dy, treat these infinitesimal rectangles as point charges, find the electric field contribution at point P for each rectangle, then do a two-dimensional integral to find the net electric field at point P. This would be quite tedious. We can instead note that there are symmetries here which allow us to exploit a solution to a previous problem. First, note that our result for the electric field at a point located a distance h from the midpoint of a line of charge q with length l is

$$E_{\text{line}} = h\frac{2kq}{l^2 + 4h^2}$$

We make use of it now by noting that the result is true for *any* point a distance h from the midpoint of the line of charge no matter what it's orientation. This is called cylindrical symmetry. The direction of the field is radially inward, i.e. toward the line of charge, if the charge is negative, but the magnitude is unaffected. That means we can use this result for the current problem since all we need do is to assume that our sheet of charge is composed of an infinite number of infinitesimal lines of charge, each with length l and width dy.

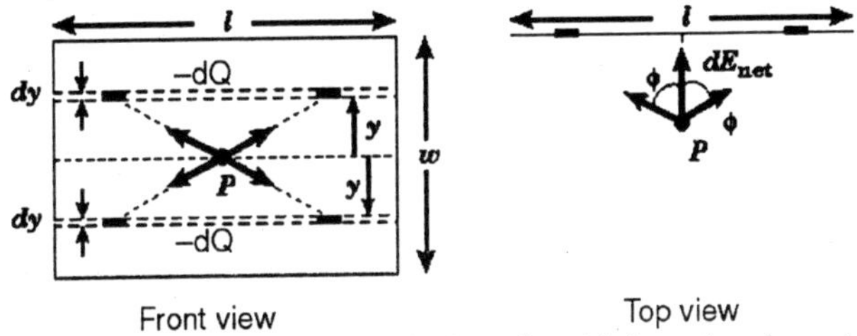

A sheet of charge -Q of length l and width w broken into lines of length l and width dy.

Note that we show only two of these lines, both a distance y from the midpoint, to point out the symmetry that proves that the net electric field of the two is directed perpendicular to the plane and toward it from point P. We define the position of each line as distance y from the centre of the sheet and let each have a charge dQ. Then the magnitude of the electric field at point P due to each line is

$$dE = h\frac{2kdQ}{l^2 + 4h^2}$$

The direction of this field contribution is towards the line. If we consider *pairs* of lines, each having charge dQ, and located y above or below the centre of the sheet, then we can exploit cylindrical symmetry again: each rectangle contributes a field whose component parallel to the sheet cancels the parallel component of the other rectangle. We are left with the perpendicular components of the electric field due to both rectangles adding.

Therefore, for two rectangles, a distance y above/below the centre of the sheet, the net electric field has magnitude where f represents the angle of the electric field vector with respect to a line perpendicular to the plane of the charge sheet. To integrate over all rectangular charges dQ, we first need to note that

$$dQ = \frac{-Q}{w}dy$$

Then the net electric field when we integrate over all pairs of rectangles is directed from point P towards the midpoint of the charged sheet.

DEFINING GAUSS' LAW

To make more progress on understanding the role of electric fields in explaining electrical phenomena, we need to specify a way in which the nature of the electric field produced from a given charge distribution manifests itself in space. Hopefully you have seen from even the few examples we have done with Coulomb's Law just how important the role of symmetry can be in reducing the complexity inherent in the problem of this "specification".

Since Coulomb's Law works for point charges and forces us to reduce every finite distribution of charge into points, it clearly does not make very powerful use of symmetry (the only symmetry available is that due to a point charge and the ability to reduce the task of the integration by choosing integration strategies that maximize the use of symmetry to reduce the work of evaluating the integral).

An indirect approach created by Karl Friedrich Gauss does intrinsically bring in symmetry in the formulation of the method for explaining how an electric field is generated from a charge distribution.

It starts by noting that a proper description of the electric field invokes the notion of a vector field.

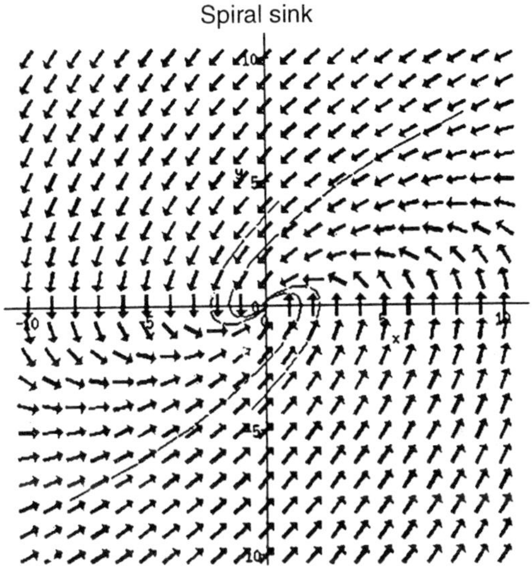

Fig. A Sink.

The arrows here depict the direction and magnitude of something that can be depicted by vectors at each point in space.

This probably reminds you of something like water flowing down a sink. In fact, it could represent a number of situations in which something is coupled to an event which leads to a progression in time which tends towards zero.

The vector field above could represent the velocity, at various points, in a pool of water flowing down a drain located at the origin of the coordinate system or any number of other things, e.g. the inflow of dust circling around a nascent star.

This tells us of the tremendous power of vectors for description. Any kind of information can be represented in vectorial form if we make the correct associations. Vector fields can also be used to represent electric fields. The field lines we draw are, in fact, a crude representation of a vector field.

Gauss realized that a more useful quantitative representation of how things *flow* throughout space could be realized by being more precise about how a vector field is defined.

We can start by being more precise about where the vectors should be drawn from, i.e. use a more definitive way of specifying how to draw the vector field so as to maximize the information we can get from it. We can start by being more orderly as to where we draw the vectors.

If, for example, we start from a given surface, we can state what the vectors look like in defining the flow of something from a specified *set* of points which can be used to define an initial position.

The flow of something can be defined at the surface A and we know we have a definitive description of the flow at the position of A. Without this convention, saying "where" the flow is at any given time is somewhat ambiguous.

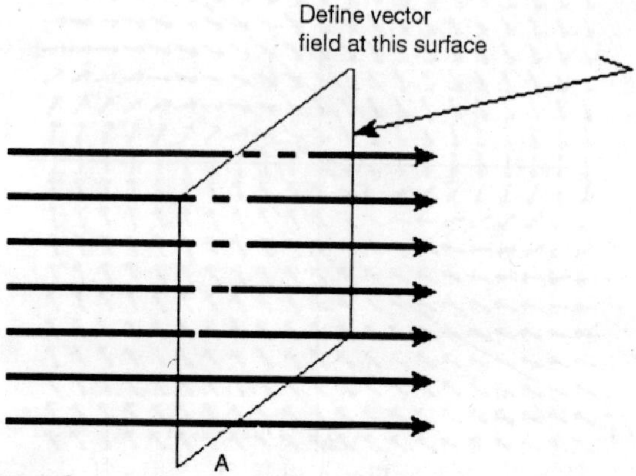

Fig. The Vector Field Defines the Flux Through Surface A.

Although it might not appear as though this type of "defining" might be useful, it already quite powerful for talking about physical observation.

For example, if you've ever noticed water flowing from a faucet set open to give a relatively small flow of water, you might have noticed that the flow narrows as the water falls away from the faucet opening.

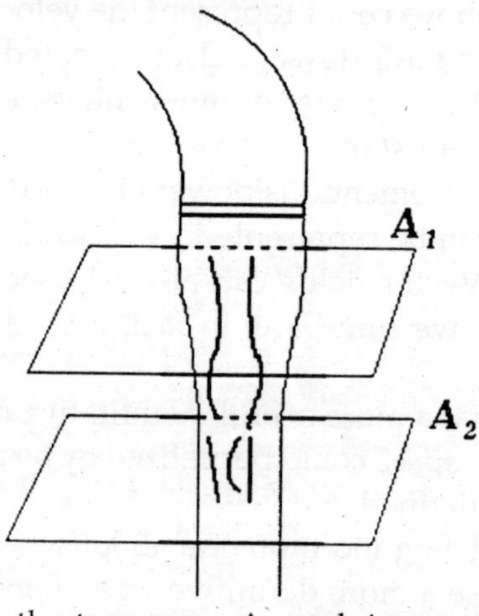

The flux through the two areas A_1 and A_2 must be the same.

If we look at the *flux* of water, defined as the amount or mass of water flowing across an area, A_1, per unit of time, then we know that the same mass of water per unit time must flow across A_2, placed at a lower position. However, since gravity causes the speed of the water to increase as it falls, the cross-sectional area of the water, that is to say, the size of the stream going through A_1 must be larger than the size of the stream going through A_2 since the volume of water (and hence the total mass since the water has constant density) per unit time crossing A_1 or A_2 is given by (cross-sectional area) × (velocity).

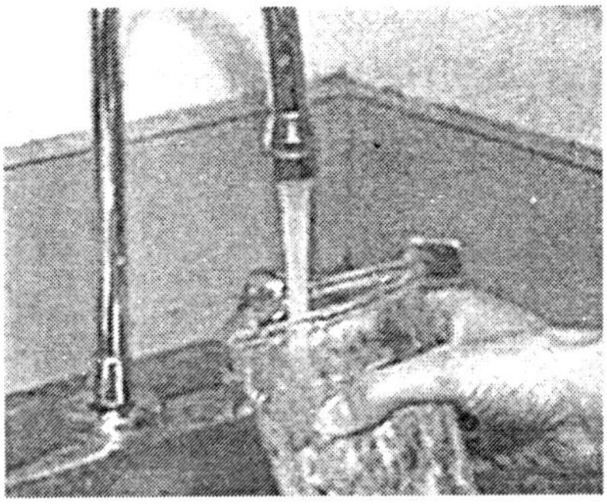

Hence, since velocity goes up, cross-sectional area must go down. Note that water reaches its terminal velocity quite quickly so that the stream reaches a constant cross-sectional area after just a small distance of travel.

To be precise about measuring flux, we should define that the flow of something through a surface is dependent on the angle between the direction of flow and the surface. Since any vector can be resolved into any set of orthogonal components, we should state that the flux is only proportional to the component of the flow vector which is *perpendicular* to the surface.

This turns out to be such a useful case that physicists generally define area as a vector! The magnitude of this vector is what we normally think of as the size of the area and the direction is always perpendicular to the surface.

Since you always have two directions in which you can be perpendicular, the convention is that the direction is chosen so as to point away from any volume which is *enclosed* by the surface whose area we are interested in. In other words, physics generally is interested in areas which confine (in fact, define) some volume.

Pointing away from the volume distinguishes inside the volume from outside the volume. The flux is then defined by the dot product of a vector representing the flow and this vector. We can state this relationship mathematically as $flux = v \cdot A$ where v is the vector describing the flow and A is a vector perpendicular to the purple area (and directed along the flow by convention) with magnitude equal to the area covered by the flow.

While v and A are parallel, the flux is indeed a maximum (the dot product is a maximum when the two vectors of the product are parallel) and equal to zero when v and A are perpendicular (in that case the purple area is parallel to the flow).

SPHERICAL SYMMETRY

The first practical application of these ideas of flux is to a mystery first explained by Newton. In his theory of Universal Gravitation, Newton speculated that his force formula for gravitational attraction between two *point* masses, m_1 and m_2, should work equally well for *spherical* masses m_1 and m_2. If we replace a spherical mass distribution by a point at the centre of the spherical distribution with the same total mass, the gravitational force looks exactly the same to the universe outside the sphere. Gauss would say that this "coincidence" works because points and spheres possess the same symmetry (spherical symmetry). Imagine a point charge, q_1. The charge emits an electric field in all directions. No direction is preferred.

If we replace the electric charge by a spherical distribution of charge q_1 with radius R, the same condition holds. The sphere looks the same from any angle so the electric field it produces looks the same from any direction. Put any other charge further away from q_1 than R and it cannot tell the difference between the point charge and the spherical charge distribution. Gauss takes advantage of this to relate his electric flux formula for the spherical charge distribution to Coulomb's Law for a point charge. Gauss states that electric flux is Φ_E = E·dA and that this is proportional to the electric charge. Some comments are in order: first, note that we use an integral to define the flux.

This accounts for the fact that we generally need to define the flux through an area which has to be broken into infinitesimally small pieces. The flux through each piece is evaluated and the total determined via integration. Gauss' Law is the following: Φ_E q_{enc} for an area which *encloses* (hence the circle through the integral) a charge q, the electric flux through this area is proportional to the charge enclosed. Assuming this to be right for the moment, we can evaluate this for a positive spherical charge distribution of radius R.

Since the electric flux extends outward, if we choose the surface of the sphere as the surface from which to evaluate the flux, then E is perpendicular to the surface at each point (since there are no preferred directions) then E · dA is just E dA since the vectors are parallel at every point on the spherical surface.

Also, since the sphere is assumed to be uniform, the E field has equal magnitude along this surface, hence the integral of E dA is just

EA = E(4πR²). Thus, if the proportionality constant is called ε_0, then

$$q = \varepsilon_0 E(4\pi R^2)$$

$$E = \frac{q}{4\pi\varepsilon_0 R^2}$$

Clearly, to be consistent with the result expected from Coulomb's Law we need

$$k = 1/4\pi\varepsilon_0,\ \varepsilon_0 = 8.854 \times 10^{-12}\ C^2/N{\cdot}m^2.$$

Thus, Gauss' Law is expressed as

$$E{\cdot}dA = \frac{q_{enc}}{\varepsilon_0}$$

with the subscript enc meaning the charge enclosed in the volume contained by the surface for which the flux is defined.

APPLICATIONS OF GAUSS' LAW

Let's find the electric field due to a line charge. As we have done this before, much of the setup of the problem is already done Let's make things a bit tougher by considering the field due to an *infinitely* long line of charge as opposed to the one of finite length which we did before. It's clear here that it's impossible for us to talk about a finite amount of charge stretched over an infinitely long distance. Instead, we state that the line has a constant linear charge density, l.

Realistically, all line charges are finite, but, first, we have done the finite length problem explicitly, and second, a line of charge which is quite long compared to the distance from it at which we would like to know the E field is not an uncommon problem. We can deal with the approximate answer as easily as with the true solution and compare the differences.

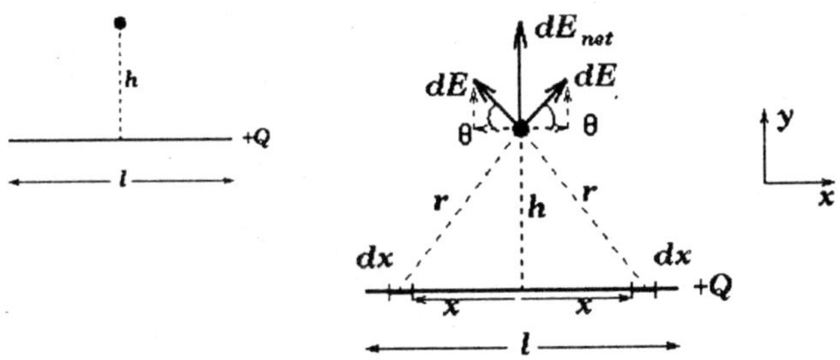

Fig. Calculation of the Electric field at the Midpoint of a line Charge of Length *l*.

Physicists will often engage in thoughts of this kind in theoretical research as it is important to know where our ideas "break down". Those ideas which can hold up under the most extreme extrapolations without delivering clearly nonsensical results indicate something deep about our understanding. The above being said, we still need to do the calculation. Consider the figure which shows a view of the line charge and a point P a distance h away from it.

We wish to find the electric field at point P. To set up the integral, we do, as before, the trick of taking infinitesimally small line segments of charge in pairs so that their horizontal components cancel and the vertical (i.e. radial) components add.

Hence we only need to change the definition of dq, the charge on an infinitesimal segment with length dx, otherwise our approach for the finite line charge is unchanged for the infinite length case. Now, $dq = \lambda dx$, so, as with the finite length charge, we use the angle, q with respect to the vertical to identify the radial component, r, as the distance from the infinitesimal charge to point P;, i.e. and thus We calculate the magnitude of E as follows. $E_{net} = \lambda/2\lambda\varepsilon_0 h$.

We can also approach this problem using Gauss' Law.

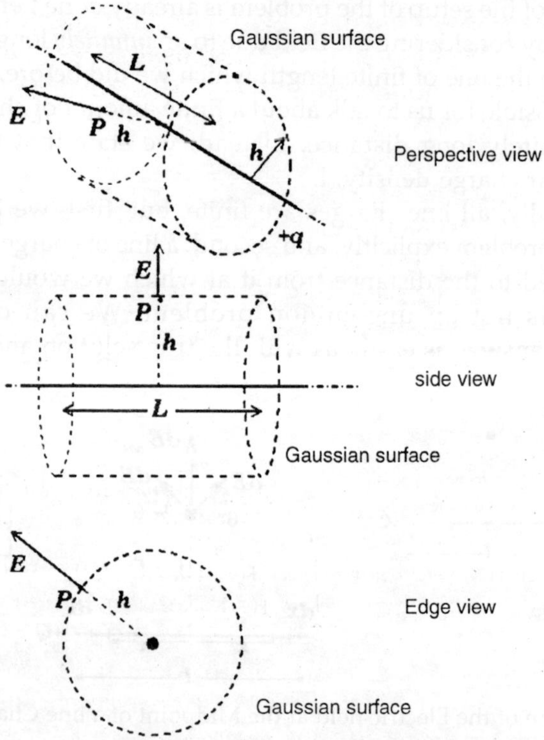

Fig. Gaussian Surface with Cylindrical Symmetry Surrounding a Line Charge.

The Gaussian surface should be a closed, hollow can. The sides of the can are perpendicular to the E field. Hence E·dA is just E dA since the electric field also only depends on h as we just showed with the explicit integration done earlier.

If we make the Gaussian can of finite length (we should since we will need to find the amount of enclosed charge), say L, then the charge enclosed is l L. The calculation of the E field proceeds in this way: $E_{net} = \lambda/2\pi\varepsilon_0 h$

We get the same result, but with somewhat less work. Gauss' Law is useful for calculation of the electric field for situations which have a high degree of symmetry. In fact, for freshman physics, we can generally only apply it in cases in which one of the following symmetries are exhibited:

- Spherical symmetry (point charges, spherical charge distributions).
- Cylindrical symmetry (infinitely long lines of charge, infinitely long cylindrical charge distributions)
- Planar symmetry (infinite sheets or blocks of charge)

PLANAR SYMMETRY

In the case of charge distributions which exhibit planar symmetry, i.e. that have a charge distribution which extends in two dimensions, we wish to find the E field due to a large, assumed infinite, sheet of charge. The infinite extent applies to only two of a possible three spatial dimensions. We talk of the sheet being infinitely wide and infinitely long, but it may be of finite or infinite thickness.

The character of the electric field depends not only on the arrangement of the charges on the sheet but on whether the sheet is composed of conducting or non-conducting material.

Before beginning this examination of planar symmetry though, it's important to understand some implications of Gauss' Law (or more importantly of the symmetries implicit in the use of Gauss' Law) for charges on conductors. First, note that conducting material has the property of allowing charges to move freely, i.e. without resistance to their motion, through them.

The mobility of the charges means that they respond to any electric field in their vicinity by moving according to F = qE. If we think about the consequences of this, you'll quickly come to the conclusion that the electric field inside the bulk of any conducting material must quickly go to and stay at zero.

The reasoning is simple: if charges move in response to electric fields, then the charges will only stop moving once the electric field goes to zero. However, since the negative and positive charges move in opposite directions they will themselves create an electric field.

A simple picture suffices to show that the electric field created always opposes the external electric field creating the charge separation.

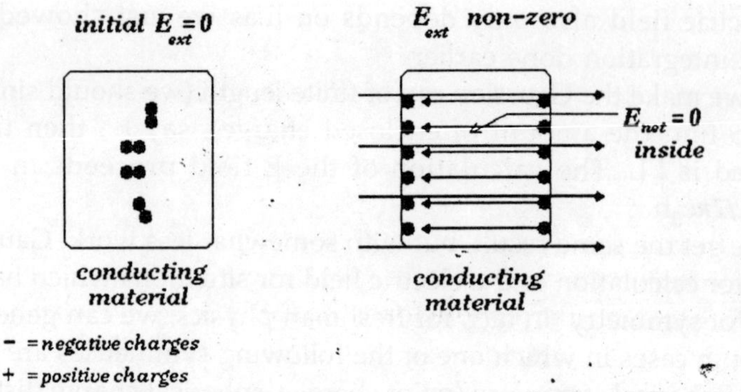

Fig. Charges Separating within a Perfectly Conducting Material.

The charges keep separating until they reach the surface of the material. More and more charges accumulate at the surface and create a net electric field that exactly cancels that of the external field. We assume that there are enough charges in the material so that any electric field can be neutralized.

Thus, the initial electric field in the material is zero since the positive and negative charges are "combined" so as to leave no net charge to create an electric field.

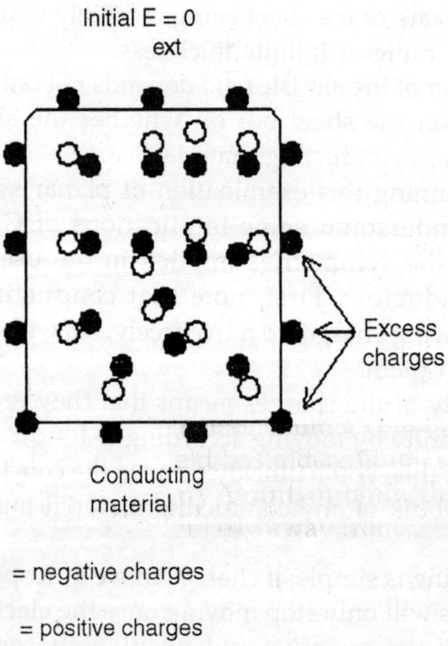

Fig. A Perfect Conductor with a Non-zero Electric Charge.

Once an external field is turned on, the net electric field inside the bulk material is zero since the charges separate to produce an electric field that opposes that of the external field. Later, we will see how to modify the above picture for "perfect" conductors so that we can determine what happens for real conductors.

What we have shown above works fine for conductors that have no net charge.

We can also apply similar reasoning to the case of a perfect conductor with a non-zero net charge. In the absence of any external electric field, the net charge on the conductor winds up on its *outside* surface. This maximizes the separation between the charges and minimizes the forces acting on them.

Note that the charges inside the conductor don't separate, but instead orient themselves so as to still generate a net field that opposes that due to the external charges, thereby neutralizing them so that net electric field inside the bulk material remains zero.

Again, it is assumed that there are enough charges in the conducting bulk to neutralize the field due to any amount of external charge placed on the conducting material. Again, we will talk more of this later, but for real materials we know that the number of internal charges is not infinite. So, the electric field can penetrate a certain very small distance into the conducting material.

We refer to this as the skin depth of the material. We can assume the skin depth to be of zero width. This is an acceptable approximation for many "good" conductors.

The picture of what happens to the net field produced by charges confined to a sheet of infinitesimal thickness. The figure shows the result of adding up the E fields from charges at various points on the sheet. As in the spherical and cylindrical cases, we see a symmetry that allows us to cancel components of the net E field that are *parallel* to the sheet and only keep those that are perpendicular to the sheet.

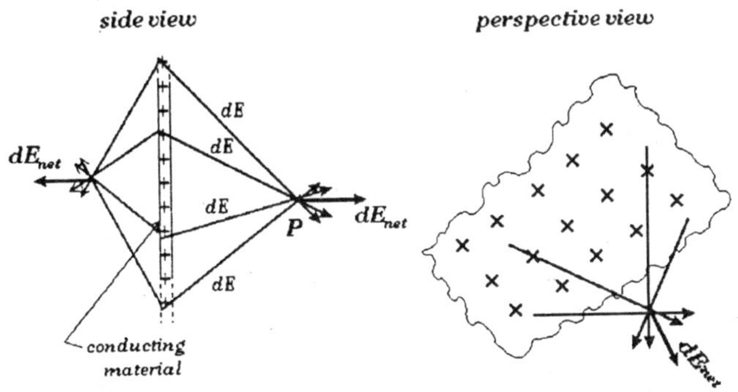

Fig. The Electric Field near an Infinite Sheet of Charge.

Before going through this problem with Gauss' Law, we can take the "old" approach of Coulomb's Law. At least we can now make use of a previous result to reduce our workload. Remember that we have already calculated the magnitude and direction of the E field due to a wire. We now make use of this result. Consider the sheet as being composed of an infinite number of infinitesimally thin wires, each with width dx as shown. We wish to find the electric field at a point P located a distance b from the sheet.

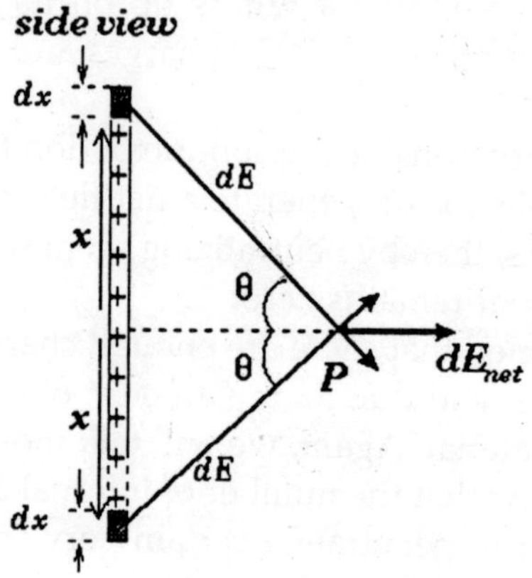

Fig. Calculating the E Field Near an Infinite Sheet of Charge.

As usual, we note that the symmetry, planar symmetry in this case, allows us to ignore components of the net E field as these cancel. The components perpendicular to the sheet add, so, since the sheet is infinite in extent width and length, the E field is perpendicular to the sheet no matter what point P we pick.

Let's choose, as a reference point, the spot on the charged sheet which is closest to the point P where we want to know the field magnitude (marked as point O in above figure.

This point lies on a line through P which is perpendicular to the sheet. If we choose "wires" which are equidistant from point P, then the contributions to the field at point P from components which are parallel to the charged sheet will exactly cancel.

The perpendicular components are each dE cosq where q is the angle between a line from the wire to point P and the line from point O to point P. The electric field at point P due to a single wire is

$$E = \frac{\sigma dx}{2\pi\varepsilon_0 b}$$

Note that we have replaced the linear charge density of the previous lecture with σdx. The distance from the wire to point P, h, has been replaced by the distance r.

Now the integration over all the wires can occur. $E_{net} = \sigma/2\varepsilon_0$, where you should note that the integral can be evaluated using Maple or by consulting a table of integrals. The above result works for an infinitesimally thin sheet of conducting material or for any finite thickness sheet of non-conducting material since the E field configuration will be the same.

If we apply Gauss' Law to the same case, we can arrive at the solution considerably faster. Imagine a Gaussian surface in the form of a block with a rectangular cross-section.

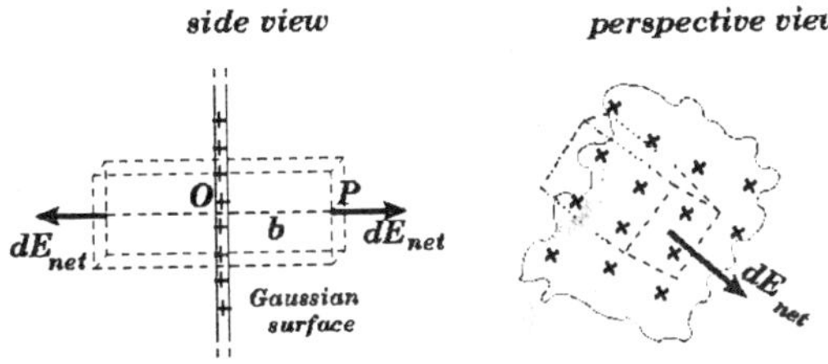

Gaussian pillbox for a finite thickness planar charge distribution.

The electric field through the Gaussian surface is parallel to the sides and perpendicular to both the endcaps.

The result is that the flux through the sides is zero while the flux through the endcaps is EA + EA.

With A being the cross-sectional area and where we take advantage of our previous calculation with Coulomb's Law for the infinitesimally thick conducting sheet (or finite thickness non-conducting sheet) to state that the electric field should be constant in direction and magnitude on either side of the sheet.

Hence, Gauss' Law states that

$$\text{flux} = \Phi = 2EA = \frac{q_{enc}}{\varepsilon_0}$$

Since the charge enclosed by the Gaussian surface is

$$q_{enc} = \sigma \cdot AE = \frac{\sigma \cdot A}{2A\varepsilon_0} = \frac{\sigma}{2\varepsilon_0}$$

This is the same result as was derived using Coulomb's Law as we expect. The difference in effort to get this result is impressive, although we did depend on having seen the symmetry that allowed us to relate the flux to the electric field in this case.

The real payoff of using Gauss' Law comes when we apply the same thought process to the case for a conducting sheet of semi-infinite thickness.

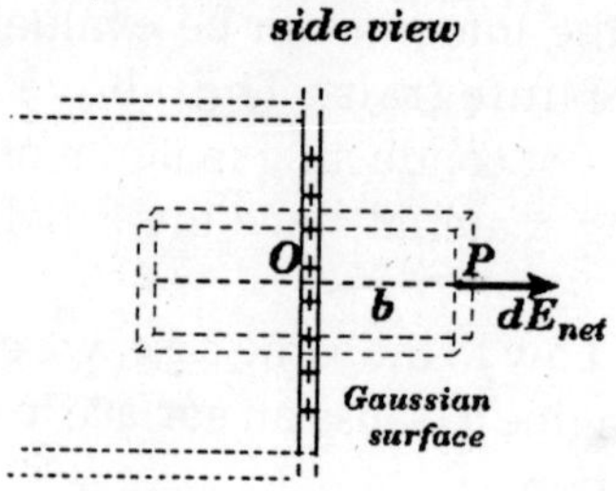

Fig. Gauss' Law Applied to a Slab of Semi-infinite Thickness.

Using Gauss' Law with the same arrangement as before, we note that now only one endcap receives a net flux since the endcap inside the conducting material sits in a zero field region. Hence the magnitude of the E field in this case is

$$\text{flux} = \Phi = EA = \frac{q_{enc}}{\varepsilon_0} = \frac{\sigma}{\varepsilon_0}$$

so the field is twice the magnitude of the non-conducting sheet of finite thickness or the conducting sheet of infinitesimal thickness.

THE FINITE THICKNESS INFINITE SHEET

Even though we've used planar symmetry to derive results for a couple of different examples of infinite sheets of charge, we should consider one more example, not for derivation, but strictly for application of the ideas learned so far. Consider the following problem from the 1991 midterm I. An infinite slab of insulating material has a uniform volume charge density of 2.0×10^{-6} C/m^3. The surfaces of the slab are located at x = +1 cm and x = –1 cm.

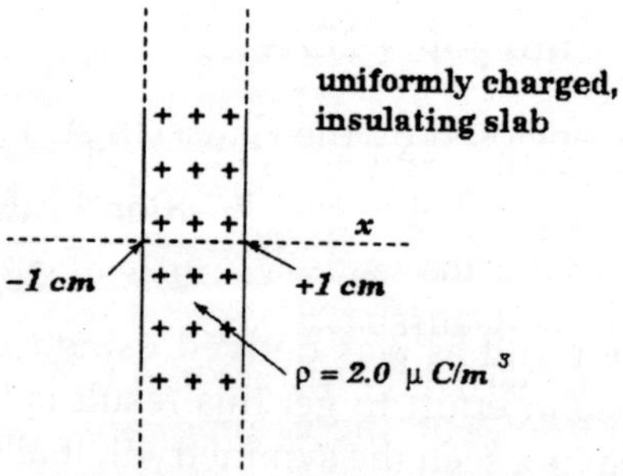

An infinite slab of insulating material with a uniform charge density throughout.

We need to draw a Gaussian surface which reflects the symmetry. From what we just covered in the last lecture, we know that the electric field should be constant in magnitude and direction *outside* the nonconducting material.

For positive charge, the electric field is directly perpendicularly away from the surface of the material. We can use a Gaussian surface of any shape we desire, but let's choose one of rectangular cross-section as we have thus far in our derivations. That means we have a situation that looks as follows:

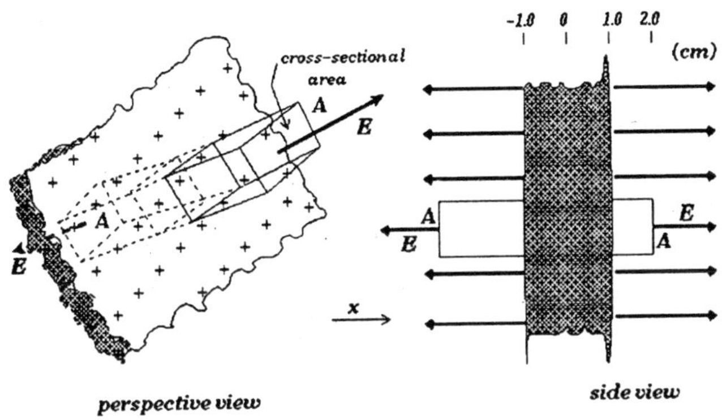

Fig. A Rectangular Gaussian Surface in the Vicinity of a Charged Slab of Infinite Extent.

Note that the Gaussian surface on the +x side extends only to the 2.0 cm mark for which we desire the field magnitude (direction is already known to be along +x, i.e. perpendicular to the surface at +1.0 cm).

On the $-x$ side we need only note that the Gaussian surface must extend the non-conductor to preserve the symmetry of the electric field.

The reason is that we know the field has constant magnitude and direction outside the nonconducting material, hence it does not matter where we place the endcap of our Gaussian surface as long as it is at x less than –1 (i.e. $x = -3$ cm or $x = -10$ cm work equally well).

Since the field has the same magnitude on either side of the nonconducting material and we have drawn the surface so the sides of the Gaussian "box" are parallel to the direction of the field for either +x or $-x$, the net flux through the Gaussian surface is 2E·A where A is the cross-sectional area of the Gaussian box.

To get the charge enclosed by the box, we use the volume charge density, which we call rho, and the volume contained by the part of the Gaussian box which is inside the nonconducting material. This total charge is $q_{enclosed} = \rho \cdot A \cdot t$ where t is the thickness of the non-conducting material. From the information given, t is obviously 2 cm. Hence, it is quick to apply Gauss' Law: $2EA = q_{enclosed}/\varepsilon_0$, E= 2260 N/C

- By symmetry, at the centre of the non-conducting material must have zero net electric field. Fields from each half of the slab would cancel.

- Now we have a choice of how to draw the Gaussian surface. We can choose a symmetric arrangement in which the two endcaps lie at -0.7 cm and +0.7 cm or one in which the two endcaps lies at 0 and +0.7 cm.

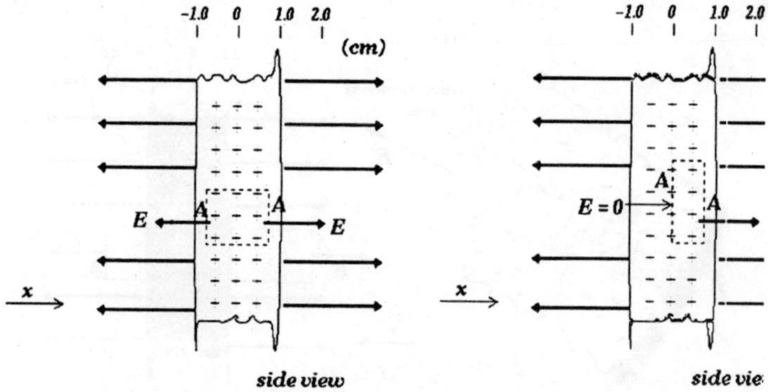

Fig. A Gaussian Surface Partially Embedded in a Charged Slab of Infinite Extent.

The reason either of these works is because we can evaluate the flux and separate out the contribution of E to the flux in either case. For the first case, the net flux through the surface is just 2EA since, by symmetry, the E field will be the same magnitude at the same distance from the symmetry point of x = 0. In the second case, the flux is just through the right endcap since the magnitude of E at x = 0 (location of the left endcap in this case) is zero. Hence the flux would be EA. Now note that we get the same result for either case. For example, in the first case $2EA = q_{enclosed}/\varepsilon_0$, E= 1582 N/C

We can easily verify that the second case gives exactly the same result. Remember that the flux is one-half of what it is in the case just shown, but that the volume enclosed and therefore charge enclosed in the second case is also half of what we get for the first case.

ELECTRIC POTENTIAL

We are still operating under the assumption that the electric field provides a convenient "mechanism" for dealing with electrical phenomena. Even though electrical forces are what we observe, the electric field is assumed to the be the "agent" by which the force is manifested on charged particles. However, just as we found with Newton's Laws, it becomes increasingly clear, with some experience, that we would prefer, for practical reasons, to adopt a way of describing the electric field in terms of scalar quantities rather than through the vector nature of the E field itself. With Newton's Laws of Motion, it turned out to be very useful to define the concepts of *work* and *kinetic energy*.

These definitions, namely $W = F \cdot ds, K = 1/2mv^2$ relate work to an integral over a path (which is broken into infinitesimal increments of displacement, ds) and the kinetic energy to the mass and velocity of a point particle. We note that these two are related according to the *work-energy theorem*, $W = K_f - K_0$. Furthermore, for work done by conservative forces, we can define a potential energy, $U(x) = F(x)\, dx$*The*

generalized work-energy theorem relates changes in potential energy to work and hence to changes in kinetic energy for the case of conservative forces

$$W = -\Delta U \ = \Delta K\ K_0 + U_0 = K_f + U_f$$

The potential energy and kinetic energy allowed us to work with scalar quantities dealing with motion and the change in motions caused by forces. We can develop an analogous scalar quantity to the electric field. It is called the electric potential and is usually represented as V. The relationships we seek then look, schematically, as follows

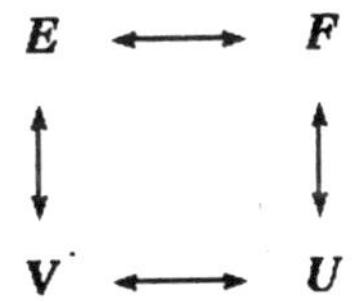

Fig. Relationships between Scalar and Vector Quantities Related to Static Electric Fields.

There's nothing especially tricky here. V becomes the scalar quantity of choice to use for problems involving the electric field. The idea is that the scalar nature of V will be generally easier, and therefore more practical, for uses than E just as potential energy is more useful for many problems than force, F. However, for V to be useful, we need a definition for it which allows us to consistently realise the pictorial relationship between force, potential energy, electric field, and electric potential that is drawn above.

The definition which works is

$$U(x) = -F(x)\, dx \ = -q_0 E_x\, dx$$

$$\Delta V - W/q_0 \ = \Delta U/q_0\ V = U/q_0$$

That is, we state that the work per unit charge is the change in potential. This is entirely consistent with thinking about V in the same way we think about height for the gravitational field. The higher the value of V with respect to some fixed reference, the more potential energy (for a positive charge) there is in the electric field, just as increasing the height of a mass raises the potential energy in the gravitational field. One consequence of this definition is the relationship between electric potential and electric field. From the generalized work-energy theorem and the definition of electric potential, we have

$$= -W/q_0$$
$$= \Delta U/q_0$$
$$= -1/q_0 \text{ F.ds}$$
$$= -1/q_0 \text{ q E.ds}$$
$$\Delta V = -\text{E.ds}$$

The s refers to a path along a electric field line. V becomes the scalar quantity of choice to use for problems involving the electric field. The idea is that the scalar nature of V will be generally easier, and therefore more practical, for uses than E just as potential energy is more useful for many problems than force, F.

The Electric Potential

We will find many uses for the electric potential in the near future. For now, we note that it is indeed easier to work with than the electric field because of its scalar nature.

The first way to calculate the potential is through the use of integral as defined above. We can apply this to a case for which we know the electric field, namely a finite length charged, rod. We wish to find the electric potential at a point P which is a distance h above the midpoint of the rod of length L.

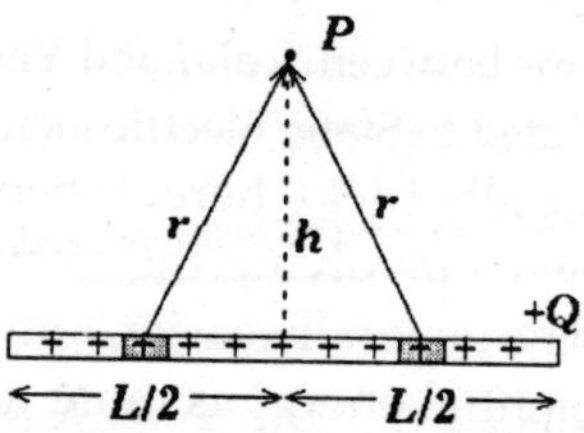

Fig. Finding the Electric Potential over the Midpoint of a Finite Length, Charged Rod.

If we wish to find the potential at point P relative to infinity, then we should integrate the dot product of E with ds over a path from infinity in to point P. The easiest path to integrate is a straight-line path as the E field is always vertically upward and opposite to a straight line path from infinity down to point P. So, noting that y is the integration variable and goes from the line charge *out* to infinity (and therefore is opposite the ds direction). $V_P = -E \cdot ds$

We can also consider doing this calculation another way. If we make use of the techniques of calculus, we can consider dividing the rod into infinitesimal lengths of charge, dq, then considering the potential at point P for each.

Adding up the potential for all the dq's by integration should yield the same result as above.

Since the infinitesimal lengths can be considered as point charges, we need to know the potential at point P, relative to infinity, for a point charge dq.

First, consider any point charge, Q, in space. Then the electric potential at a point a distance r away can be calculated using y as the integration variable as follows.

The electric field due to a point charge is radial, so we can integrate over a straight line path along a radius from infinity to r. with x being the distance of dq from the midpoint of the line segment charge. Even though we are not dealing with vector components, we can still take advantage of symmetry in that we see that the right and left halfs of the line segment (i.e. to the right or left of the midpoint of the line segment) are completely symmetrical: they give equal contributions to the potential.

As expected, this gives the same result. This example brings up something of a lesson here. If you wish to find the potential of a charge distribution of finite size, you can always break the distribution into infinitesimal pieces, treat each as a point charge for getting the potential at the desired point, then integrate over the whole distribution to get the net potential.

If you know the electric field distribution of a charge distribution or can easily calculate it, you can find the potential relative to some reference point by integration of the electric field over some path. The fact that you can use any path means you always want to choose the path that minimizes the difficulty of the integral.

Electric Potential for a Finite Thickness, Infinite Sheet

We now want to apply what we developed at the end of the last lecture in deciding which of our two methods to use for calculating the electric potential due to a charge distribution. Let's consider the case of the infinite, charged, non-conducting sheet. If we assume the sheet has finite thickness, t, and a volume charge density, r, then we know,that the electric field has a magnitude $E = \rho t/2\varepsilon_0$.

At any point outside the sheet. The electric field is directed perpendicularly away from the sheet (for a positively charged sheet). So, if we want to know the potential at point P, *relative to the right surface of the sheet,* then we can evaluate the potential via a straight line path from the right face of the sheet out to point P. $V_P = -Ed$

If we considered trying to break the infinite sheet into infinitesimal charges and integrating, we would get the same answer, but the effort we would need to expend is significantly greater.

Clearly this problem is best done using our previous knowledge of the properties of the electric field.

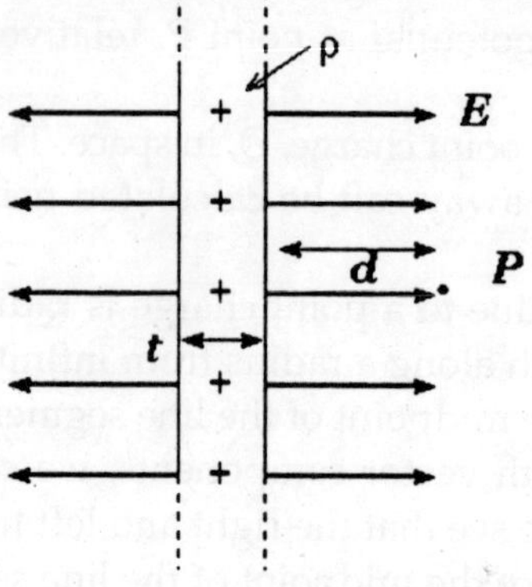

Fig. An Infinite Plane with Volume Charge Density Rand Finite Thickness.

Potential of a Charged Ring

Consider a thin ring of plastic with a radius R and a uniformly distributed net charge –Q.

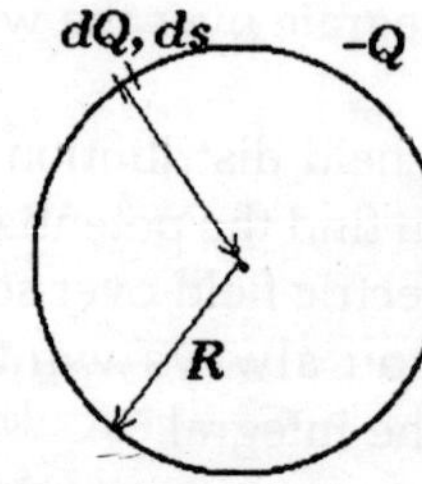

Fig. A Thin Ring of Radius R and Uniformly Distributed Charge -Q.

We can quickly surmise the answer to the first part of the problem. By symmetry, the electric field at point P must be zero. Since we can break the circle into infinitesimal pieces, dQ, with arc-length ds, we can quickly see that every piece has another piece, diametrically opposite it, which delivers a contribution to the net field which has equal magnitude and opposite direction.

We would be unwise to be so hasty with the second part of this question however. The electric potential, after all, is evaluated by integrating over a path from infinity to point P. Even though the electric field at point P is zero, it is definitely not zero for any arbitrary position which is not at point P.

Hence we have no reason to expect that we would get zero when integrating E · ds, and, in fact, we don't. In this case calculating using the E field is tedious compared to just integrating over the contributions due to each infinitesimal piece of the ring, dQ. To derive dQ, use the linear charge density method, $l = Q/(2\pi R)$, where $2\pi R$ is the circumference, or total length, of the circle. So, $dQ = \lambda ds$. $V_P = -kQ/R$, The answer is distinctly not zero.

The lesson is, be careful! Knowing E at a point says nothing about the value of V at that point. Although we can certainly say that, in regions in which E = 0, the potential does not change.

Electric Potential Inside a Spherical Thin Shell Conductor

Consider a hollow metal sphere of radius R which has a net charge +Q. Find the electric potential, relative to infinity, at the centre of the sphere.

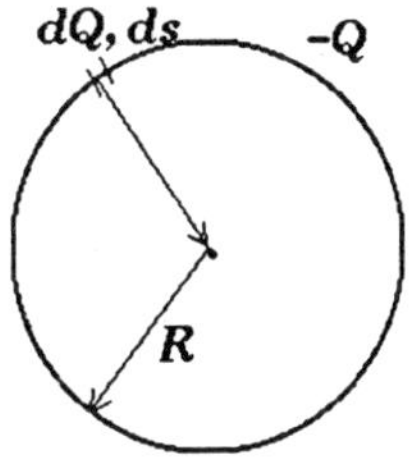

Fig. A Thin Spherical Shell of Radius R and Charge +Q.

We know that a conductor must have all of its net charge sitting on a surface. In this case, its the outer surface of the sphere. Furthermore, we know the electric field is the same as for a point charge sitting at the centre of the sphere for points which are *outside* the sphere.

Inside the sphere, the electric field is zero since any Gaussian surface we draw which is completely contained inside the sphere would contain no net charge.

If we set the origin of our coordinate system as being at the sphere centre, then we need to integrate the electric field over a path (it might as well be straight-line since that's the easiest to integrate) from infinity to the sphere and from the sphere surface to 0 (since the electric field is different inside than outside). So, remembering that the path is opposite to the direction of the E field (radially inward for the path, radially outward for E). $V_P = -kQ/R$.

Electric Potential Inside a Hollow Metal Sphere

Suppose we allowed the shell from the previous problem to have a finite thickness? How does that change our result for the electric potential inside the shell?

Assume that we have a spherical shell with outer radius b and inner radius a. What is the electric potential, relative to zero at infinity, for a point P at radius $r < a$?

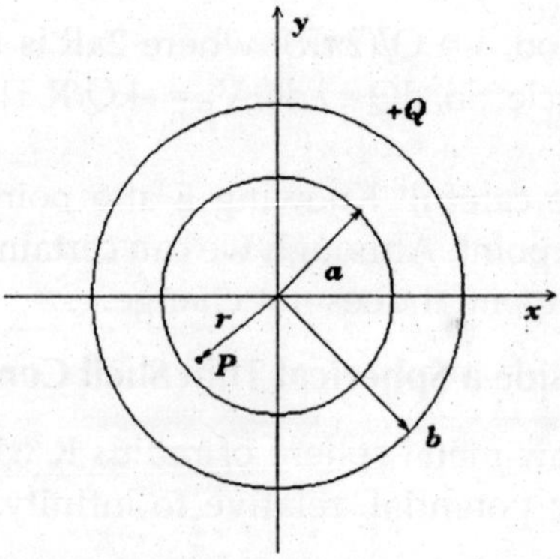

A spherical, conducting shell with inner radius a and outer radius b. We wish to find the electric potential for a point located at r < a.

The approach to the solution is identical to our previous example for r > b. That is to say, the potential of the outer radius of the conducting shell is

$$V_b = \int \vec{E} \cdot d\vec{s} = -\int_{\infty}^{b} \frac{kQ}{r^2} dr = \frac{kQ}{b}$$

How about the region a < r < b? The electric field in the bulk of a conductor must be zero, so the electric potential cannot change for this region. Hence,

$$V_{a<r<b} = V_b = \frac{kQ}{b}$$

Finally, for the region r < a we also have no electric field, hence, $V_{r<a} = V_{a<r<b} = V_b = kQ/b$,Therefore, for any point inside the shell (or within the conducting material of the shell), the potential is the same as the outside surface. This includes the potential at the centre of the shell, hence the result is unchanged!

Equipotential

We take from the previous example a potent lesson: any path for which conducting material provides a continuous link must be at the *same* electric potential.

We can make use of this to work out problems that might seem intractable at first.

For example, consider two conducting spheres with differing radii R_a and R_b sitting on insulating stands far apart. The sphere with radius R_a has an electric charge +Q.

If we connect a thin, conducting line between the spheres, then disconnect it, what are the charges on the two spheres?

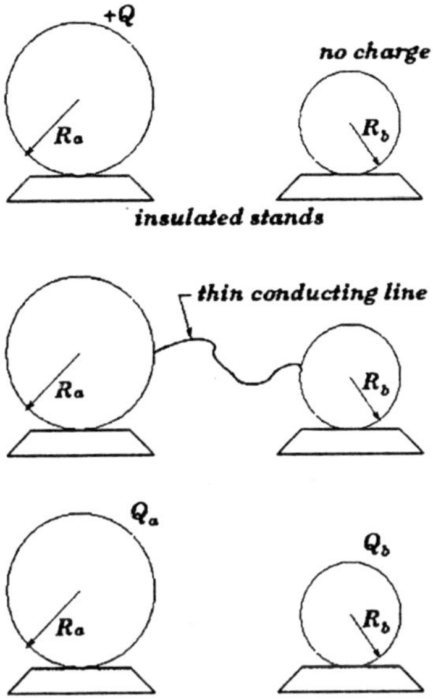

Fig. Two Shells on Insulated Stands are Briefly Connected by a Conducting Line, then Disconnected.

The conducting line brings the two spheres to the same electric potential. Since the sphere with radius R_b was initially uncharged, bringing the two spheres to the same electric potential requires some of the charge on the R_a sphere to go to the R_b sphere. This reduces the potential on R_a and increases it on R_b until the two come to the same electric potential relative to infinity. We can calculate what this charge is by noting that the final charges must satisfy charge conservation such that $Q_a + Q_b = Q \Rightarrow Q_a = Q - Q_b$.

The final condition of the potentials being equal is easily represented provided we have the spheres separated by a considerable distance. This ensures that the electric field of one sphere does not have much influence on the charge distribution of the other sphere. For the potentials to be equal, we must have

CAPACITANCE

Methods of storing energy for the purpose of doing work and directly studying electric fields and their effect on matter was an important focus of research in the mid-1700's. Methods of storing electrical charges for long periods of time were developed, the most important being the creation of the Leydan jar.

Methods of improving the Leydan jar by simplifying its construction and increasing its *capacity* for storing charge soon resulted in devices consisting of just two flat metal sheets separated by air. One of the sheets can be charged, then disconnected from all electrically conducting material.

The charged sheet induces an equal charge on the second sheet. In its simplest model, any two oppositely charged metal objects separated by a distance forms a "condenser" of charge or a capacitor, as we now say. In order to provide a "figure of merit" for how effective any type of capacitor is at holding charge, we define the capacitance through the relationship $C = Q/V$

The higher the capacitance, the more effective the capacitor is in the following sense: if we look at the relationship between C, Q, and V as defined, we note that the more charge we store per unit of V, the higher the value of C.

Since V relates to the amount of work needed to produce the charge Q on the plates of the capacitor, having a high capacitance means we store more charge for less work done in charging.

Capacitance also is a measure of how effective the capacitor is at storing energy, and hence how effective the capacitor can be at *doing* work after it is charged.

First, let's consider the parallel-plate capacitor.

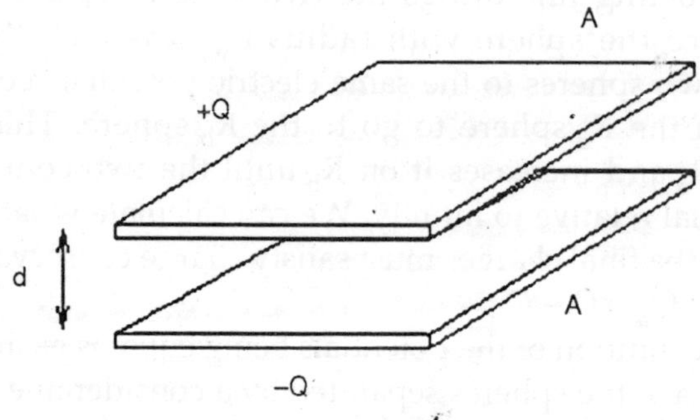

Fig. Two thin Plate Conductors with Fixed Distance d, Plate Area A, and Charges +Q on one Plate and –Q on the other form a Parallel-plate Capacitor.

The two conducting plates each have an area A and are separated by a distance d.

By using a source of EMF or ElectroMotive Force (a device which moves charges from one point to another), we can drive electrons onto the bottom plate while removing them from the top plate. After the plates are charged, we can, if we wish, remove the EMF source and the plates, if isolated from any other conducting material will hold their charges for a considerable period of time; years in some cases.

We can evaluate the capacitance for the parallel-plate case by noting that the electric field between the plates is, according to Gauss's Law as applied in previous lectures, constant in magnitude and direction.

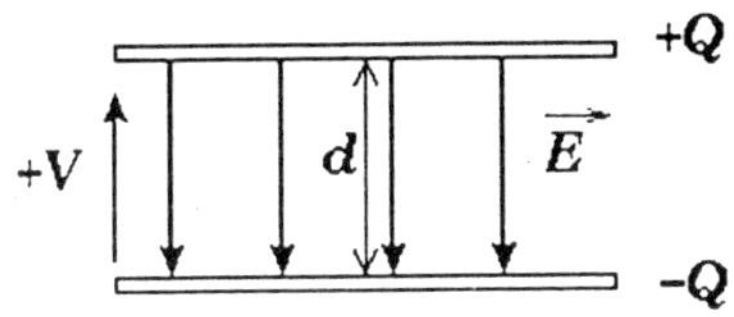

Fig. A parallel-plate capacitor with plate area A and charge Q.

Hence, we know that $E = Q/e_0A$. The voltage between the plates increases in going from the negative to the positive plate and has a magnitude V = Ed. Therefore,

$$C = Q/V = \frac{Q}{Qd/\varepsilon_0 A} = \frac{\varepsilon_0 A}{d}$$

Therefore, for the parallel-plate capacitor, the capacitance is proportional only to geometric factors (the area of the plates and the separation distance between them) and the natural constant, ε_0.

Batteries and Capacitor

We note that we are more familiar with the concept of storing electrical energy in *batteries*. Hence, it makes some sense to distinguish between batteries and capacitors.

The purpose of a battery is to use chemical (or other) processes to produce the EMF needed to move charges around. This is done by using the chemical or other process to maintain a constant potential between the *poles* (i.e. the positive and negative parts of the battery). This constant potential provides the EMF for moving charges from one conductor of a capacitor to another, for example.

Hence, this is the principal distinguishing feature of a battery and a capacitor: a battery maintains a *constant* potential difference between it's poles whereas a capacitor can maintain a potential difference between its positive and negatively charged conductors, but there is no need for that potential difference to be constant - the potential difference depends on the magnitude of the charge on the conductors!

Thus, while batteries store energy, capacitors store both energy *and* charge. The motion of charges can be quite rapid and readily controlled hence capacitors can be used for electrical energy storage that requires timing as part of its function. Batteries are generally not good for this purpose.

Energy Storage in Capacitors

We can now talk about the energy stored in the electric field of or, equivalently, stored in the arrangement of electric charges within the capacitor. Again, we use the parallel-plate capacitor because it is reasonably simple. We will use *schematic* representations of circuit elements to save drawing time. For a source of EMF, or voltage that can move charges, we have the symbol in figure .

Fig. Symbol for a Source of EMF within a Circuit.

To calculate the energy stored, we hook the EMF source to a capacitor, e.g. by closing a switch that allows conducting lines to link the EMF source and a capacitor together.

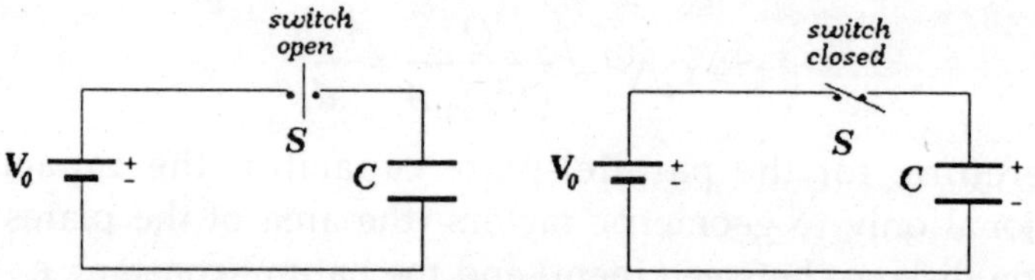

Fig. Closing a Switch to Charge a Capacitor.

As soon as the EMF source is able to push charge onto the capacitor plates, it does so by stealing electrons from the top plate of the capacitor and pushing them onto the bottom plate.

As soon as the first infinitesimal bit of charge hits the capacitor, an electric field is created between the plates. Although the field is also infinitesimally small, it still produces a voltage (again, infinitesimally small). This voltage operates *opposite* to the direction that charge is flowing, i.e. the capacitor voltage tends to push positive charge in the counter-clockwise direction while the EMF source pushes it in the clockwise direction.

Since the EMF source has a much higher voltage, it succeeds in pushing more charge onto the plates of the capacitor. As the charge continues to build though, the electric field strength, and hence the voltage across the plates of the capacitor, continues to build. When the capacitor voltage equals that of the EMF source, charge flow stops and the situation is static. At this point, the voltage across C is just V_0, the same as that across the EMF source.

To find the amount of work done by the battery to bring this situation about, let's translate the above description into calc-speak, i.e say it mathematically. The first bit of charge to arrives takes no work on the part of the EMF source assuming that C is initially uncharged. This bit of charge on C causes a potential V to appear across C. The next infinitesimal bit of charge, dq, coming through the EMF source must have work done on it to push it against the potential from C. This work is just dW = dq V. Now we can integrate to see what the total work is to get V up to V_0.

$$W = \frac{1}{2} C V_0^2$$

We can write this in a couple of other forms if we note that $C = Q/V_0$ for this case:

$$W = \frac{1}{2}CV_0^2 = \frac{1}{2}QV_0 = \frac{1}{2}Q^2/C$$

These results were derived for the parallel-plate case but are general for a capacitor with any geometry. Since the work done is also equal to the potential energy stored for a system subject only to conservative forces, we note that the work done on the charges to get them to the capacitor plates also represents the potential energy of the charged capacitor, i.e. its ability to do work.

Another task of importance will be understanding the connection between the potential energy in the arrangement of charges on the capacitor plates and its connection to the electric field. Again, we can use the parallel-plate to provide a solution which holds generally. Assume we were interested not in knowing the potential energy of C, but of the energy density or potential energy per unit volume between its plates. The potential energy divided by the volume provides this, so $u = ½\, \varepsilon_0 E^2$

This represents the potential energy per unit volume of free space (i.e. vacuum) when an electric field of magnitude E is created in that unit volume. It also represents the work done per unit volume to establish a field of magnitude E in that space. Hence we see the role of ε_0 as signifying the effort needed to establish an electric field in a volume of empty space. Hence the name for ε_0, the permittivity of free space.

DIELECTRICS

Capacitors by noting their importance as a means for storing charge (and hence electrical energy) that could be used to do work or make studies of the electric field. As such, there was a continual push to develop capacitors of larger charge-storing capability. There are practical limitations to storing charge in that the electric field between the parallel plates of a capacitor, and hence the electric potential, goes up as the charge goes up.

At a certain point, the air gap between the plates breaks down, i.e. the nitrogen in the air is ionized, becomes conducting, and therefore provides a path that enables the plates to short out. One can also decrease the distance between the plates. Since the potential across the plates goes as $Ed = Qd/\varepsilon_0 A$, reducing d decreases V for a given value of Q on the plates. This has practical limitations in keeping the distance between the plates uniform over the area.

To see why keeping the plates apart is a problem, let's calculate the force between the plates. First, look at figure showing a parallel-plate capacitor with a separation distance x between the plates. To pull the plates apart by an additional infinitesimal distance dx, we need to exert a force because the positive and negative charge plates attract one another.

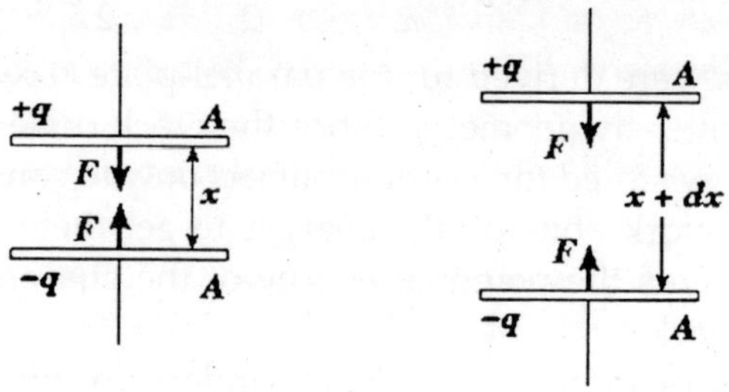

Fig. The Mutual Forces on Capacitor Plates Means that it takes Work to Separate Them.

To be formal about this, note that the potential energy of the original capacitor is

$$U = \frac{1}{2}qV = \frac{1}{2}q\frac{qx}{\varepsilon_0 A} = \frac{1}{2}\frac{q^2 x}{\varepsilon_0 A}$$

After the separation, the new potential energy is (note that the charge on the plates cannot change since there is no where for the charges to go)

$$U + dU = \frac{1}{2}qV = \frac{1}{2}q\frac{q(x+dx)}{\varepsilon_0 A} = \frac{1}{2}\frac{q^2(x+dx)}{\varepsilon_0 A}$$

The negative sign on the right-hand side indicates that the force is indeed attractive as we thought.

The fact that it is non-zero indicates that plates have to be held apart, especially if the charge on them is to be large. Note also that the force increases as the square of the potential difference between the plates and inversely as the distance between them, hence attempting to increase the capacitance by reducing the distance between the plates only increases the force needed to keep them apart.

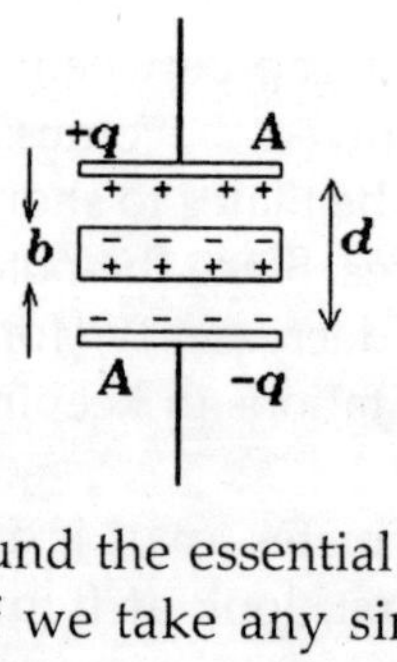

So, how do we get around the essential constraints on capacitance? First, note what happens if we take any single, charged parallel-plate capacitor and insert a slab of conducting material into the space between the plates, but do it such that the conducting material does not touch either plate. Conducting material placed inside a capacitor effectively reduces the distance between conductors and thereby increases capacitance for a given charge on the conducting plates.

The conducting material assumes the charge separation. The electric field above and below the conductor remains the same as before the conducting material was introduced, $E = q/\varepsilon_0 A$ because the electric field in the vicinity of the charged plates is unaffected by the presence of the conducting material. The electric field in the conducting material is zero, so the potential from the bottom to the top of the conducting material is the same. Therefore, the potential difference between the bottom and top plates are (note that the spaces between the conductor and the plates are both $(d - b)/2$)

$$V = \frac{q(d-b)}{2\varepsilon_0 A} + \frac{q(d-b)}{2\varepsilon_0 A} = \frac{d(d-b)}{\varepsilon_0 A}$$

If we find the capacitance for the arrangement with the conductor inside, we get

$$C = \frac{q}{V} = \frac{\varepsilon_0 A}{d-b}$$

If we compare this to the original capacitance without the conductor in the middle, $C = q/V = \varepsilon_0 A/d$, we see that the capacitance has gone up with the introduction of the conductor. Effectively, we have reduced the gap between conducting surfaces.

This still doesn't help, technologically speaking, with making better capacitances, because we still have to keep the conductor from touching the plates of the capacitor.

To get around this, we need something that reduces the electric field magnitude in the volume of space between the plates without allowing for conduction between the plates. The solution is to fill the space between the plates with a dielectric material. Dielectrics are non-conductors that are formulated from molecules which either have permanent electric dipole moments or can be induced to have dipole moments in the presence of an external electric field.

These dielectric molecules align partially with any external electric field. They generate a field of their own within the material. This field opposes that of the external field and therefore reduces the magnitude of E within the material. While not quite as effective as a conductor, this material does not conduct charge between the plates of a conductor and still reduces the voltage as a function of charge on the plates.

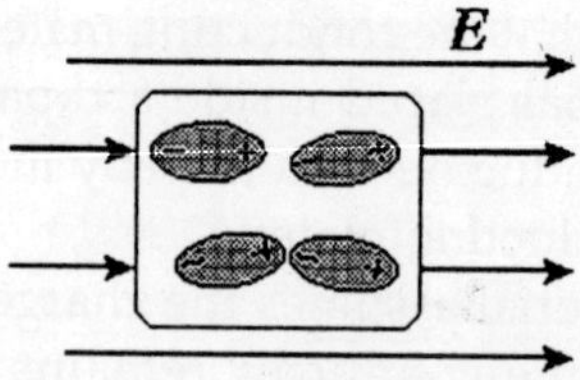

Fig. A Dielectric in an External Electric Field.

The reduction factor is referred to as the dielectric constant of the material and is usually represented by the letter K. The electric field produced by the charge on the plates of the capacitor remains $E_0 = q/\varepsilon_0 A$ in any region between the plates which is not occupied by dielectric. In the region which is occupied by dielectric material, the electric field is E} = E_0/K with $K > 1$, i.e. the electric field magnitude is reduced.

Since the potential difference between the plates goes as the electric field times the distance, the lower E} (and therefore the higher K is), the more charge can be stored on the plates for a given potential difference. One way to think about the proper use of K is to note that any formula derived for capacitance can be adopted to the case in which a dielectric completely fills the space between the plates of the capacitor by simply replacing every instance of the permittivity, e_0, with the value Ke_0. Hence, the voltage between the plates of a parallel-plate capacitor in which a material with dielectric constant K fills the volume between the plates with separation d is

$$V = \frac{qd}{K\varepsilon_0 A}$$

The capacitance is $C = K\varepsilon_0 A/d$, etc. If we apply the above consideration of calculating quantities by replacing e_0 with Ke_0 to the case in which an empty capacitor with charge q on its plates has a dielectric introduced into it, we find an anomaly.

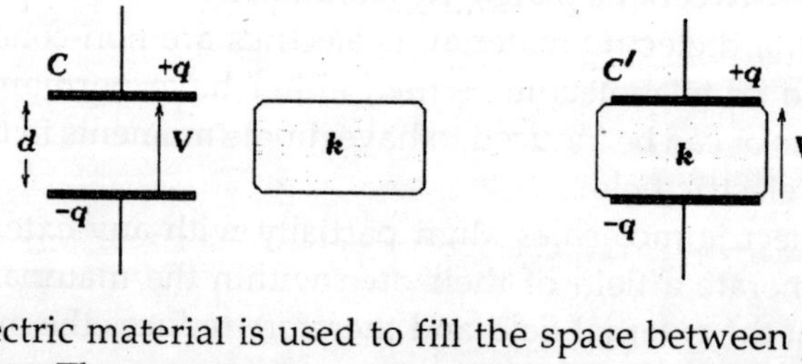

A dielectric material is used to fill the space between parallel plates of a capacitor. The capacitance and voltage are modified but the electric charge on the conducting plates themselves is unchanged. The net surface charge density is reduced due to the dielectric.

Before and after the dielectric is inserted, we have, if we let $\varepsilon = K\varepsilon_0$,

Before Insertion	*After Insertion*
$C = (\varepsilon_0 A)/d$ $V = qd/(\varepsilon_0 A)$ $U = {}^1/_2 CV^2 = (q^2 d)/(\varepsilon_0 A)$	C} $= (\varepsilon A)/d$ V} $= qd/(\varepsilon A)$ U} $= (q^2 d)/(\varepsilon A)$

Notice that the energy stored has decreased by a factor of K! Where did this energy go? To understand, we again have to turn to forces exerted by the capacitor plates. For any real capacitor we must have fields which bow out at the edges, the so-called fringe fields.

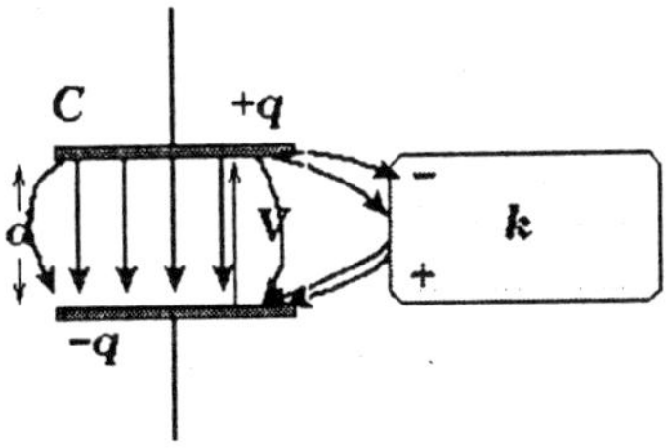

The fringe fields at the edge of the capacitor attract the dielectric material and do work to pull the dielectric into the space between the capacitor plates.

As the dielectric is introduced into the fringe field, the dielectric molecules start to align with the external field. This creates an attractive force as the plus charges of the dielectrics are aligned so that they are closest to the negative plate of the capacitor (and of course the same process occurs for the negative dielectric charges and positive plate of the capacitor). The horizontal components of the fringe field yield a horizontal, attractive force which pulls the dielectric into the space between the capacitor plates.

The work done by this force in accelerating the dielectric into the capacitor volume decreases the energy stored in the capacitor, thereby lowering the energy stored in the capacitor. Some energy is also converted into electromagnetic radiation and into heat.

COMBINATIONS OF CAPACITORS

Parallel Combinations

Having determined the basics for determing capacitance for various capacitor geometries, it is now time to consider cases in which more than one capacitor may be involved in a circuit.

We'll restrict the derivations to examples using parallel-plate capacitors for mathematical ease although the results will be quite general.Let's start by looking at the parallel combination. In this case, we have capacitors which have the same potential difference across their plates.

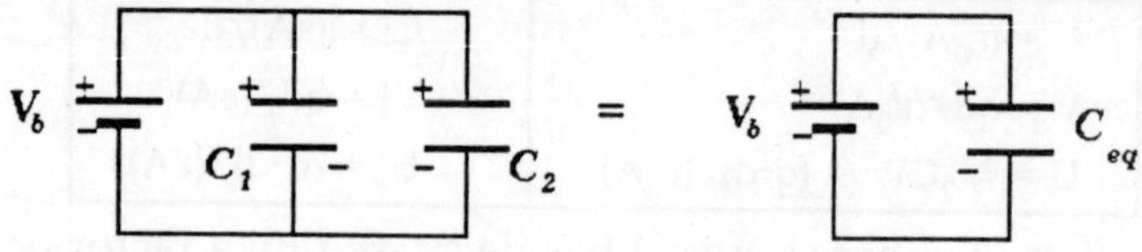

Fig. Two Capacitors in Parallel and a Circuit with an Equivalent Capacitance.

For such combinations, we can simplify matters a bit by finding an *equivalent* capacitor whose characteristics make exactly the same demands from the EMF source in charging up the capacitors (presuming they had the same initial charge before the EMF source was connected).

For this to be the case, the EMF source will have to push the same amount of charge onto the equivalent capacitor as it does for the two original capacitors it replaces.

The work done by the emf source should be the same as well. This means that the potential difference created on the equivalent capacitor should be the same as for the two capacitors it replaces. We can now ask what the capacitance of the C_{eq} must be for these conditions to hold. Since the potentials across C_1, C_2, and C_{eq} must all be the same and the charge on the equivalent capacitor must be the same as on the two individual capacitors in the original circuit, we have $Q_{eq} = Q_1 + Q_2$, $C_{eq}V_b = C_1V_1 + C_2V_2 = C_1V_b + C_2V_b$, $C_{eq} = C_1 + C_2$

So the equivalent capacitor needs to have a capacitance which is the sum of the two capacitors it replaces. This result can easily be extended to any number of capacitors so that the result of n capacitors connected in parallel is $C_{eq} = C_1 + C_2 + C_3 + ... + C_n$

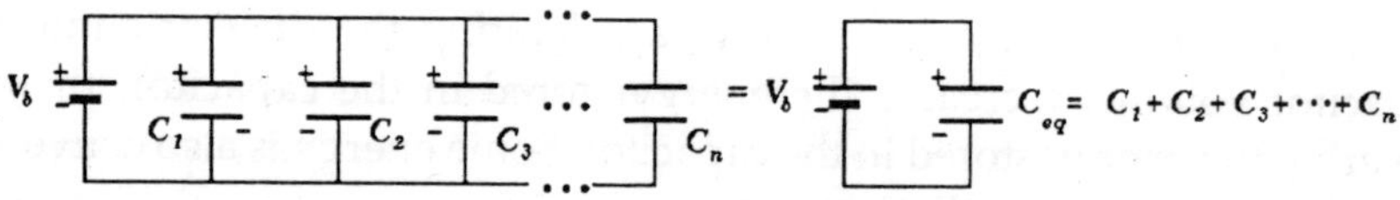

Fig. Extending the Parallel Capacitor Formula to any Number of Capacitors.

Series Combinations

We can also consider capacitors arranged so that the potential across the combination is equal to the sum of the potential difference across each.

As with the parallel combination, we want to find the equivalent capacitance that can replace the pair of capacitors here without changing anything that the EMF source must do to produce full charge on the combination. In this case, it is the charges that must be equal magnitude on the combination since we note that the positive pole of the EMF source contacts one plate of one capacitor in the pair while the negative pole contacts one plate of the other capacitor.

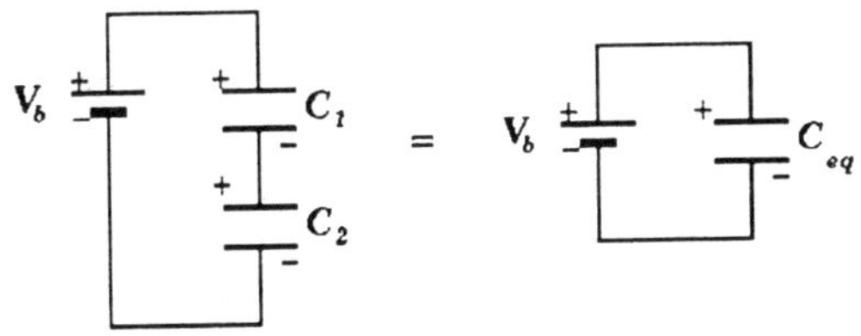

Fig. A Series Arrangement of Capacitors and a Circuit with Equivalent Capacitance.

In each case, the plate connected to the battery pole *induces* the equal magnitude charge on the other plate for that capacitor. Since the inner plates are electrically isolated from the world, they maintain a net charge of zero (assuming all capacitors were initially uncharged before the EMF source was connected). Thus, $q_1 = q_2 = q_{eq}$ for the combination. The potential in going from the bottom plate of C_2 to the top plate of C_1 is $V_2 + V_1$, hence $V_b = V_1 + V_2$, $q_{eq}/C_{eq} = q_1/C_1 + q_2/C_2$, $1/C_{eq} = 1/C_1 + 1/C_2$,

This can also be extended to any number of capacitors in parallel. For example, for n capacitors in series,

$$\frac{1}{C_{eq}} = \frac{1}{C_1} + \frac{1}{C_2} + \frac{1}{C_3} + \ldots + \frac{1}{C_n}$$

CURRENT AND RESISTANCE

Much of our determination of the properties of systems involving charges has been with the implicit assumption that those charges were *static*. Just as in mechanics, we can learn a lot by understanding the static conditions of objects. Note that static here did not imply that objects didn't move. The equation for motion apply to any object that has constant acceleration, i.e. the only thing that had to be static about the system was its acceleration.

We could determine a number of things about the world just from this. To truly understand motion, however, we needed Newton's Laws as a way of describing dynamics.

This opened up a whole new universe of understanding which went well beyond Galileo's equations for motion. Phenomena that were complete mysteries before, e.g. the orbits of planets, suddenly had precise descriptions in terms of Newton's Equation of Motion.We have the same situation with electricity. To get to the point of understanding this though, we have to start from the beginning. We begin by formally defining current as the time rate of change of charge through a conductor. Thus, if we have a conductor material of any cross-section, we can define the current in that material as the amount of charge that passes through a hypothetical plane that cuts all the way through the material.

Each of the hypothetical planes above sees the same current, i = dq/dt, go by it. We define the unit of current as the Ampere and abbreviate it as A. 1 A = 1 coulomb/sec.

Since charge is conserved, note that current, the time rate of change of charge, must also be a conserved quantity. If were not, then charges could accumulate or disappear at some point in the material.

This is assumed not to happen for conducting material nor for electrical devices. Currents flow as the result of potential differences between positions.

RESISTANCE

Currents are not dependent solely on potential differences. An object capable of carrying current can be characterized by its resistance where resistance is defined as

$$R = \frac{V}{i}$$

The resistance of an object depends on its geometry and its resistivity. Resistivity is a property of the material itself and is the same no matter what the shape or dimension of the material. The symbol for resistivity is the Greek letter rho. For an object made from material with resistivity rho,

$$R = \frac{\rho L}{A}$$

where L is the length of the object and A is its cross-sectional area which is assumed to be constant throughout the object. If A it is not constant, then we need to do an integration over infinitesimally small distances of the material where we can assume that the cross-sectional area is a constant.

For this course, unless explicitly stated otherwise, you can always assume that there is a linear connection between potential difference, current, and resistance in which V = i R

Although this appears to be identical to the definition of R, keep in mind that this relationship assumes that potential difference and current are linearly related over some reasonably large set of possible V's and i'. Semiconductor materials, as an example, display remarkably (and most usefully) non-linear relationships between potential difference and current.

The above equation of $V = i*R$ is called Ohm's Law because it holds reasonably well for a large number of objects used as electrical devices.

ELECTRIC POWER DEFINED

One final definition is needed. When we talk about rates of energy change in an electrical circuit, we deal with two notions of the term, *power*. The power provided by sources of electromotive force, emf's, is given by the formula derived as follows. EMF's provide potential differences that move charges. The energy provided by the emf can be calculated by looking at infinitesimal charges moved through the potential difference, $dU = dq\ V = i\ V\ dt$ $dU/dt = i\ V =$ Power.

Thus the power provided by an emf is the current times the potential difference it maintains between its positive and negative poles. In any material or device that has electrical resistance, the rate of change of energy for current flowing through it is lost or rather transformed into thermal energy.

To determine the relationship for this, note that Ohm's Law tells us that the potential difference across a material or device with resistance R is $V_R = iR$. Therefore, the power necessary to maintain current i going through the material or device is $P = i\ V_R = i^2R = V_R^2/R$.As stated before, this power is the rate at which electrical energy is turned into thermal energy in the resistance.

Chapter 2

Ohm's and Kirchhoff's Laws

An electric circuit is formed when a conductive path is created to allow free electrons to continuously move. This continuous movement of free electrons through the conductors of a circuit is called a *current,* and it is often referred to in terms of "flow," just like the flow of a liquid through a hollow pipe. The force motivating electrons to "flow" in a circuit is called *voltage.* Voltage is a specific measure of potential energy that is always relative between two points.

When we speak of a certain amount of voltage being present in a circuit, we are referring to the measurement of how much *potential* energy exists to move electrons from one particular point in that circuit to another particular point. Without reference to *two* particular points, the term "voltage" has no meaning. Free electrons tend to move through conductors with some degree of friction, or opposition to motion. This opposition to motion is more properly called *resistance.*

The amount of current in a circuit depends on the amount of voltage available to motivate the electrons, and also the amount of resistance in the circuit to oppose electron flow. Just like voltage, resistance is a quantity relative between two points. For this reason, the quantities of voltage and resistance are often stated as being "between" or "across" two points in a circuit. To be able to make meaningful statements about these quantities in circuits, we need to be able to describe their quantities in the same way that we might quantify mass, temperature, volume, length, or any other kind of physical quantity. For mass we might use the units of "pound" or "gram." For temperature we might use degrees Fahrenheit or degrees Celsius. Here are the standard units of measurement for electrical current, voltage, and resistance:

Quanity Abbreviation	*Symbol*	*Unit of Measurement*	*Unit*
Current	I	Ampere("Amp")	A
Voltage	E or V	Volt	V
Resistance	R	Ohm	Ω

The "symbol" given for each quantity is the standard alphabetical letter used to represent that quantity in an algebraic equation. Standardized letters like these are common in the disciplines of physics and engineering, and are internationally recognized. The "unit abbreviation" for each quantity represents the alphabetical symbol used as a shorthand notation for its particular unit of measurement. And, yes, that strange-looking "horseshoe" symbol is the capital Greek letter Ω, just a character in a *foreign* alphabet.

The mathematical symbol for each quantity is meaningful as well. The "R" for resistance and the "V" for voltage are both self-explanatory, whereas "I" for current seems a bit weird.

The "I" is thought to have been meant to represent "Intensity" (of electron flow), and the other symbol for voltage, "E," stands for "Electromotive force."

There seems to be some dispute over the meaning of "I." The symbols "E" and "V" are interchangeable for the most part, although some texts reserve "E" to represent voltage across a source (such as a battery or generator) and "V" to represent voltage across anything else. All of these symbols are expressed using capital letters, except in cases where a quantity (especially voltage or current) is described in terms of a brief period of time (called an "instantaneous" value).

For example, the voltage of a battery, which is stable over a long period of time, will be symbolized with a capital letter "E," while the voltage peak of a lightning strike at the very instant it hits a power line would most likely be symbolized with a lower-case letter "e" (or lower-case "v") to designate that value as being at a single moment in time.

This same lower-case convention holds true for current as well, the lower-case letter "i" representing current at some instant in time. Most direct-current (DC) measurements, however, being stable over time, will be symbolized with capital letters.

One foundational unit of electrical measurement, often taught in the beginnings of electronics courses but used infrequently afterwards, is the unit of the *coulomb,* which is a measure of electric charge proportional to the number of electrons in an imbalanced state. One coulomb of charge is equal to 6,250,000,000,000,000,000 electrons. The symbol for electric charge quantity is the capital letter "Q," with the unit of coulombs abbreviated by the capital letter "C."

It so happens that the unit for electron flow, the amp, is equal to 1 coulomb of electrons passing by a given point in a circuit in 1 second of time. Cast in these terms, current is the *rate of electric charge motion* through a conductor.

Voltage is the measure of *potential energy per unit charge* available to *charge* available to motivate electrons from one point to another. Before we can precisely define what a "volt" is, we must understand how to measure this quantity we call "potential energy." The general metric unit for energy of any kind is the *joule*, equal to the amount of work performed by a force of 1 newton exerted through a motion of 1 meter (in the same direction). In British units, this is slightly less than 3/4 pound of force exerted over a distance of 1 foot.

Put in common terms, it takes about 1 joule of energy to lift a 3/4 pound weight 1 foot off the ground, or to drag something a distance of 1 foot using a parallel pulling force of 3/4 pound. Defined in these scientific terms, 1 volt is equal to 1 joule of electric potential energy per (divided by) 1 coulomb of charge. Thus, a 9 volt battery releases 9 joules of energy for every coulomb of electrons moved through a circuit.

These units and symbols for electrical quantities will become very important to know as we begin to explore the relationships between them in circuits. The first, and perhaps most important, relationship between current, voltage, and resistance is called Ohm's Law, discovered by Georg Simon Ohm and published in his 1827 paper, *The Galvanic Circuit Investigated Mathematically*.

Ohm's principal discovery was that the amount of electric current through a metal conductor in a circuit is directly proportional to the voltage impressed across it, for any given temperature. Ohm expressed his discovery in the form of a simple equation, describing how voltage, current, and resistance interrelate:

$$E = IR$$

In this algebraic expression, voltage (V) is equal to current (I) multiplied by resistance (R). Using algebra techniques, we can manipulate this equation into two variations, solving for I and for R, respectively:

$$1 = \frac{E}{R}$$

$$R = \frac{E}{I}$$

Let's see how these equations might work to help us analyse simple circuits:

In the above circuit, there is only one source of voltage (the battery, on the left) and only one source of resistance to current (the lamp, on the right). This makes it very easy to apply Ohm's Law. If we know the values of any two of the three quantities (voltage, current, and resistance) in this circuit, we can use Ohm's Law to determine the third.

In this first example, we will calculate the amount of current (I) in a circuit, given values of voltage (E) and resistance (R):

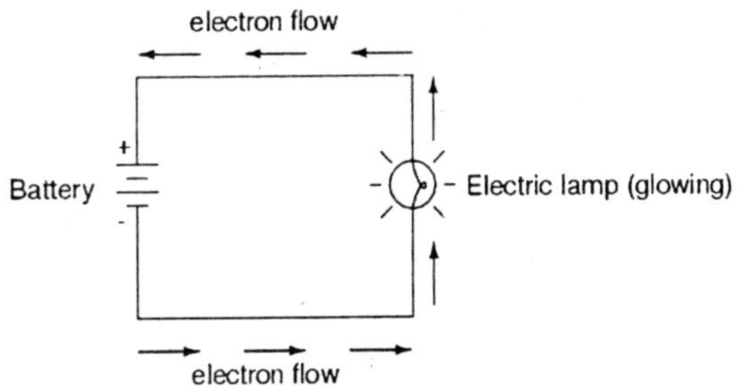

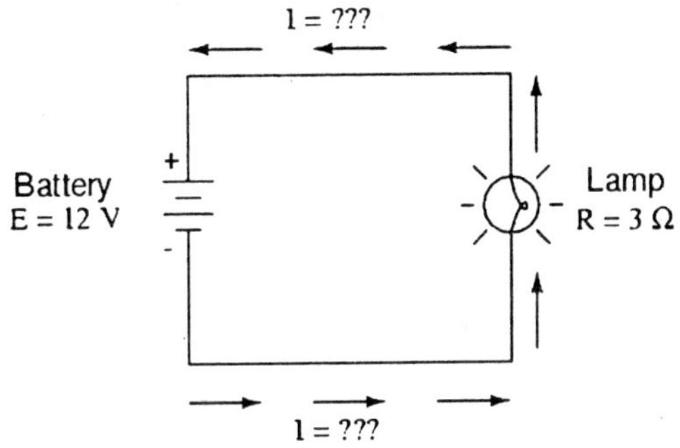

What is the amount of current (I) in this circuit?

$$I = \frac{E}{R} = \frac{12V}{3\Omega} = 4A$$

In this second example, we will calculate the amount of resistance (R) in a circuit, given values of voltage (E) and current (I):

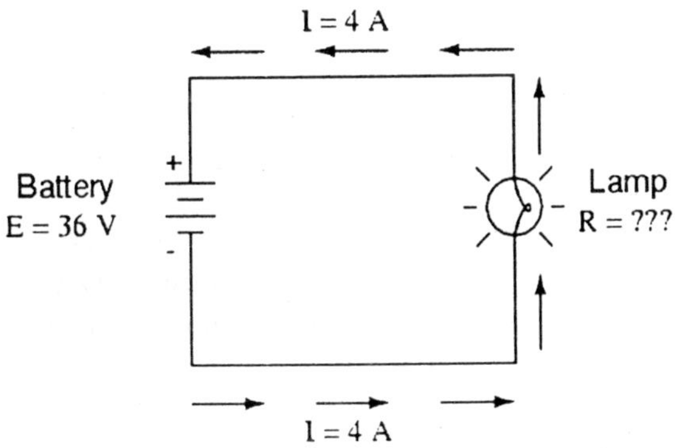

What is the amount of resistance (R) offered by the lamp?

$$R = \frac{E}{I} = \frac{36V}{4A} = 9\Omega$$

In the last example, we will calculate the amount of voltage supplied by a battery, given values of current (I) and resistance (R):

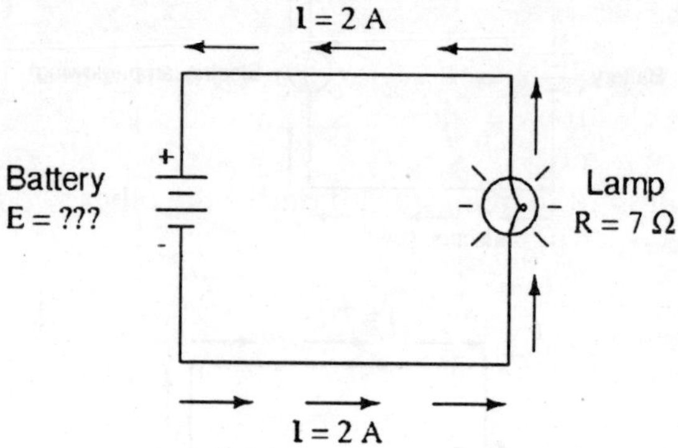

What is the amount of voltage provided by the battery?

$$E = 1R = (2A)(7\ \Omega) = 14V$$

Ohm's Law is a very simple and useful tool for analyzing electric circuits. It is used so often in the study of electricity and electronics that it needs to be committed to memory by the serious student. For those who are not yet comfortable with algebra, there's a trick to remembering how to solve for any one quantity, given the other two. First, arrange the letters E, I, and R in a triangle like this:

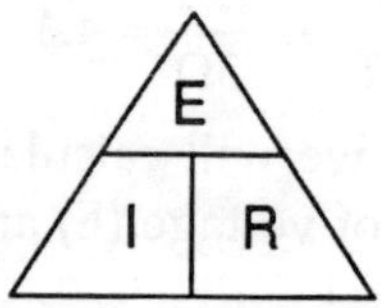

If you know E and I, and wish to determine R, just eliminate R from the picture and see what's left:

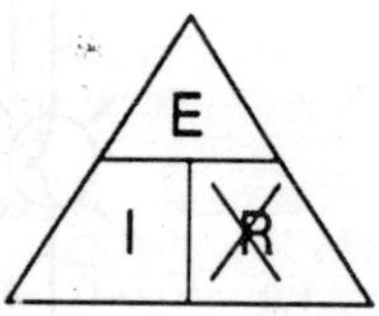

$$R = \frac{E}{I}$$

If you know E and R, and wish to determine I, eliminate I and see what's left:

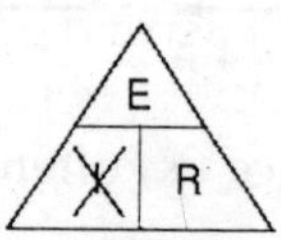

$$I = \frac{E}{R}$$

Lastly, if you know I and R, and wish to determine E, eliminate E and see what's left:

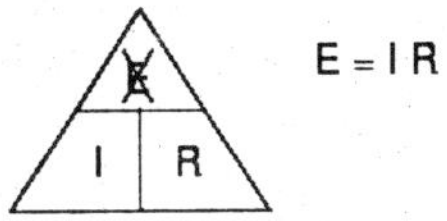

Eventually, you'll have to be familiar with algebra to seriously study electricity and electronics, but this tip can make your first calculations a little easier to remember. If you are comfortable with algebra, all you need to do is commit E=IR to memory and derive the other two formulae from that when you need them!

Ohm's Law also makes intuitive sense if you apply it to the water-and-pipe analogy. If we have a water pump that exerts pressure (voltage) to push water around a "circuit" (current) through a restriction (resistance), we can model how the three variables interrelate. If the resistance to water flow stays the same and the pump pressure increases, the flow rate must also increase.

If the pressure stays the same and the resistance increases (making it more difficult for the water to flow), then the flow rate must decrease: If the flow rate were to stay the same while the resistance to flow decreased, the required pressure from the pump would necessarily decrease. As odd as it may seem, the actual mathematical relationship between pressure, flow, and resistance is actually more complex for fluids like water than it is for electrons.

POWER IN ELECTRIC CIRCUITS

In addition to voltage and current, there is another measure of free electron activity in a circuit: *power*. First, we need to understand just what power is before we analyse it in any circuits.

Power is a measure of how much work can be performed in a given amount of time. *Work* is generally defined in terms of the lifting of a weight against the pull of gravity. The heavier the weight and/or the higher it is lifted, the more work has been done. *Power* is a measure of how rapidly a standard amount of work is done.

For American automobiles, engine power is rated in a unit called "horsepower," invented initially as a way for steam engine manufacturers to quantify the working ability of their machines in terms of the most common power source of their day: horses. One horsepower is defined in British units as 550 ft-lbs of work per second of time. The power of a car's engine won't indicate how tall of a hill it can climb or how much weight it can tow, but it will indicate how *fast* it can climb a specific hill or tow a specific weight.

The power of a mechanical engine is a function of both the engine's speed and it's torque provided at the output shaft. Speed of an engine's output shaft is measured in revolutions per minute, or RPM. Torque is the amount of twisting force produced by the engine, and it is usually measured in pound-feet, or lb-ft (not to be confused with foot-pounds or ft-lbs, which is the unit for work). Neither speed nor torque alone is a measure of an engine's power.

A 100 horsepower diesel tractor engine will turn relatively slowly, but provide great amounts of torque. A 100 horsepower motorcycle engine will turn very fast, but provide relatively little torque. Both will produce 100 horsepower, but at different speeds and different torques. The equation for shaft horsepower is simple:

$$\text{Horsepower} = \frac{2\pi ST}{33{,}000}$$

where,

S = shaft speed in r.p.m.
T = shaft torque in lb-ft.

Notice how there are only two variable terms on the right-hand side of the equation, S and T. All the other terms on that side are constant: 2, pi, and 33,000 are all constants (they do not change in value). The horsepower varies only with changes in speed and torque, nothing else. We can re-write the equation to show this relationship:

Horsepower α ST

α This symbol means "proportional to"

Because the unit of the "horsepower" doesn't coincide exactly with speed in revolutions per minute multiplied by torque in pound-feet, we can't say that horsepower *equals* ST. However, they are *proportional* to one another. As the mathematical product of ST changes, the value for horsepower will change by the same proportion.

In electric circuits, power is a function of both voltage and current. Not surprisingly, this relationship bears striking resemblance to the "proportional" horsepower 0formula above:

$$P = IE$$

In this case, however, power (P) is exactly equal to current (I) multiplied by voltage (E), rather than merely being proportional to IE. When using this formula, the unit of measurement for power is the *watt*, abbreviated with the letter "W."

It must be understood that neither voltage nor current by themselves constitute power. Rather, power is the combination of both voltage *and* current in a circuit. Remember that voltage is the specific work (or potential energy) per unit charge, while current is the rate at which electric charges move through a conductor. Voltage (specific work) is analogous to the work done in lifting a weight against the pull of gravity. Current (rate) is analogous to the speed at which that weight is lifted. Together as a product (multiplication), voltage (work) and current (rate) constitute power.

Just as in the case of the diesel tractor engine and the motorcycle engine, a circuit with high voltage and low current may be dissipating the same amount of power as a circuit with low voltage and high current. Neither the amount of voltage alone nor the amount of current alone indicates the amount of power in an electric circuit.

In an open circuit, where voltage is present between the terminals of the source and there is zero current, there is *zero* power dissipated, no matter how great that voltage may be. Since P=IE and I=0 and anything multiplied by zero is zero, the power dissipated in any open circuit must be zero.

Likewise, if we were to have a short circuit constructed of a loop of superconducting wire (absolutely zero resistance), we could have a condition of current in the loop with zero voltage, and likewise no power would be dissipated. Since P=IE and E=0 and anything multiplied by zero is zero, the power dissipated in a superconducting loop must be zero.

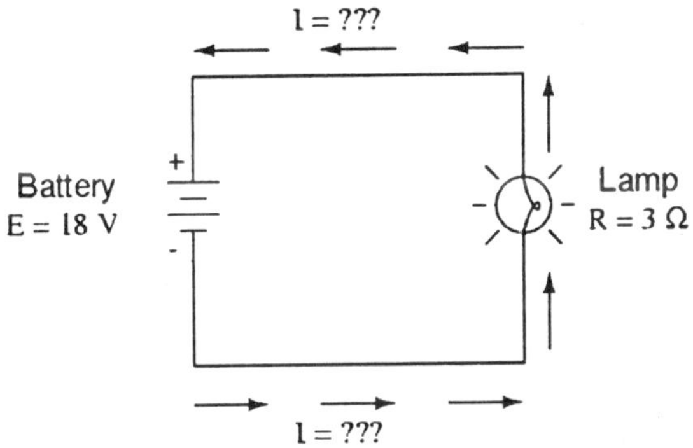

Whether we measure power in the unit of "horsepower" or the unit of "watt," we're still talking about the same thing: how much work can be done in a given amount of time. The two units are not numerically equal, but they express the same kind of thing. In fact, European automobile manufacturers typically advertise their engine power in terms of kilowatts (kW), or thousands of watts, instead of horsepower! These two units of power are related to each other by a simple conversion formula:

$$1 \text{ Horsepower} = 745.7 \text{ Watts}$$

So, our 100 horsepower diesel and motorcycle engines could also be rated as "74570 watt" engines, or more properly, as "74.57 kilowatt" engines. In European engineering specifications, this rating would be the norm rather than the exception. We've seen the formula for determining the power in an electric circuit: by multiplying the voltage in "volts" by the current in "amps" we arrive at an answer in "watts." Let's apply this to a circuit example: In the above circuit, we know we have a battery voltage of 18 volts and a lamp resistance of 3 Ω. Using Ohm's Law to determine current, we get:

$$I = \frac{E}{R} = \frac{18V}{3\Omega} = 6A$$

Now that we know the current, we can take that value and multiply it by the voltage to determine power:

$$P = IE = (6A)\ (18V) = 108V$$

Let's try taking that same circuit and increasing the battery voltage to see what happens. Intuition should tell us that the circuit current will increase as the voltage increases and the lamp resistance stays the same. Likewise, the power will increase as well:

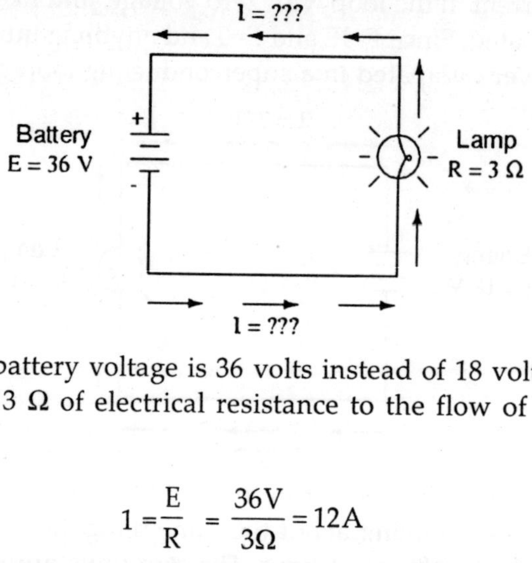

Now, the battery voltage is 36 volts instead of 18 volts. The lamp is still providing 3 Ω of electrical resistance to the flow of electrons. The current is now:

$$I = \frac{E}{R} = \frac{36V}{3\Omega} = 12A$$

This stands to reason: if I = E/R, and we double E while R stays the same, the current should double. Indeed, it has: we now have 12 amps of current instead of 6. Now, what about power?

$$P = IE = (12A)\ (36V) = 432\ W$$

Notice that the power has increased just as we might have suspected, but it increased quite a bit more than the current. Why is this? Because power is a function of voltage multiplied by current, and *both* voltage and current doubled from their previous values, the power will increase by a factor of 2 × 2, or 4.

You can check this by dividing 432 watts by 108 watts and seeing that the ratio between them is indeed 4.

If we only know voltage (E) and resistance (R):

If,

$$1 = \frac{E}{R} \text{ and } P = IE$$

Then,

$$P = \frac{E}{R}E \text{ or } P = \frac{E^2}{R}$$

If we only know current (I) and resistance (R):

$$\text{If, } E = IR \text{ and } P = IE$$

Then,

$$P = I(IR) \text{ or } P = I^2R$$

An historical note: it was James Prescott Joule, not Georg Simon Ohm, who first discovered the mathematical relationship between power dissipation and current through a resistance. This discovery, published in 1841, followed the form of the last equation ($P = I^2R$), and is properly known as Joule's Law. However, these power equations are so commonly associated with the Ohm's Law equations relating voltage, current, and resistance (E=IR; I=E/R; and R=E/I) that they are frequently credited to Ohm.

Power equations

$$P = IE \quad P = \frac{E^2}{R} \quad P = 1^2R$$

RESISTORS

Because the relationship between voltage, current, and resistance in any circuit is so regular, we can reliably control any variable in a circuit simply by controlling the other two. Perhaps the easiest variable in any circuit to control is its resistance. This can be done by changing the material, size, and shape of its conductive components.

Special components called *resistors* are made for the express purpose of creating a precise quantity of resistance for insertion into a circuit. They are typically constructed of metal wire or carbon, and engineered to maintain a stable resistance value over a wide range of environmental conditions.

Unlike lamps, they do not produce light, but they do produce heat as electric power is dissipated by them in a working circuit. Typically, though, the purpose of a resistor is not to produce usable heat, but simply to provide a precise quantity of electrical resistance.

The most common schematic symbol for a resistor is a zig-zag line:

Resistor values in ohms are usually shown as an adjacent number, and if several resistors are present in a circuit, they will be labeled with a unique identifier number such as R_1, R_2, R_3, etc. Resistor symbols can be shown either horizontally or vertically:

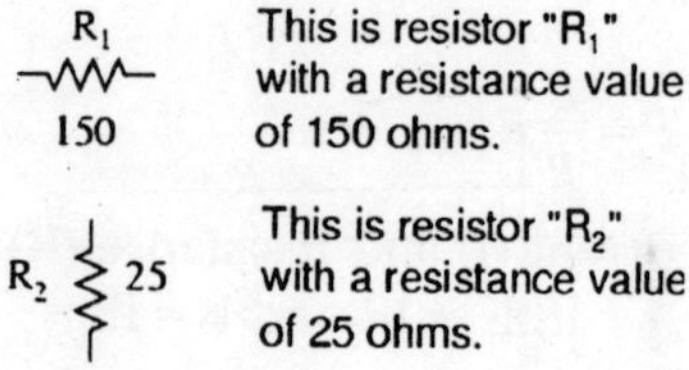

Real resistors look nothing like the zig-zag symbol. Instead, they look like small tubes or cylinders with two wires protruding for connection to a circuit. Here is a sampling of different kinds and sizes of resistors:

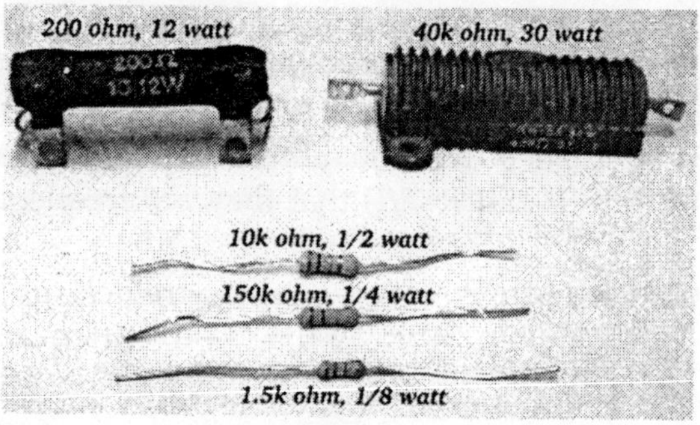

In keeping more with their physical appearance, an alternative schematic symbol for a resistor looks like a small, rectangular box:

Resistors can also be shown to have varying rather than fixed resistances. This might be for the purpose of describing an actual physical device designed for the purpose of providing an adjustable resistance, or it could be to show some component that just happens to have an unstable resistance:

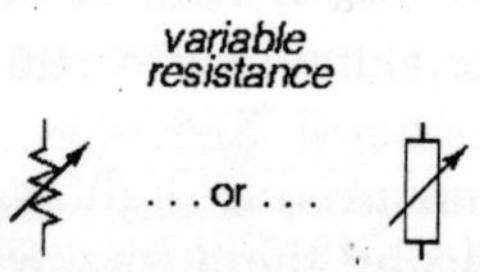

In fact, any time you see a component symbol drawn with a diagonal arrow through it, that component has a variable rather than a fixed value. This symbol "modifier" (the diagonal arrow) is standard electronic symbol convention.

Variable resistors must have some physical means of adjustment, either a rotating shaft or lever that can be moved to vary the amount of electrical resistance. Here is a photograph showing some devices called *potentiometers,* which can be used as variable resistors:

Because resistors dissipate heat energy as the electric currents through them overcome the "friction" of their resistance, resistors are also rated in terms of how much heat energy they can dissipate without overheating and sustaining damage.

Naturally, this power rating is specified in the physical unit of "watts." Most resistors found in small electronic devices such as portable radios are rated at 1/4 (0.25) watt or less. The power rating of any resistor is roughly proportional to its physical size. Note in the first resistor photograph how the power ratings relate with size: the bigger the resistor, the higher its power dissipation rating.

Although it may seem pointless now to have a device doing nothing but resisting electric current, resistors are extremely useful devices in circuits.

Because they are simple and so commonly used throughout the world of electricity and electronics, we'll spend a considerable amount of time analyzing circuits composed of nothing but resistors and batteries.

For a practical illustration of resistors' usefulness, It is a picture of a *printed circuit board,* or *PCB*: an assembly made of sandwiched layers of insulating phenolic fiber-board and conductive copper strips, into which components may be inserted and secured by a low-temperature welding process called "soldering." The various components on this circuit board are identified by printed labels. Resistors are denoted by any label beginning with the letter "R".

This particular circuit board is a computer accessory called a "modem," which allows digital information transfer over telephone lines. There are at least a dozen resistors (all rated at 1/4 watt power dissipation) that can be seen on this modem's board.

Every one of the black rectangles (called "integrated circuits" or "chips") contain their own array of resistors for their internal functions, as well.

Another circuit board example shows resistors packaged in even smaller units, called "surface mount devices."

This particular circuit board is the underside of a personal computer hard disk drive, and once again the resistors soldered onto it are designated with labels beginning with the letter "R": There are over one hundred surface-mount resistors on this circuit board, and this count of course does not include the number of resistors internal to the black "chips." These two photographs should convince anyone that resistors — devices that "merely" oppose the flow of electrons — are very important components in the realm of electronics!

In schematic diagrams, resistor symbols are sometimes used to illustrate any general type of device in a circuit doing something useful with electrical energy. Any non-specific electrical device is generally called a *load*, so if you see a schematic diagram showing a resistor symbol labeled "load," especially in a tutorial circuit diagram explaining some concept unrelated to the actual use of electrical power, that symbol may just be a kind of shorthand representation of something else more practical than a resistor.

To summarize what we've learned in this lesson, let's analyse the following circuit, determining all that we can from the information given:

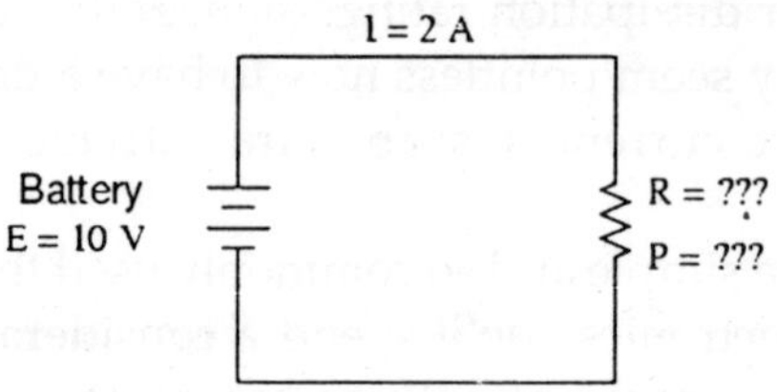

All we've been given here to start with is the battery voltage (10 volts) and the circuit current (2 amps).

We don't know the resistor's resistance in ohms or the power dissipated by it in watts. Surveying our array of Ohm's Law equations, we find two equations that give us answers from known quantities of voltage and current:

$$R = \frac{E}{I} \quad \text{and } P = IE$$

Inserting the known quantities of voltage (E) and current (I) into these two equations, we can determine circuit resistance (R) and power dissipation (P):

$$R = \frac{10V}{2A} = 5\Omega$$

$$P = (2A)\ (10V) = 20W$$

For the circuit conditions of 10 volts and 2 amps, the resistor's resistance must be 5 Ω. If we were designing a circuit to operate at these values, we would have to specify a resistor with a minimum power rating of 20 watts, or else it would overheat and fail.

NONLINEAR CONDUCTION

Ohm's Law is a simple and powerful mathematical tool for helping us analyse electric circuits, but it has limitations, and we must understand these limitations in order to properly apply it to real circuits. For most conductors, resistance is a rather stable property, largely unaffected by voltage or current.

For this reason we can regard the resistance of many circuit components as a constant, with voltage and current being directly related to each other.

For instance, our previous circuit example with the 3 Ω lamp, we calculated current through the circuit by dividing voltage by resistance (I=E/R). With an 18 volt battery, our circuit current was 6 amps. Doubling the battery voltage to 36 volts resulted in a doubled current of 12 amps. All of this makes sense, of course, so long as the lamp continues to provide exactly the same amount of friction (resistance) to the flow of electrons through it: 3 Ω.

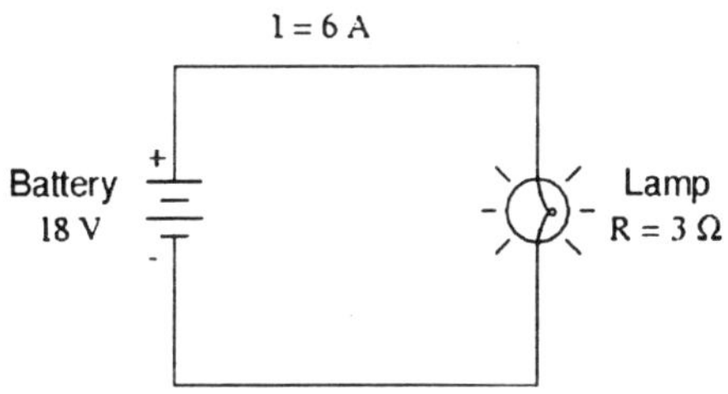

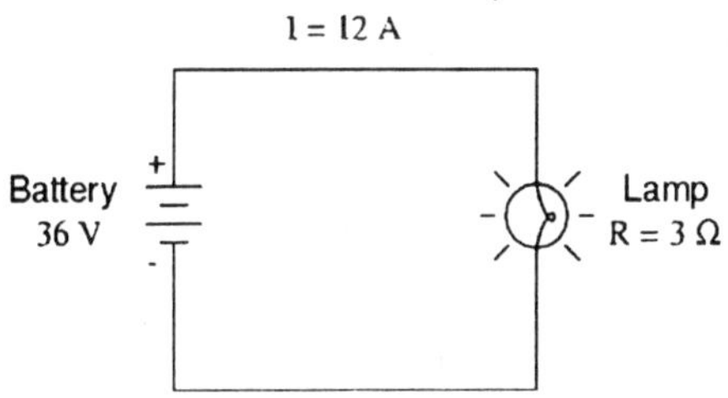

However, reality is not always this simple.In an incandescent l amp (the kind employing the principle of electric current heating a thin filament of wire to the point that it glows white-hot), the resistance of the filament wire will increase dramatically as it warms from room temperature to operating temperature.

If we were to increase the supply voltage in a real lamp circuit, the resulting increase in current would cause the filament to increase temperature, which would in turn increase its resistance, thus preventing further increases in current without further increases in battery voltage. Consequently, voltage and current do not follow the simple equation "I=E/R" (with R assumed to be equal to 3 Ω) because an incandescent lamp's filament resistance does not remain stable for different currents.

The phenomenon of resistance changing with variations in temperature is one shared by almost all metals, of which most wires are made. For most applications, these changes in resistance are small enough to be ignored. In the application of metal lamp filaments, the change happens to be quite large.

This is just one example of "nonlinearity" in electric circuits. It is by no means the only example. A "linear" function in mathematics is one that tracks a straight line when plotted on a graph. The simplified version of the lamp circuit with a constant filament resistance of 3 Ω generates a plot like this:

The straight-line plot of current over voltage indicates that resistance is a stable, unchanging value for a wide range of circuit voltages and currents. In an "ideal" situation, this is the case. Resistors, which are manufactured to provide a definite, stable value of resistance, behave very much like the plot of values seen above. A mathematician would call their behaviour "linear."

A more realistic analysis of a lamp circuit, however, over several different values of battery voltage would generate a plot of this shape:

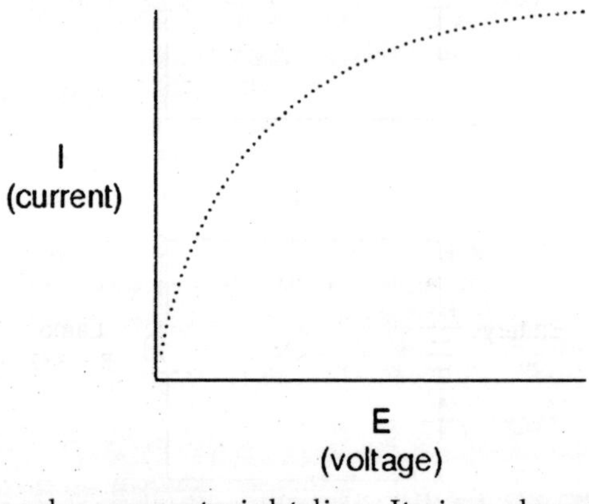

The plot is no longer a straight line. It rises sharply on the left, as voltage increases from zero to a low level. As it progresses to the right we see the line flattening out, the circuit requiring greater and greater increases in voltage to achieve equal increases in current.

If we try to apply Ohm's Law to find the resistance of this lamp circuit with the voltage and current values plotted above, we arrive at several different values. We could say that the resistance here is *nonlinear*, increasing with increasing current and voltage. The nonlinearity is caused by the effects of high temperature on the metal wire of the lamp filament.

Another example of nonlinear current conduction is through gases such as air. At standard temperatures and pressures, air is an effective insulator. However, if the voltage between two conductors separated by an air gap is increased greatly enough, the air molecules between the gap will become "ionized," having their electrons stripped off by the force of the high voltage between the wires.

Once ionized, air (and other gases) become good conductors of electricity, allowing electron flow where none could exist prior to ionization. If we were to plot current over voltage on a graph as we did with the lamp circuit, the effect of ionization would be clearly seen as nonlinear:

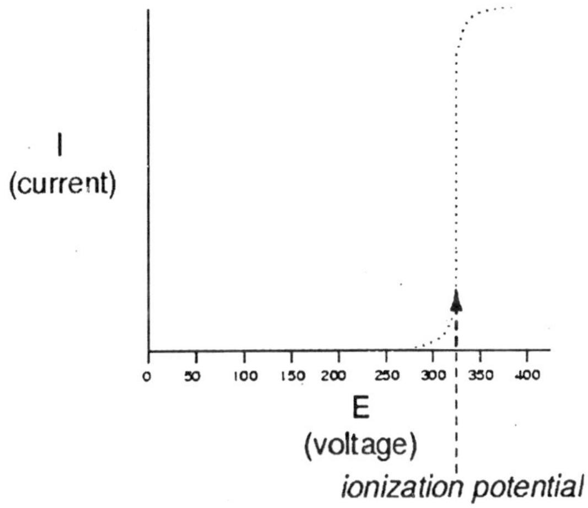

The graph shown is approximate for a small air gap (less than one inch). A larger air gap would yield a higher ionization potential, but the shape of the I/E curve would be very similar: practically no current until the ionization potential was reached, then substantial conduction after that. Incidentally, this is the reason lightning bolts exist as momentary surges rather than continuous flows of electrons.

The voltage built up between the earth and clouds must increase to the point where it overcomes the ionization potential of the air gap before the air ionizes enough to support a substantial flow of electrons. Once it does, the current will continue to conduct through the ionized air until the static charge between the two points depletes. Once the charge depletes enough so that the voltage falls below another threshold point, the air de-ionizes and returns to its normal state of extremely high resistance.

Many solid insulating materials exhibit similar resistance properties:

extremely high resistance to electron flow below some critical threshold voltage, then a much lower resistance at voltages beyond that threshold. Once a solid insulating material has been compromised by high-voltage *breakdown*, as it is called, it often does not return to its former insulating state, unlike most gases.

It may insulate once again at low voltages, but its breakdown threshold voltage will have been decreased to some lower level, which may allow breakdown to occur more easily in the future.

This is a common mode of failure in high-voltage wiring: insulation damage due to breakdown. Such failures may be detected through the use of special resistance meters employing high voltage(1000 volts or more).

There are circuit components specifically engineered to provide nonlinear resistance curves, one of them being the *varistor*. Commonly manufactured from compounds such as zinc oxide or silicon carbide, these devices maintain high resistance across their terminals until a certain "firing" or "breakdown" voltage (equivalent to the "ionization potential" of an air gap) is reached, at which point their resistance decreases dramatically.

Unlike the breakdown of an insulator, varistor breakdown is repeatable: that is, it is designed to withstand repeated breakdowns without failure. A picture of a varistor is shown here:

There are also special gas-filled tubes designed to do much the same thing, exploiting the very same principle at work in the ionization of air by a lightning bolt.

Other electrical components exhibit even stranger current/voltage curves than this. Some devices actually experience a *decrease* in current as the applied voltage *increases*. Because the slope of the current/voltage for this phenomenon is negative (angling down instead of up as it progresses from left to right), it is known as *negative resistance*.

Most notably, high-vacuum electron tubes known as *tetrodes* and semiconductor diodes known as *Esaki* or *tunnel* diodes exhibit negative resistance for certain ranges of applied voltage.

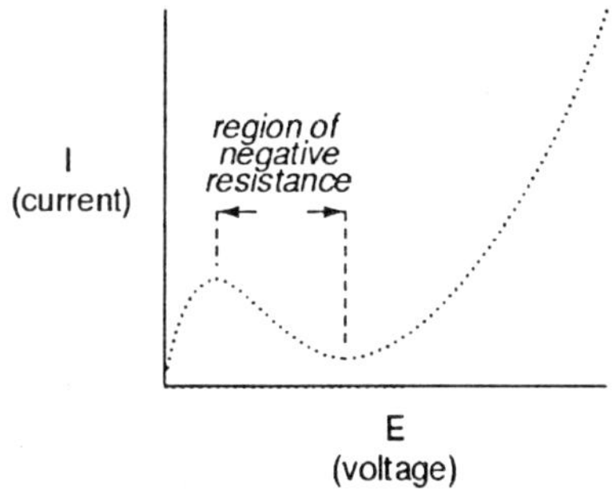

Ohm's Law is not very useful for analyzing the behaviour of components like these where resistance varies with voltage and current. Some have even suggested that "Ohm's Law" should be demoted from the status of a "Law" because it is not universal.

It might be more accurate to call the equation (R=E/I) a *definition of resistance,* befitting of a certain class of materials under a narrow range of conditions.

CIRCUIT WIRING

So far, we've been analyzing single-battery, single-resistor circuits with no regard for the connecting wires between the components, so long as a complete circuit is formed. Does the wire length or circuit "shape" matter to our calculations? Let's look at a couple of circuit configurations and find out: When we draw wires connecting points in a circuit, we usually assume those wires have negligible resistance. As such, they contribute no appreciable effect to the overall resistance of the circuit, and so the only resistance we have to contend with is the resistance in the components. In the above circuits, the only resistance comes from the 5 Ω resistors, so that is all we will consider in our calculations.

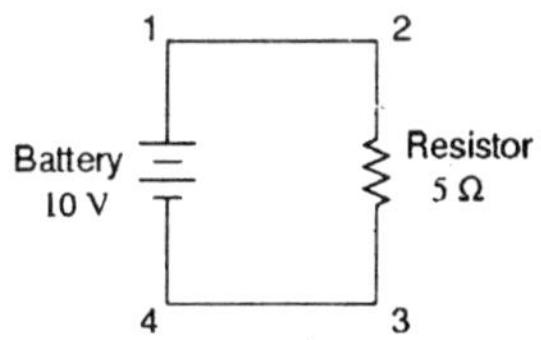

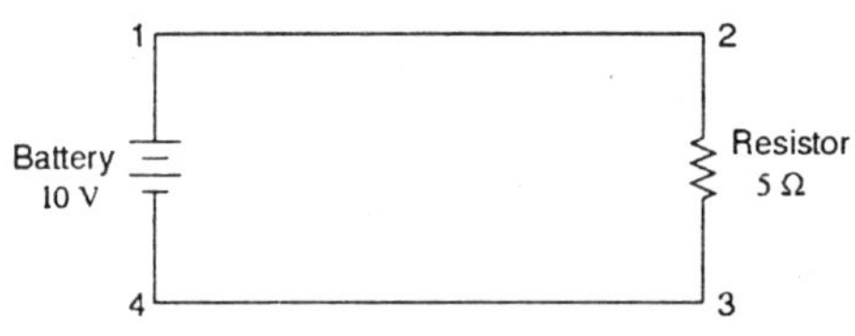

In real life, metal wires actually *do* have resistance (and so do power sources!), but those resistances are generally so much smaller than the resistance present in the other circuit components that they can be safely ignored. Exceptions to this rule exist in power system wiring, where even very small amounts of conductor resistance can create significant voltage drops given normal (high) levels of current.

If connecting wire resistance is very little or none, we can regard the connected points in a circuit as being *electrically common*. That is, points 1 and 2 in the above circuits may be physically joined close together or far apart, and it doesn't matter for any voltage or resistance measurements relative to those points. The same goes for points 3 and 4. It is as if the ends of the resistor were attached directly across the terminals of the battery, so far as our Ohm's Law calculations and voltage measurements are concerned.

This is useful to know, because it means you can re-draw a circuit diagram or re-wire a circuit, shortening or lengthening the wires as desired without appreciably impacting the circuit's function. All that matters is that the components attach to each other in the same sequence. It also means that voltage measurements between sets of "electrically common" points will be the same.

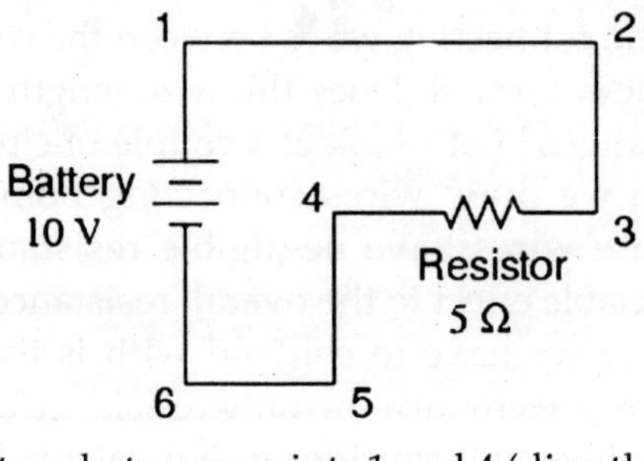

That is, the voltage between points 1 and 4 (directly across the battery) will be the same as the voltage between points 2 and 3 (directly across the resistor). Take a close look at the following circuit, and try to determine which points are common to each other: Here, we only have 2 components excluding the wires: the battery and the resistor.

Though the connecting wires take a convoluted path in forming a complete circuit, there are several electrically common points in the electrons' path. Points 1, 2, and 3 are all common to each other, because they're directly connected together by wire. The same goes for points 4, 5, and 6.

This makes sense mathematically, too. With a 10 volt battery and a 5 Ω resistor, the circuit current will be 2 amps. With wire resistance being zero, the voltage drop across any continuous stretch of wire can be determined through Ohm's Law as such:

$$E = IR$$
$$E = (2A)\ (0W)$$
$$E = 0V$$

It should be obvious that the calculated voltage drop across any uninterrupted length of wire in a circuit where wire is assumed to have zero resistance will always be zero, no matter what the magnitude of current, since zero multiplied by anything equals zero.

Because common points in a circuit will exhibit the same relative voltage and resistance measurements, wires connecting common points are often labeled with the same designation. This is not to say that the *terminal* connection points are labeled the same, just the connecting wires. Take this circuit as an example:

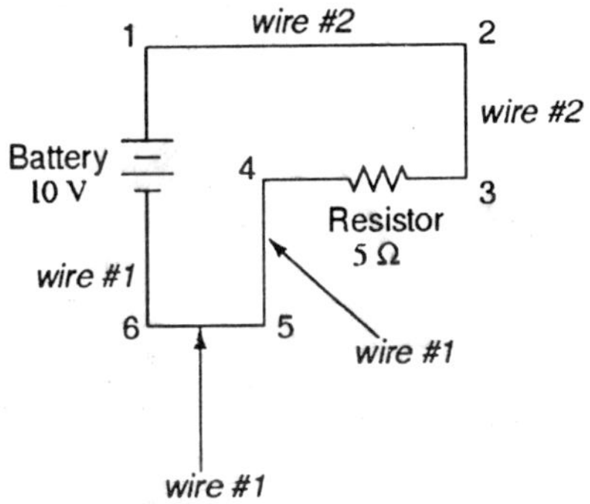

Points 1, 2, and 3 are all common to each other, so the wire connecting point 1 to 2 is labeled the same (wire 2) as the wire connecting point 2 to 3 (wire 2). In a real circuit, the wire stretching from point 1 to 2 may not even be the same colour or size as the wire connecting point 2 to 3, but they should bear the exact same label. The same goes for the wires connecting points 6, 5, and 4.

Knowing that electrically common points have zero voltage drop between them is a valuable troubleshooting principle. If I measure for voltage between points in a circuit that are supposed to be common to each other, I should read zero.

If, however, I read substantial voltage between those two points, then I know with certainty that they cannot be directly connected together. If those points are *supposed* to be electrically common but they register otherwise, then I know that there is an "open failure" between those points. One final note: for most practical purposes, wire conductors can be assumed to possess zero resistance from end to end.

In reality, however, there will always be some small amount of resistance encountered along the length of a wire, unless it's a superconducting wire. Knowing this, we need to bear in mind that the principles learned here about electrically common points are all valid to a large degree, but not to an *absolute* degree. That is, the rule that electrically common points are guaranteed to have zero voltage between them is more accurately stated as such: electrically common points will have *very little* voltage dropped between them.

That small, virtually unavoidable trace of resistance found in any piece of connecting wire is bound to create a small voltage across the length of it as current is conducted through.

So long as you understand that these rules are based upon *ideal* conditions, you won't be perplexed when you come across some condition appearing to be an exception to the rule.

POLARITY OF VOLTAGE DROPS

We can trace the direction that electrons will flow in the same circuit by starting at the negative (-) terminal and following through to the positive (+) terminal of the battery, the only source of voltage in the circuit. From this we can see that the electrons are moving counter-clockwise, from point 6 to 5 to 4 to 3 to 2 to 1 and back to 6 again.

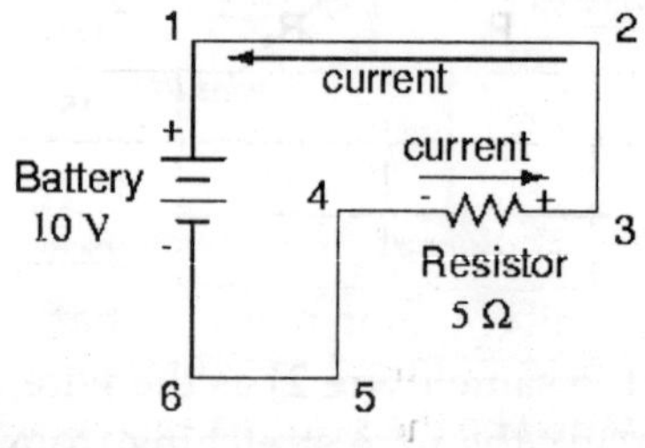

As the current encounters the 5 Ω resistance, voltage is dropped across the resistor's ends.

The polarity of this voltage drop is negative (-) at point 4 with respect to positive (+) at point 3. We can mark the polarity of the resistor's voltage drop with these negative and positive symbols, in accordance with the direction of current (whichever end of the resistor the current is *entering* is negative with respect to the end of the resistor it is *exiting*:

While it might seem a little silly to document polarity of voltage drop in this circuit, it is an important concept to master. It will be critically important in the analysis of more complex circuits involving multiple resistors and/or batteries.

It should be understood that polarity has nothing to do with Ohm's Law: there will never be negative voltages, currents, or resistance entered into any Ohm's Law equations! There are other mathematical principles of electricity that do take polarity into account through the use of signs (+ or -), but not Ohm's Law.

VOLTAGE DIVIDER CIRCUITS

Let's analyse a simple series circuit, determining the voltage drops across individual resistors:

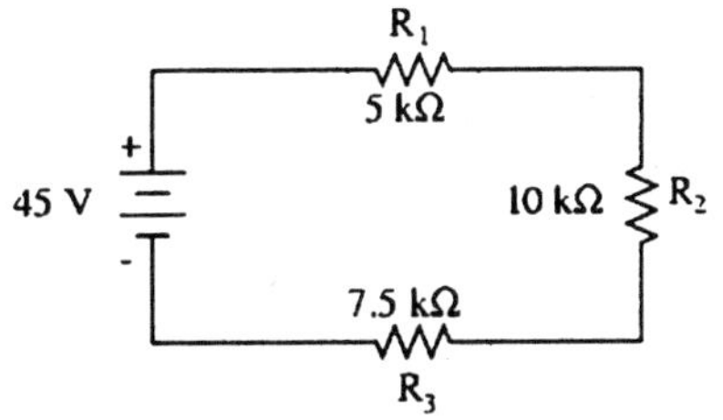

	R_1	R_2	R_3	Total	
E				45	Volts
I					Amps
R	5k	10k	7.5k		Ohms

From the given values of individual resistances, we can determine a total circuit resistance, knowing that resistances add in series:

	R_1	R_2	R_3	Total	
E				45	Volts
I					Amps
R	5k	10k	7.5k	**22.5k**	Ohms

From here, we can use Ohm's Law (I=E/R) to determine the total current, which we know will be the same as each resistor current, currents being equal in all parts of a series circuit:

	R_1	R_2	R_3	Total	
E				45	Volts
I	**2m**	**2m**	**2m**	**2m**	Amps
R	5k	10k	7.5k	22.5k	Ohms

Now, knowing that the circuit current is 2 mA, we can use Ohm's Law (E=IR) to calculate voltage across each resistor:

	R_1	R_2	R_3	Total	
E	**10**	**20**	**15**	45	Volts
I	2m	2m	2m	2m	Amps
R	5k	10k	7.5k	22.5k	Ohms

It should be apparent that the voltage drop across each resistor is proportional to its resistance, given that the current is the same through all resistors. Notice how the voltage across R_2 is double that of the voltage across R_1, just as the resistance of R_2 is double that of R_1.

If we were to change the total voltage, we would find this proportionality of voltage drops remains constant:

	R_1	R_2	R_3	Total	
E	40	80	60	180	Volts
I	8m	8m	8m	8m	Amps
R	5k	10k	7.5k	22.5k	Ohms

The voltage across R_2 is still exactly twice that of R_1's drop, despite the fact that the source voltage has changed. The proportionality of voltage drops (ratio of one to another) is strictly a function of resistance values.

With a little more observation, it becomes apparent that the voltage drop across each resistor is also a fixed proportion of the supply voltage. The voltage across R_1, for example, was 10 volts when the battery supply was 45 volts.

When the battery voltage was increased to 180 volts (4 times as much), the voltage drop across R_1 also increased by a factor of 4 (from 10 to 40 volts). The *ratio* between R_1's voltage drop and total voltage, however, did not change:

$$\frac{E_{R1}}{E_{total}} = \frac{10\text{ v}}{45\text{ v}} = \frac{40\text{ v}}{180\text{ v}} = 0.22222$$

Likewise, none of the other voltage drop ratios changed with the increased supply voltage either:

$$\frac{E_{R2}}{E_{total}} = \frac{20\text{ v}}{45\text{ v}} = \frac{80\text{ v}}{180\text{ v}} = 0.44444$$

$$\frac{E_{R3}}{E_{total}} = \frac{15\text{ v}}{45\text{ v}} = \frac{60\text{ v}}{180\text{ v}} = 0.33333$$

For this reason a series circuit is often called a *voltage divider* for its ability to proportion — or divide — the total voltage into fractional portions of constant ratio.

With a little bit of algebra, we can derive a formula for determining series resistor voltage drop given nothing more than total voltage, individual resistance, and total resistance:

Voltage drop across any resistor· $E_n = I_n R_n$

Current in a series ciruit $I_{total} = \dfrac{E_{total}}{R_{total}}$

... Substituting $\dfrac{E_{total}}{R_{total}}$ for I_n in the first equation...

Voltage drop across any series resitor $E_n = \frac{E_{total}}{R_{total}} R_n$ ... or ...

$$E_n = E_{total} \frac{R_n}{R_{total}}$$

The ratio of individual resistance to total resistance is the same as the ratio of individual voltage drop to total supply voltage in a voltage divider circuit. This is known as the *voltage divider formula,* and it is a short-cut method for determining voltage drop in a series circuit without going through the current calculation(s) of Ohm's Law.

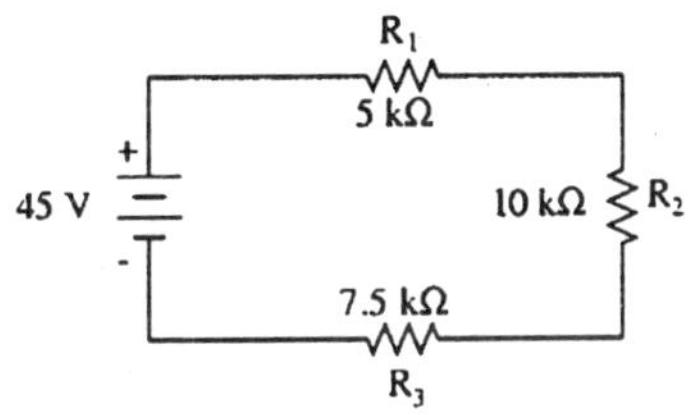

$$E_{R1} = 45\text{ V} \frac{5\text{ k}\Omega}{22.5\text{ k}\Omega} = 10\text{ V}$$

$$E_{R2} = 45\text{ V} \frac{10\text{ k}\Omega}{22.5\text{ k}\Omega} = 20\text{ V}$$

$$E_{R3} = 45\text{ V} \frac{7.5\text{ k}\Omega}{22.5\text{ k}\Omega} = 15\text{ V}$$

Using this formula, we can re-analyse the example circuit's voltage drops in fewer steps: Voltage dividers find wide application in electric meter circuits, where specific combinations of series resistors are used to "divide" a voltage into precise proportions as part of a voltage measurement device. One device frequently used as a voltage-dividing component is the *potentiometer,* which is a resistor with a movable element positioned by a manual knob or lever.

The movable element, typically called a *wiper,* makes contact with a resistive strip of material (commonly called the *slidewire* if made of resistive metal wire) at any point selected by the manual control:

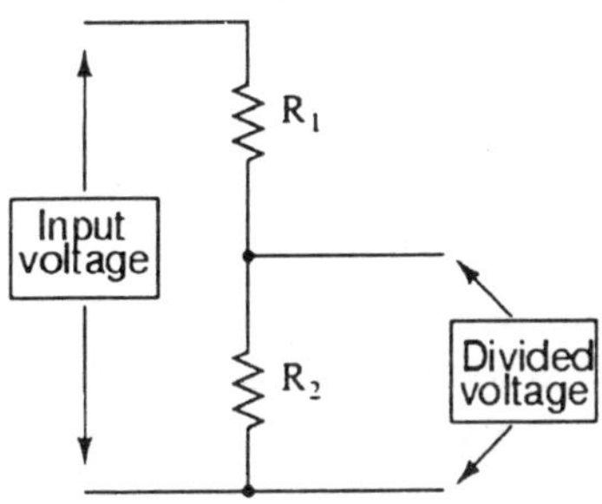

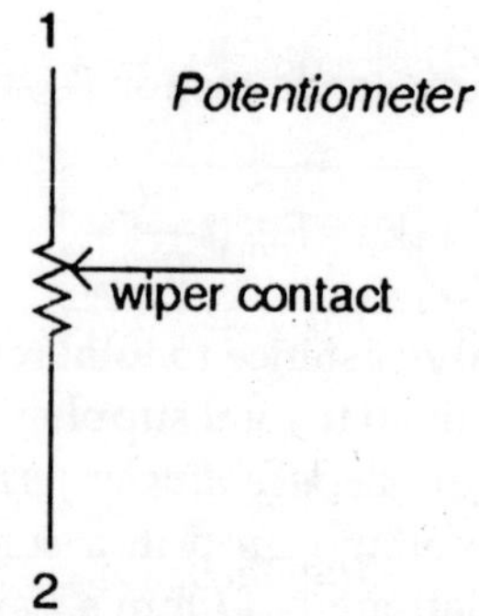

The wiper contact is the left-facing arrow symbol drawn in the middle of the vertical resistor element. As it is moved up, it contacts the resistive strip closer to terminal 1 and further away from terminal 2, lowering resistance to terminal 1 and raising resistance to terminal 2. As it is moved down, the opposite effect results. The resistance as measured between terminals 1 and 2 is constant for any wiper position.

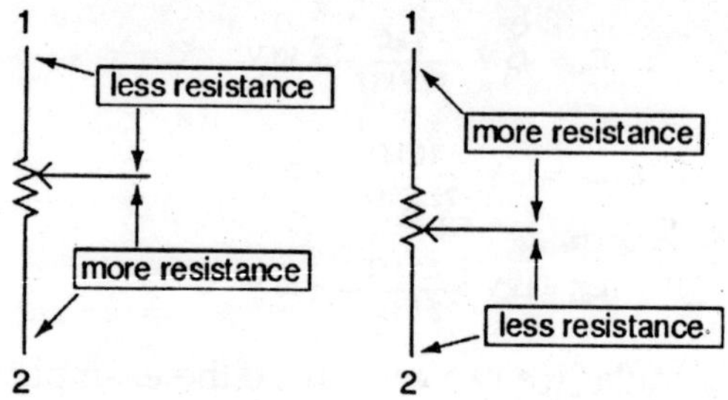

Shown here are internal illustrations of two potentiometer types, rotary and linear:

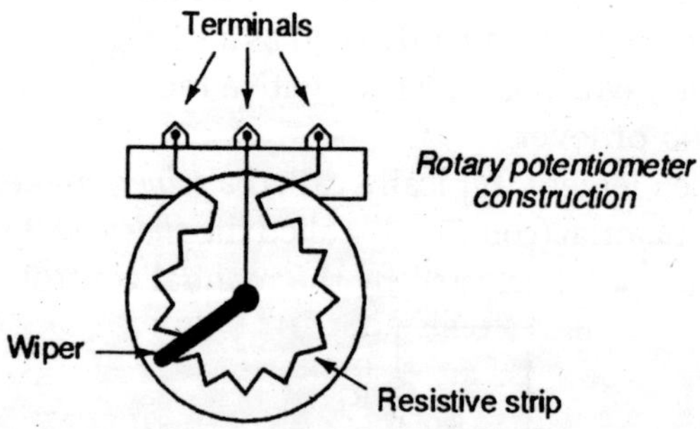

Some linear potentiometers are actuated by straight-line motion of a lever or slide button. Others, like the one depicted in the previous illustration, are actuated by a turn-screw for fine adjustment ability. The latter units are sometimes referred to as *trimpots*, because they work well for applications requiring a variable resistance to be "trimmed" to some precise value.

Linear potentiometer construction

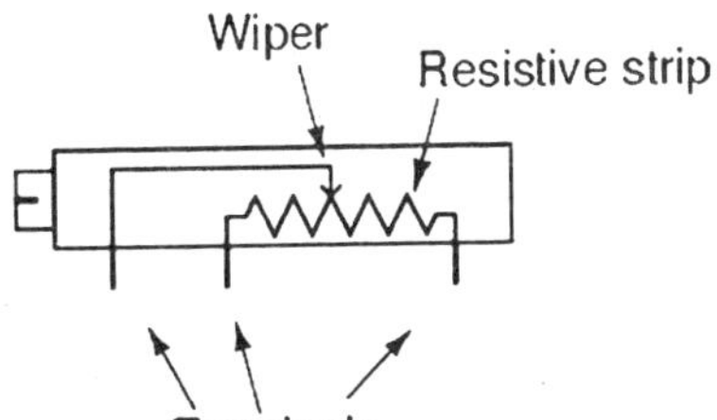

It should be noted that not all linear potentiometers have the same terminal assignments. With some, the wiper terminal is in the middle, between the two end terminals.

The following photograph shows a real, rotary potentiometer with exposed wiper and slidewire for easy viewing. The shaft which moves the wiper has been turned almost fully clockwise so that the wiper is nearly touching the left terminal end of the slidewire:

Here is the same potentiometer with the wiper shaft moved almost to the full-counterclockwise position, so that the wiper is near the other extreme end of travel:

If a constant voltage is applied between the outer terminals (across the length of the slidewire), the wiper position will tap off a fraction of the applied voltage, measurable between the wiper contact and either of the other two terminals. The fractional value depends entirely on the physical position of the wiper:

Using a potentiometer as a variable voltage divider

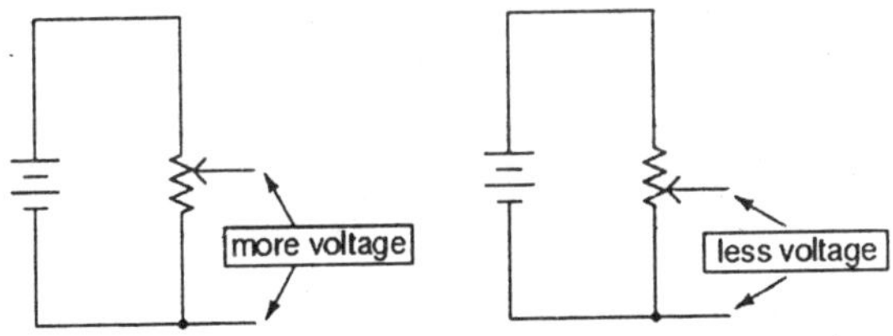

Just like the fixed voltage divider, the potentiometer's voltage *division ratio* is strictly a function of resistance and not of the magnitude of applied voltage. In other words, if the potentiometer knob or lever is moved to the 50 per cent (exact centre) position, the voltage dropped between wiper and either outside terminal would be exactly 1/2 of the applied voltage, no matter what that voltage happens to be, or what the end-to-end resistance of the potentiometer is. In other words, a potentiometer functions as a variable voltage divider where the voltage division ratio is set by wiper position.

This application of the potentiometer is a very useful means of obtaining a variable voltage from a fixed-voltage source such as a battery. If a circuit you're building requires a certain amount of voltage that is less than the value of an available battery's voltage, you may connect the outer terminals of a potentiometer across that battery and "dial up" whatever voltage you need between the potentiometer wiper and one of the outer terminals for use in your circuit:

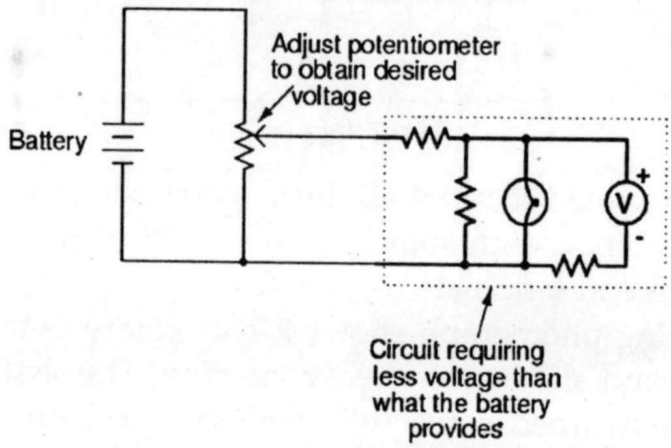

When used in this manner, the name potentiometer makes perfect sense: they meter (control) the potential (voltage) applied across them by creating a variable voltage-divider ratio.

This use of the three-terminal potentiometer as a variable voltage divider is very popular in circuit design.

Shown here are several small potentiometers of the kind commonly used in consumer electronic equipment and by hobbyists and students in constructing circuits:

The smaller units on the very left and very right are designed to plug into a solderless breadboard or be soldered into a printed circuit board. The middle units are designed to be mounted on a flat panel with wires soldered to each of the three terminals. Here are three more potentiometers, more specialized than the set just shown:

The large "Helipot" unit is a laboratory potentiometer designed for quick and easy connection to a circuit. The unit in the lower-left corner of the photograph is the same type of potentiometer, just without a case or 10-turn counting dial.

Both of these potentiometers are precision units, using multi-turn helical-track resistance strips and wiper mechanisms for making small adjustments. The unit on the lower-right is a panel-mount potentiometer, designed for rough service in industrial applications.

KIRCHHOFF'S VOLTAGE LAW (KVL)

Let's take another look at our example series circuit, this time numbering the points in the circuit for voltage reference:

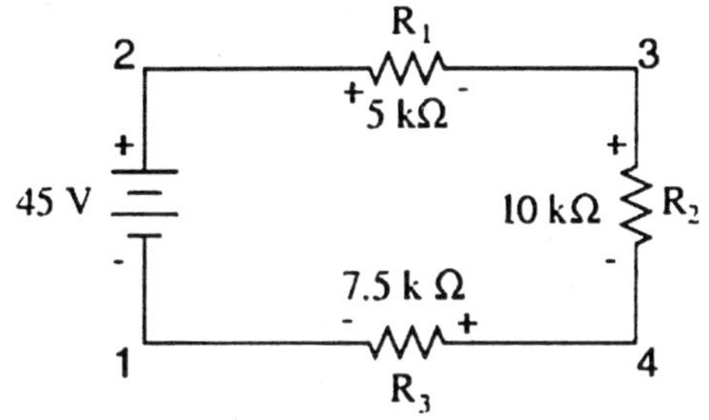

If we were to connect a voltmeter between points 2 and 1, red test lead to point 2 and black test lead to point 1, the meter would register +45 volts. Typically the "+" sign is not shown, but rather implied, for positive readings in digital meter displays. However, for this lesson the polarity of the voltage reading is very important:

$$E_{2-1} = +45V$$

When a voltage is specified with a double subscript (the characters "2-1" in the notation "E_{2-1}"), it means the voltage at the first point (2) as measured in reference to the second point (1). A voltage specified as "E_{cg}" would mean the voltage as indicated by a digital meter with the red test lead on point "c" and the black test lead on point "g": the voltage at "c" in reference to "g".

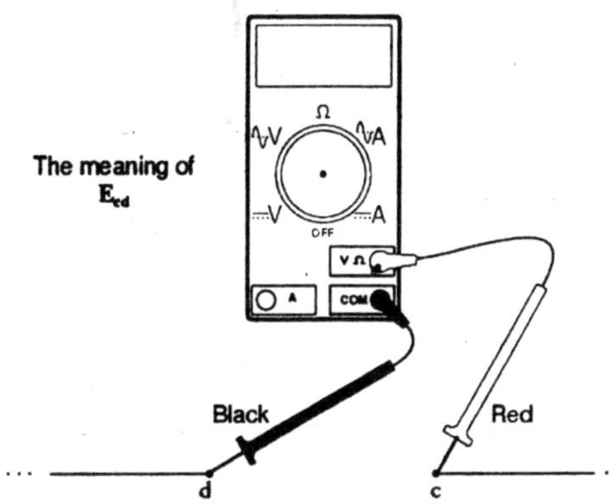

If we were to take that same voltmeter and measure the voltage drop across each resistor, stepping around the circuit in a clockwise direction with the red test lead of our meter on the point ahead and the black test lead on the point behind, we would obtain the following readings:

We should already be familiar with the general principle for series circuits stating that individual voltage drops add up to the total applied voltage, but measuring voltage drops in this manner and paying.

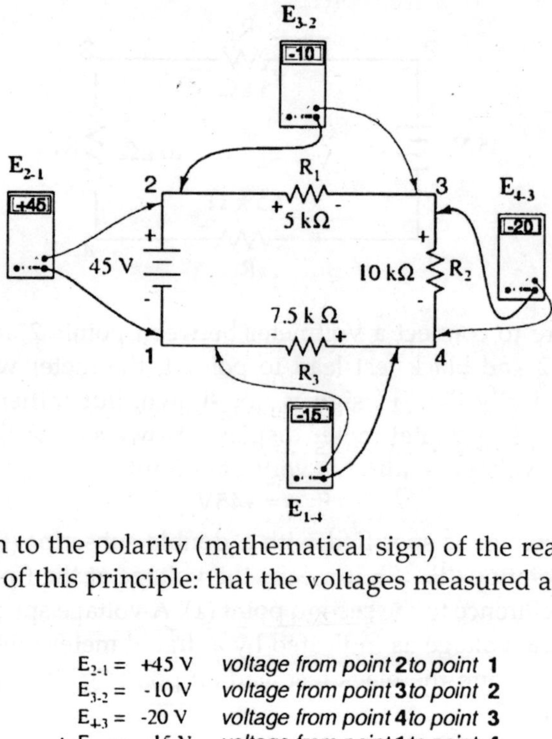

Attention to the polarity (mathematical sign) of the readings reveals another facet of this principle: that the voltages measured as such all add up to zero:

E_{2-1} =	+45 V	*voltage from point* **2** *to point* **1**
E_{3-2} =	- 10 V	*voltage from point* **3** *to point* **2**
E_{4-3} =	-20 V	*voltage from point* **4** *to point* **3**
+ E_{1-4} =	- 15 V	*voltage from point* **1** *to point* **4**
	0 V	

This principle is known as *Kirchhoff's Voltage Law* (discovered in 1847 by Gustav R. Kirchhoff, a German physicist), and it can be stated as such:

"The algebraic sum of all voltages in a loop must equal zero"

By *algebraic,* I mean accounting for signs (polarities) as well as magnitudes. By *loop,* I mean any path traced from one point in a circuit around to other points in that circuit, and finally back to the initial point. In the above example the loop was formed by following points in this order: 1-2-3-4-1.

It doesn't matter which point we start at or which direction we proceed in tracing the loop; the voltage sum will still equal zero. To demonstrate, we can tally up the voltages in loop 3-2-1-4-3 of the same circuit:

$E_{2-3} = +10$ V *voltage from point* **2** *to point* **3**
$E_{1-2} = -45$ V *voltage from point* **1** *to point* **2**
$E_{4-1} = +15$ V *voltage from point* **4** *to point* **1**
$+ E_{3-4} = +20$ V *voltage from point* **3** *to point* **4**

0 V

This may make more sense if we re-draw our example series circuit so that all components are represented in a straight line:

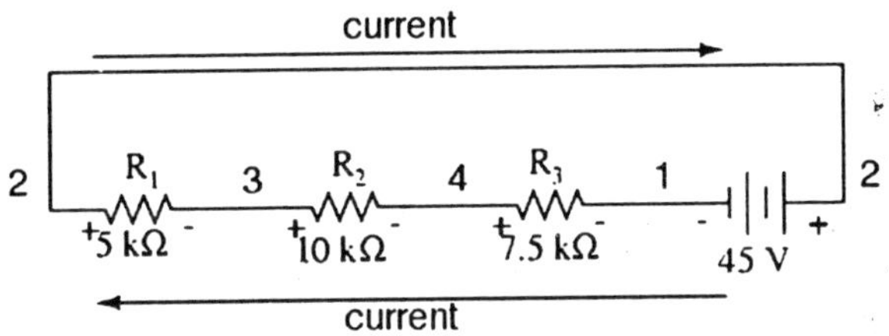

It's still the same series circuit, just with the components arranged in a different form. Notice the polarities of the resistor voltage drops with respect to the battery: the battery's voltage is negative on the left and positive on the right, whereas all the resistor voltage drops are oriented the other way: positive on the left and negative on the right.

This is because the resistors are *resisting* the flow of electrons being pushed by the battery. In other words, the "push" exerted by the resistors *against* the flow of electrons *must* be in a direction opposite the source of electromotive force. Here we see what a digital voltmeter would indicate across each component in this circuit, black lead on the left and red lead on the right, as laid out in horizontal fashion:

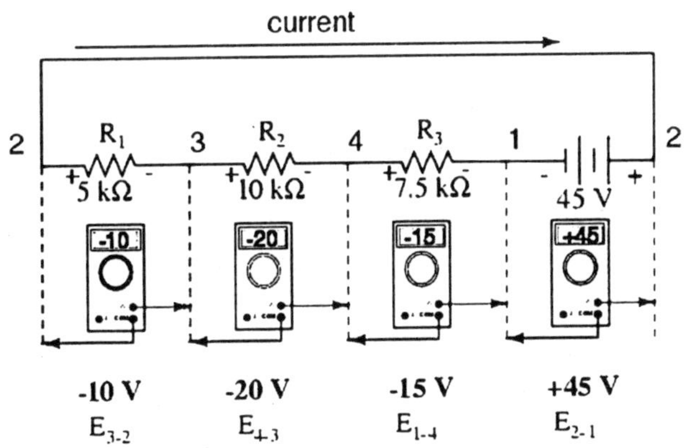

If we were to take that same voltmeter and read voltage across combinations of components, starting with only R_1 on the left and progressing across the whole string of components, we will see how the voltages add algebraically (to zero):

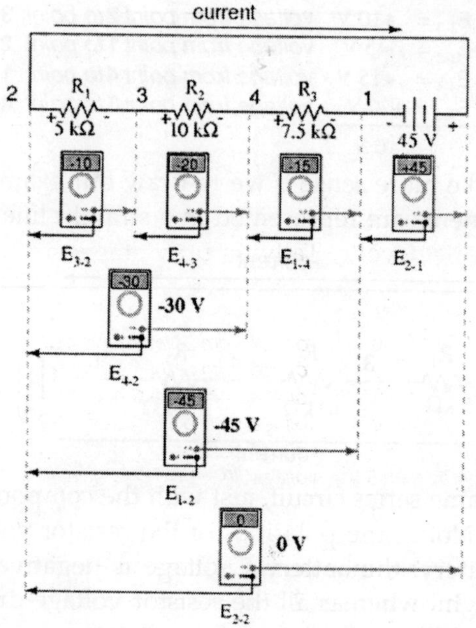

The fact that series voltages add up should be no mystery, but we notice that the *polarity* of these voltages makes a lot of difference in how the figures add.

While reading voltage across R_1, R_1-R_2, and $R_1-R_2-R_3$, we see how the voltages measure successively larger (albeit negative) magnitudes, because the polarities of the individual voltage drops are in the same orientation (positive left, negative right).

The sum of the voltage drops across R_1, R_2, and R_3 equals 45 volts, which is the same as the battery's output, except that the battery's polarity is opposite that of the resistor voltage drops (negative left, positive right), so we end up with 0 volts measured across the whole string of components.

That we should end up with exactly 0 volts across the whole string should be no mystery, either.

Looking at the circuit, we can see that the far left of the string (left side of R_1: point number 2) is directly connected to the far right of the string (right side of battery: point number 2), as necessary to complete the circuit.

Since these two points are directly connected, they are *electrically common* to each other. And, as such, the voltage between those two electrically common points *must* be zero.

Kirchhoff's Voltage Law (sometimes denoted as *KVL* for short) will work for *any* circuit configuration at all, not just simple series. Note how it works for this parallel circuit:

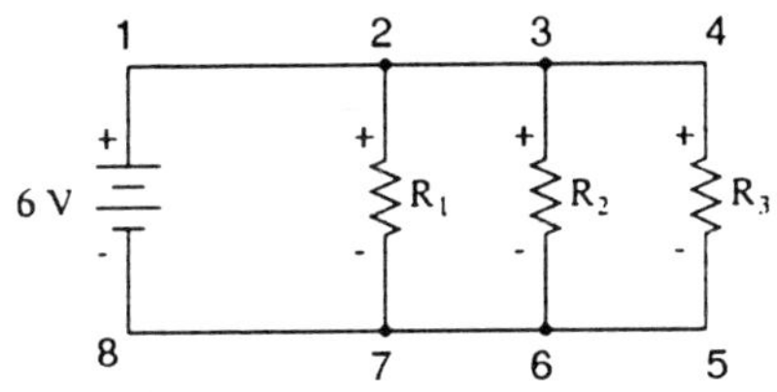

Being a parallel circuit, the voltage across every resistor is the same as the supply voltage: 6 volts. Tallying up voltages around loop 2-3-4-5-6-7-2, we get:

E_{3-2} =	0 V	*voltage from point* **3** *to point* **2**
E_{4-3} =	0 V	*voltage from point* **4** *to point* **3**
E_{5-4} =	-6 V	*voltage from point* **5** *to point* **4**
E_{6-5} =	0 V	*voltage from point* **6** *to point* **5**
E_{7-6} =	0 V	*voltage from point* **7** *to point* **6**
+ E_{2-7} =	+6 V	*voltage from point* **2** *to point* **7**
E_{2-2} =	0 V	

Note how I label the final (sum) voltage as E_{2-2}. Since we began our loop-stepping sequence at point 2 and ended at point 2, the algebraic sum of those voltages will be the same as the voltage measured between the same point (E_{2-2}), which of course must be zero.

The fact that this circuit is parallel instead of series has nothing to do with the validity of Kirchhoff's Voltage Law.

For that matter, the circuit could be a "black box" — its component configuration completely hidden from our view, with only a set of exposed terminals for us to measure voltage between — and KVL would still hold true:

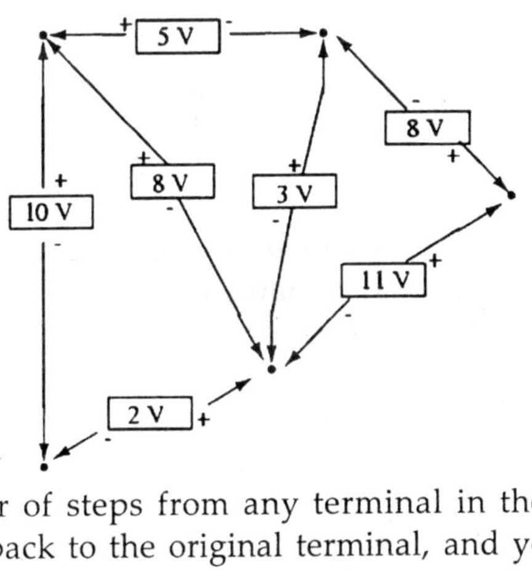

Try any order of steps from any terminal in the above diagram, stepping around back to the original terminal, and you'll find that the algebraic sum of the voltages *always* equals zero.

Furthermore, the "loop" we trace for KVL doesn't even have to be a real current path in the closed-circuit sense of the word. All we have to do to comply with KVL is to begin and end at the same point in the circuit, tallying voltage drops and polarities as we go between the next and the last point.

Consider this absurd example, tracing "loop" 2-3-6-3-2 in the same parallel resistor circuit:

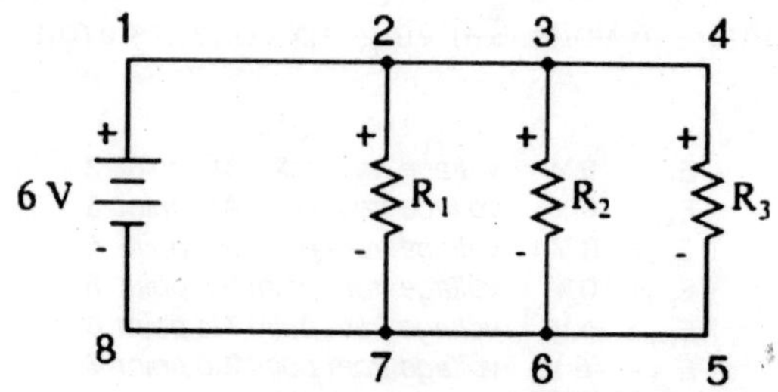

$E_{3\text{-}2} = 0$ V *voltage from point* **3** *to point* **2**
$E_{6\text{-}3} = -6$ V *voltage from point* **6** *to point* **3**
$E_{3\text{-}6} = +6$ V *voltage from point* **3** *to point* **6**
$+\ E_{2\text{-}3} = 0$ V *voltage from point* **2** *to point* **3**

$E_{2\text{-}2} = 0$ V

KVL can be used to determine an unknown voltage in a complex circuit, where all other voltages around a particular "loop" are known. Take the following complex circuit (actually two series circuits joined by a single wire at the bottom) as an example:

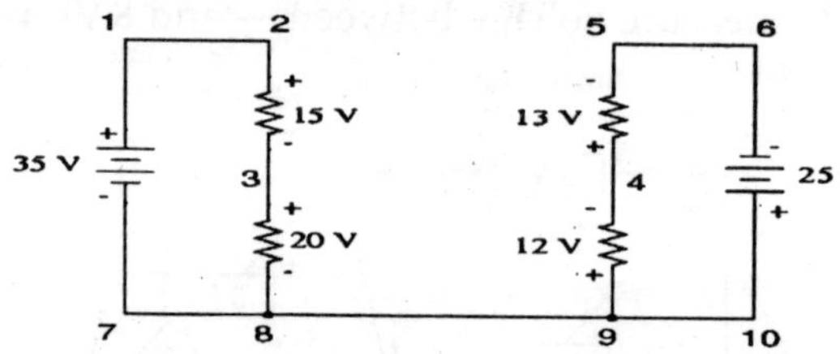

To make the problem simpler, I've omitted resistance values and simply given voltage drops across each resistor. The two series circuits share a common wire between them (wire 7-8-9-10), making voltage measurements *between* the two circuits possible. If we wanted to determine the voltage between points 4 and 3, we could set up a KVL equation with the voltage between those points as the unknown:

$$E_{4-3} + E_{9-4} + E_{8-9} + E_{3-8} = 0$$
$$E_{4-3} + 12 + 0 + 20 = 0$$
$$E_{4-3} + 32 = 0$$
$$E_{4-3} = -32\text{V}$$

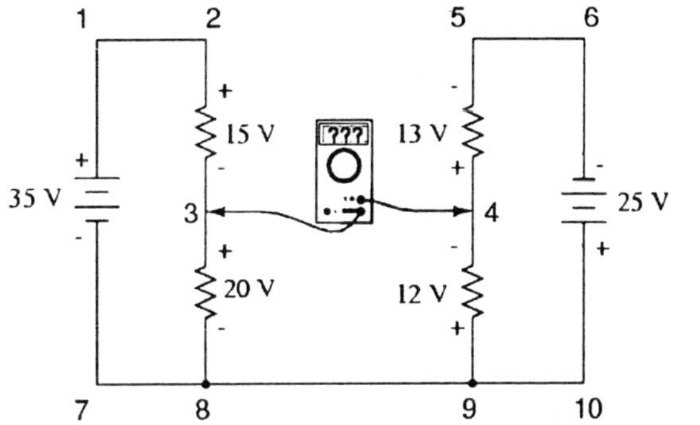

Measuring voltage from point 4 to point 3 (unknown amount)

E_{4-3}

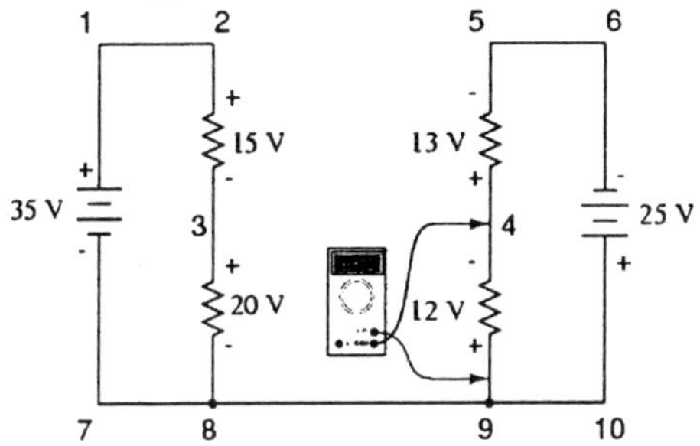

Measuring voltage from point 9 to point 4 (+12 volts)

$E_{4-3} + 12$

Stepping around the loop 3-4-9-8-3, we write the voltage drop figures as a digital voltmeter would register them, measuring with the red test lead on the point ahead and black test lead on the point behind as we progress around the loop.

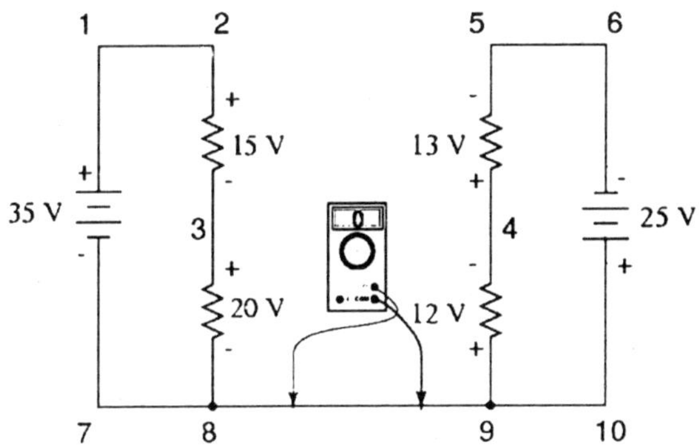

Measuring voltage from point 8 to point 9 (0 volts)

$E_{4-3} + 12 + 0$

Therefore, the voltage from point 9 to point 4 is a positive (+) 12 volts because the "red lead" is on point 9 and the "black lead" is on point 4. The voltage from point 3 to point 8 is a positive (+) 20 volts because the "red lead" is on point 3 and the "black lead" is on point 8. The voltage from point 8 to point 9 is zero, of course, because those two points are electrically common.

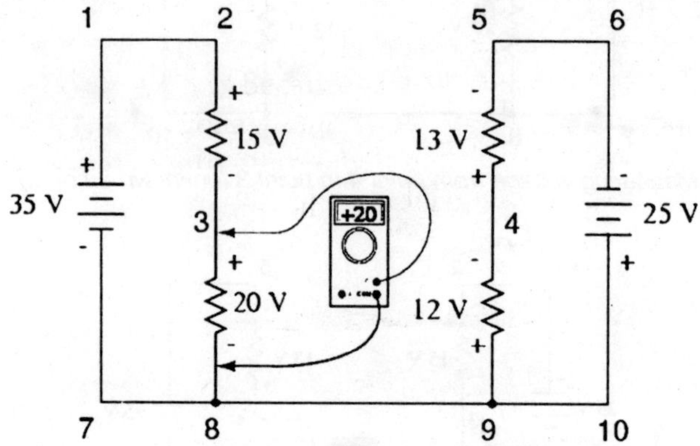

Measuring voltage from point 3 to point 8 (+20 volts)

$$E_{4-3} + 12 + 0 + 20 = 0$$

Our final answer for the voltage from point 4 to point 3 is a negative (-) 32 volts, telling us that point 3 is actually positive with respect to point 4, precisely what a digital voltmeter would indicate with the red lead on point 4 and the black lead on point 3: In other words, the initial placement of our "meter leads" in this KVL problem was "backwards."

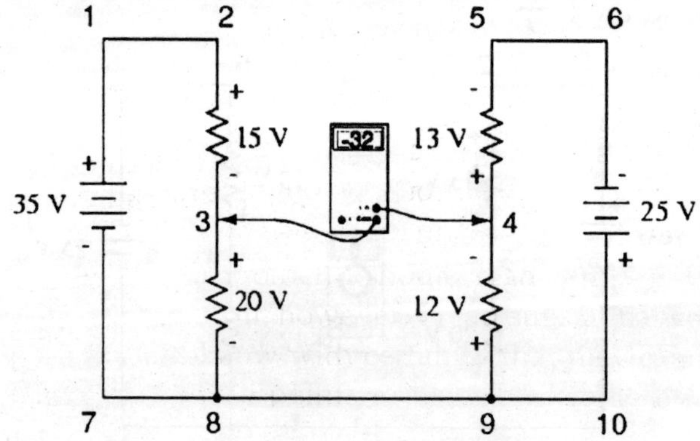

$$E_{4-3} = -32$$

Had we generated our KVL equation starting with E_{3-4} instead of E_{4-3}, stepping around the same loop with the opposite meter lead orientation, the final answer would have been E_{3-4} = +32 volts: It is important to realise that neither approach is "wrong." In both cases, we arrive at the correct assessment of voltage between the two points, 3 and 4: point 3 is positive with respect to point 4, and the voltage between them is 32 volts.

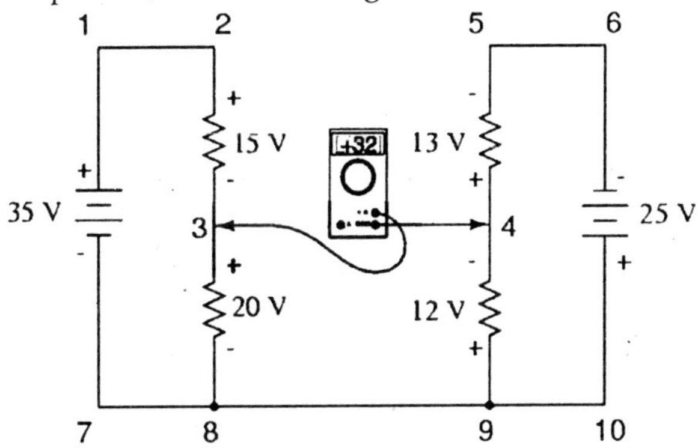

$E_{3-4} = +32$

Current Divider Circuits

Let's analyse a simple parallel circuit, determining the branch currents through individual resistors:

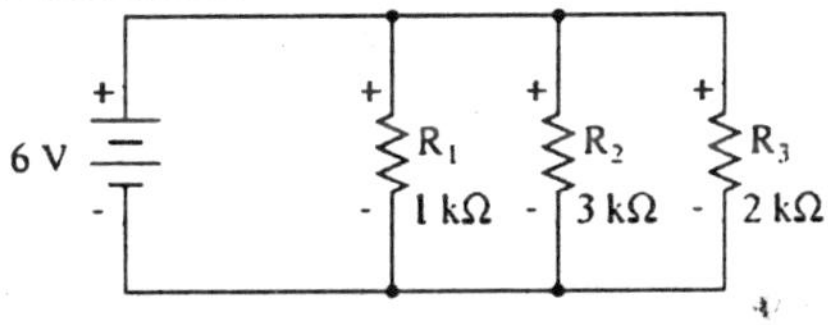

Knowing that voltages across all components in a parallel circuit are the same, we can fill in our voltage/current/resistance table with 6 volts across the top row:

	R_1	R_2	R_3	Total	
E	6	6	6	6	Volts
I					Amps
R	1k	3k	2k		Ohms

Using Ohm's Law (I=E/R) we can calculate each branch current:

	R_1	R_2	R_3	Total	
E	6	6	6	6	Volts
I	**6m**	**2m**	**3m**		Amps
R	1k	3k	2k		Ohms

Knowing that branch currents add up in parallel circuits to equal the total current, we can arrive at total current by summing 6 mA, 2 mA, and 3 mA:

	R_1	R_2	R_3	Total	
E	6	6	6	6	Volts
I	6m	2m	3m	**11m**	Amps
R	1k	3k	2k		Ohms

The final step, of course, is to figure total resistance. This can be done with Ohm's Law (R=E/I) in the "total" column, or with the parallel resistance formula from individual resistances. Either way, we'll get the same answer:

	R_1	R_2	R_3	Total	
E	6	6	6	6	Volts
I	6m	2m	3m	11m	Amps
R	1k	3k	2k	**545.45**	Ohms

Once again, it should be apparent that the current through each resistor is related to its resistance, given that the voltage across all resistors is the same. Rather than being directly proportional, the relationship here is one of inverse proportion.

For example, the current through R_1 is twice as much as the current through R_3, which has twice the resistance of R_1.

If we were to change the supply voltage of this circuit, we find that (surprise!) these proportional ratios do not change:

	R_1	R_2	R_3	Total	
E	24	24	24	24	Volts
I	24m	8m	12m	44m	Amps
R	1k	3k	2k	545.45	Ohms

i The current through R_1 is still exactly twice that of R_3, despite the fact that the source voltage has changed. The proportionality between different branch currents is strictly a function of resistance.

Also reminiscent of voltage dividers is the fact that branch currents are fixed proportions of the total current. Despite the fourfold increase in supply voltage, the ratio between any branch current and the total current remains unchanged:

$$\frac{I_{R1}}{I_{total}} = \frac{5\,mA}{11\,mA} = \frac{24\,mA}{44\,mA} = 0.54545$$

$$\frac{I_{R2}}{I_{total}} = \frac{2\,mA}{11\,mA} = \frac{8\,mA}{44\,mA} = 0.18182$$

$$\frac{I_{R3}}{I_{total}} = \frac{3\,mA}{11\,mA} = \frac{12\,mA}{44\,mA} = 0.27273$$

For this reason a parallel circuit is often called a *current divider* for its ability to proportion — or divide — the total current into fractional parts. With a little bit of algebra, we can derive a formula for determining parallel resistor current given nothing more than total current, individual resistance, and total resistance:

The ratio of total resistance to individual resistance is the same ratio as individual (branch) current to total current. This is known as the *current divider formula,* and it is a short-cut method for determining branch currents in a parallel circuit when the total current is known.

Using the original parallel circuit as an example, we can re-calculate the branch currents using this formula, if we start by knowing the total current and total resistance:

$$I_{R1} = 11\text{ mA}\,\frac{545.45\ \Omega}{1\text{ k}\Omega} = 6\text{ mA}$$

$$I_{R2} = 11\text{ mA}\,\frac{545.45\ \Omega}{3\text{ k}\Omega} = 2\text{ mA}$$

$$I_{R3} = 11\text{ mA}\,\frac{545.45\ \Omega}{2\text{ k}\Omega} = 3\text{ mA}$$

If you take the time to compare the two divider formulae, you'll see that they are remarkably similar. Notice, however, that the ratio in the voltage divider formula is R_n (individual resistance) divided by R_{Total}, and how the ratio in the current divider formula is R_{Total} divided by R_n:

Voltage divider formula

$$E_n = E_{total}\frac{R_n}{R_{total}}$$

Current divider formula

$$I_n = I_{total}\frac{R_{total}}{R_n}$$

It is quite easy to confuse these two equations, getting the resistance ratios backwards. One way to help remember the proper form is to keep in mind that both ratios in the voltage and current divider equations must equal less than one.

After all these are *divider* equations, not *multiplier* equations! If the fraction is upside-down, it will provide a ratio greater than one, which is incorrect.

Knowing that total resistance in a series (voltage divider) circuit is always greater than any of the individual resistances, we know that the fraction for that formula must be R_n over R_{Total}. Conversely, knowing that total resistance in a parallel (current divider) circuit is always less then any of the individual resistances, we know that the fraction for that formula must be R_{Total} over R_n.

Current divider circuits also find application in electric meter circuits, where a fraction of a measured current is desired to be routed through a sensitive detection device. Using the current divider formula, the proper shunt resistor can be sized to proportion just the right amount of current for the device in any given instance:

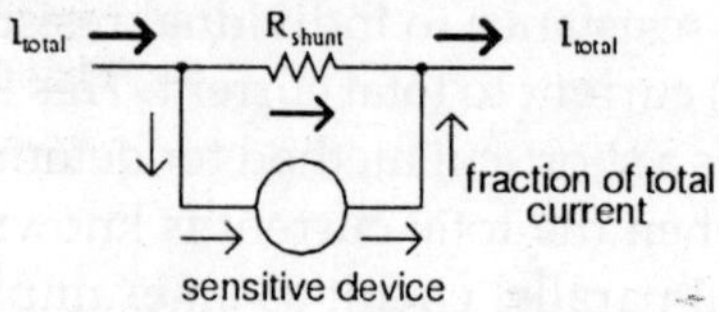

Kirchhoff's Current Law (KCL)

Let's take a closer look at that last parallel example circuit:

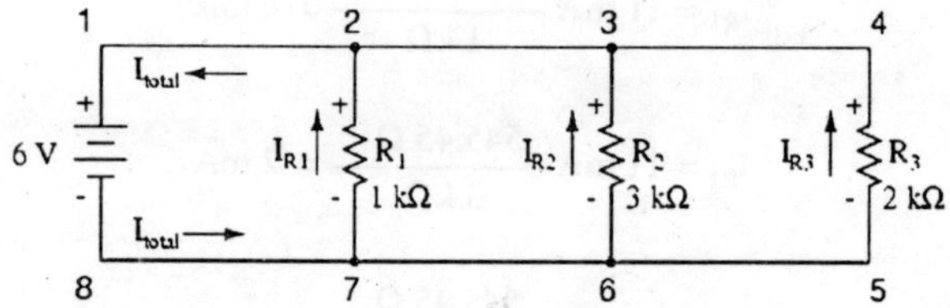

Solving for all values of voltage and current in this circuit:

	R_1	R_2	R_3	Total	
E	6	6	6	6	Volts
I	6m	2m	3m	11m	Amps
R	1k	3k	2k	545.45	Ohms

At this point, we know the value of each branch current and of the total current in the circuit. We know that the total current in a parallel circuit must equal the sum of the branch currents, but there's more going on in this circuit than just that.

Taking a look at the currents at each wire junction point (node) in the circuit, we should be able to see something else:

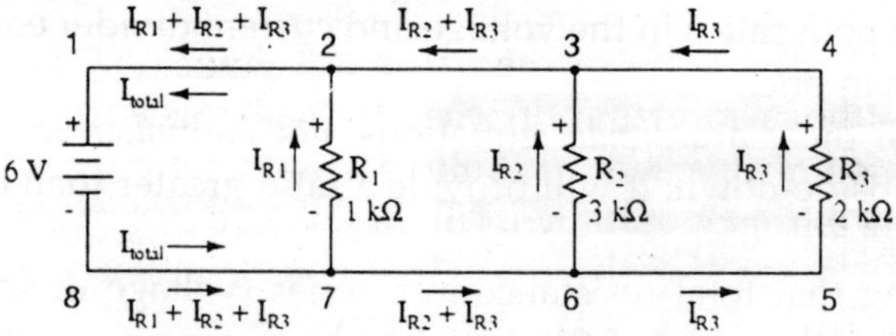

At each node on the negative "rail" (wire 8-7-6-5) we have current splitting off the main flow to each successive branch resistor. At each node on the positive "rail" (wire 1-2-3-4) we have current merging together to form the main flow from each successive branch resistor. This fact should be fairly obvious if you think of the water pipe circuit analogy with every branch node acting as a "tee" fitting, the water flow splitting or merging with the main piping as it travels from the output of the water pump toward the return reservoir or sump.

If we were to take a closer look at one particular "tee" node, such as node 3, we see that the current entering the node is equal in magnitude to the current exiting the node:

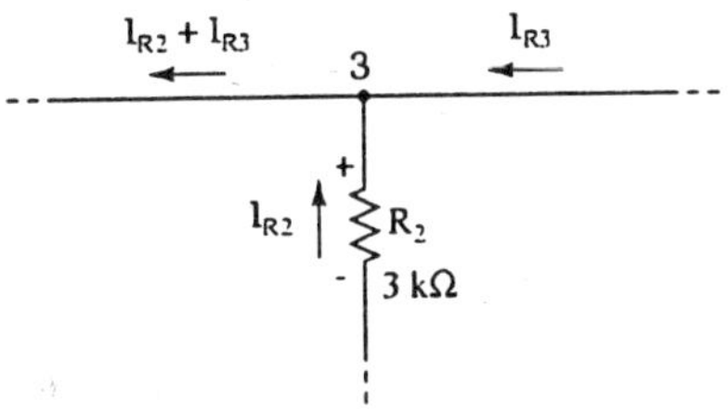

From the right and from the bottom, we have two currents entering the wire connection labeled as node 3.

To the left, we have a single current exiting the node equal in magnitude to the sum of the two currents entering. To refer to the plumbing analogy: so long as there are no leaks in the piping, what flow enters the fitting must also exit the fitting. This holds true for any node ("fitting"), no matter how many flows are entering or exiting. Mathematically, we can express this general relationship as such:

$$I_{exiting} = I_{entering}$$

Mr. Kirchhoff decided to express it in a slightly different form (though mathematically equivalent), calling it *Kirchhoff's Current Law* (KCL):

$$_{entering} + (-I_{exiting}) = 0$$

Summarized in a phrase, Kirchhoff's Current Law reads as such:

"The algebraic sum of all currents entering and exiting a node must equal zero"

That is, if we assign a mathematical sign (polarity) to each current, denoting whether they enter (+) or exit (-) a node, we can add them together to arrive at a total of zero, guaranteed.

The negative (-) sign on the value of 5 milliamps tells us that the current is *exiting* the node, as opposed to the 2 milliamp and 3 milliamp currents, which must were both positive (and therefore *entering* the node). Whether negative or positive denotes current entering or exiting is entirely arbitrary, so long as they are opposite signs for opposite directions and we stay consistent in our notation, KCL will work.

Together, Kirchhoff's Voltage and Current Laws are a formidable pair of tools useful in analyzing electric circuits.

Chapter 3

Series Circuits and Parallel Circuits

"SERIES" AND "PARALLEL" CIRCUITS

Circuits consisting of just one battery and one load resistance are very simple to analyse, but they are not often found in practical applications. Usually, we find circuits where more than two components are connected together. There are two basic ways in which to connect more than two circuit components: *series* and *parallel*. First, an example of a series circuit:

Series

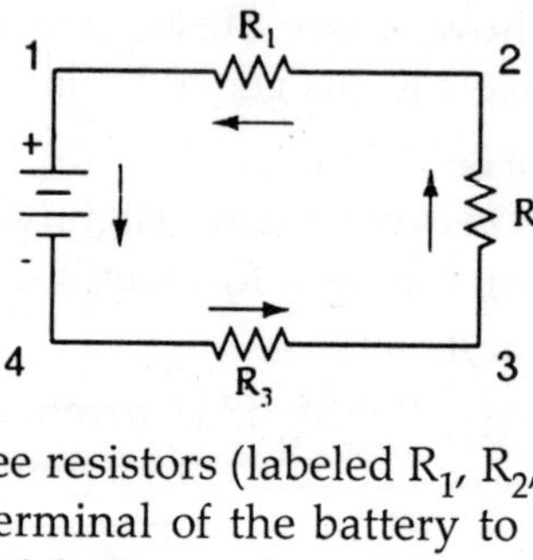

Here, we have three resistors (labeled R_1, R_2, and R_3), connected in a long chain from one terminal of the battery to the other. (It should be noted that the subscript labeling — those little numbers to the lower-right of the letter "R" — are unrelated to the resistor values in ohms. They serve only to identify one resistor from another.)

The defining characteristic of a series circuit is that there is only one path for electrons to flow. In this circuit the electrons flow in a counter-clockwise direction, from point 4 to point 3 to point 2 to point 1 and back around to 4. Now, let's look at the other type of circuit, a parallel configuration: Again, we have three resistors, but this time they form more than one continuous path for electrons to flow. There's one path from 8 to 7 to 2 to 1 and back to 8 again.

There's another from 8 to 7 to 6 to 3 to 2 to 1 and back to 8 again. And then there's a third path from 8 to 7 to 6 to 5 to 4 to 3 to 2 to 1 and back to 8 again. Each individual path (through R_1, R_2, and R_3) is called a *branch*.

Parallel

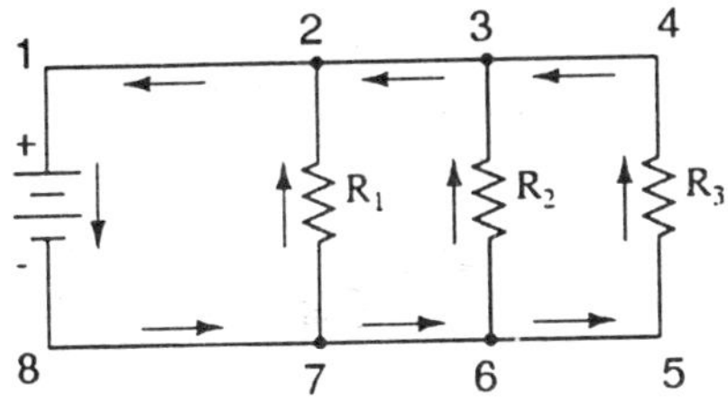

The defining characteristic of a parallel circuit is that all components are connected between the same set of electrically common points. Looking at the schematic diagram, we see that points 1, 2, 3, and 4 are all electrically common. So are points 8, 7, 6, and 5. Note that all resistors as well as the battery are connected between these two sets of points.

And, of course, the complexity doesn't stop at simple series and parallel either! We can have circuits that are a combination of series and parallel, too:

Series-parallel

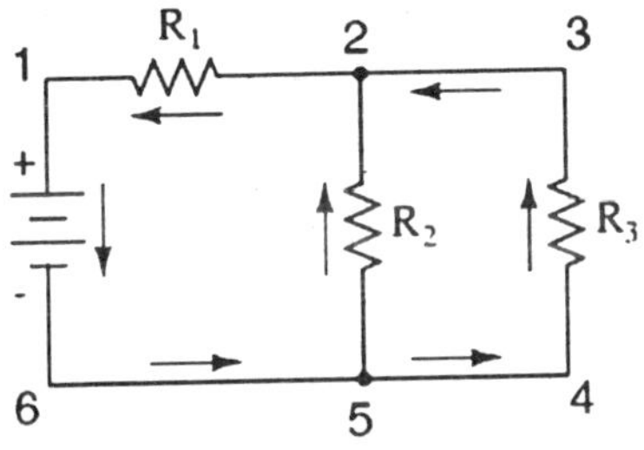

In this circuit, we have two loops for electrons to flow through: one from 6 to 5 to 2 to 1 and back to 6 again, and another from 6 to 5 to 4 to 3 to 2 to 1 and back to 6 again. Notice how both current paths go through R_1 (from point 2 to point 1). In this configuration, we'd say that R_2 and R_3 are in parallel with each other, while R_1 is in series with the parallel combination of R_2 and R_3.

The basic idea of a "series" connection is that components are connected end-to-end in a line to form a single path for electrons to flow:

Series connection

only one path for electrons to flow!

The basic idea of a "parallel" connection, on the other hand, is that all components are connected across each other's leads. In a purely parallel circuit, there are never more than two sets of electrically common points, no matter how many components are connected. There are many paths for electrons to flow, but only one voltage across all components:

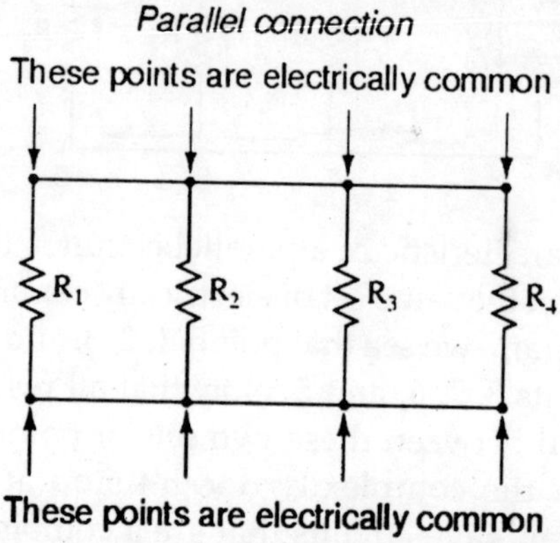

Series and parallel resistor configurations have very different electrical properties. We'll explore the properties of each configuration in the sections to come.

SIMPLE SERIES CIRCUITS

Let's start with a series circuit consisting of three resistors and a single battery:

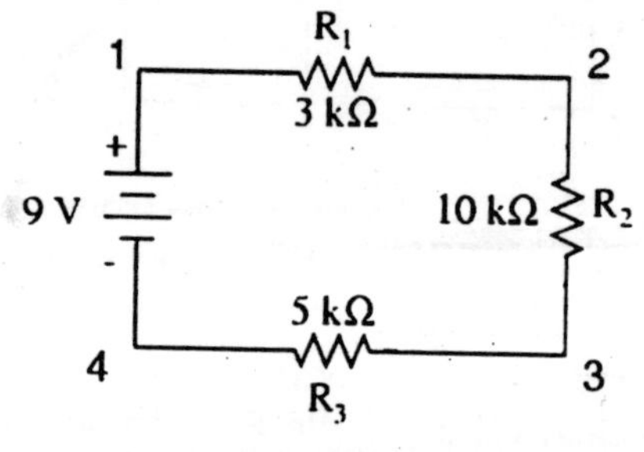

The first principle to understand about series circuits is that the amount of current is the same through any component in the circuit. This is because there is only one path for electrons to flow in a series circuit, and because free electrons flow through conductors like marbles in a tube, the rate of flow (marble speed) at any point in the circuit (tube) at any specific point in time must be equal.

From the way that the 9 volt battery is arranged, we can tell that the electrons in this circuit will flow in a counter-clockwise direction, from point 4 to 3 to 2 to 1 and back to 4. However, we have one source of voltage and three resistances. How do we use Ohm's Law here? An important caveat to Ohm's Law is that all quantities (voltage, current, resistance, and power) must relate to each other in terms of the same two points in a circuit. For instance, with a single-battery, single-resistor circuit, we could easily calculate any quantity because they all applied to the same two points in the circuit:

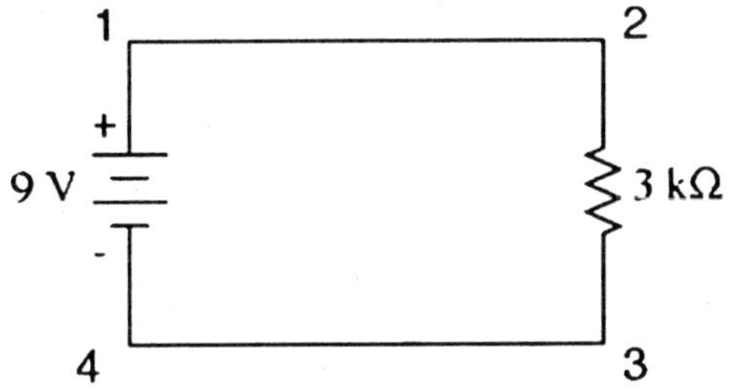

$$I = \frac{E}{R}$$

$$I = \frac{9 \text{ volts}}{3 \text{ k}\Omega} = 3 \text{ mA}$$

Since points 1 and 2 are connected together with wire of negligible resistance, as are points 3 and 4, we can say that point 1 is electrically common to point 2, and that point 3 is electrically common to point 4. Since we know we have 9 volts of electromotive force between points 1 and 4 (directly across the battery), and since point 2 is common to point 1 and point 3 common to point 4, we must also have 9 volts between points 2 and 3 (directly across the resistor).

Therefore, we can apply Ohm's Law (I = E/R) to the current through the resistor, because we know the voltage (E) across the resistor and the resistance (R) of that resistor.

All terms (E, I, R) apply to the same two points in the circuit, to that same resistor, so we can use the Ohm's Law formula with no reservation. However, in circuits containing more than one resistor, we must be careful in how we apply Ohm's Law. In the three-resistor example circuit, we know that we have 9 volts between points 1 and 4, which is the amount of electromotive force trying to push electrons through the series combination of R_1, R_2, and R_3.

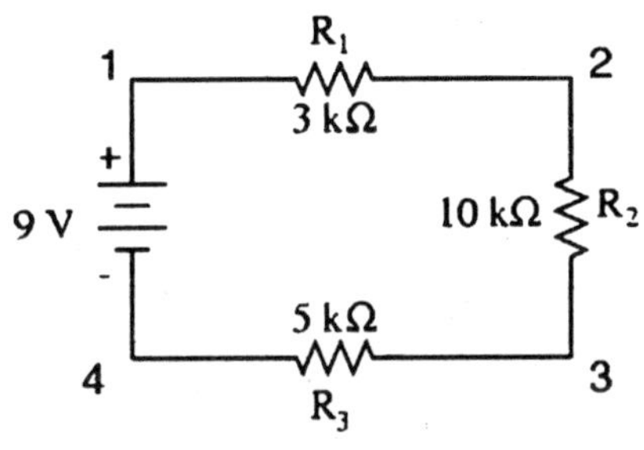

However, we cannot take the value of 9 volts and divide it by 3k, 10k or 5k Ω to try to find a current value, because we don't know how much voltage is across any one of those resistors, individually.

The figure of 9 volts is a *total* quantity for the whole circuit, whereas the figures of 3k, 10k, and 5k Ω are *individual* quantities for individual resistors.

If we were to plug a figure for total voltage into an Ohm's Law equation with a figure for individual resistance, the result would not relate accurately to any quantity in the real circuit.

For R_1, Ohm's Law will relate the amount of voltage across R_1 with the current through R_1, given R_1's resistance, 3kΩ:

$$I_{R1} = \frac{E_{R1}}{3\ k\Omega} \qquad\qquad E_{R1} = I_{R1}(3\ k\Omega)$$

But, since we don't know the voltage across R_1 (only the total voltage supplied by the battery across the three-resistor series combination) and we don't know the current through R_1, we can't do any calculations with either formula.

The same goes for R_2 and R_3: we can apply the Ohm's Law equations if and only if all terms are representative of their respective quantities between the same two points in the circuit. So what can we do? We know the voltage of the source (9 volts) applied across the series combination of R_1, R_2, and R_3, and we know the resistances of each resistor, but since those quantities aren't in the same context, we can't use Ohm's Law to determine the circuit current.

If only we knew what the *total* resistance was for the circuit: then we could calculate *total* current with our figure for *total* voltage (I=E/R).

This brings us to the second principle of series circuits: the total resistance of any series circuit is equal to the sum of the individual resistances.

This should make intuitive sense: the more resistors in series that the electrons must flow through, the more difficult it will be for those electrons to flow. In the example problem, we had a 3 kΩ, 10 kΩ, and 5 kΩ resistor in series, giving us a total resistance of 18 kΩ:

$$R_{total} = R_1 + R_2 + R_3$$

$$R_{total} = 3\ k\Omega + 10\ k\Omega + 5\ k\Omega$$

$$R_{total} = 18\ k\Omega$$

In essence, we've calculated the equivalent resistance of R_1, R_2, and R_3 combined. Knowing this, we could re-draw the circuit with a single equivalent resistor representing the series combination of R_1, R_2, and R_3:

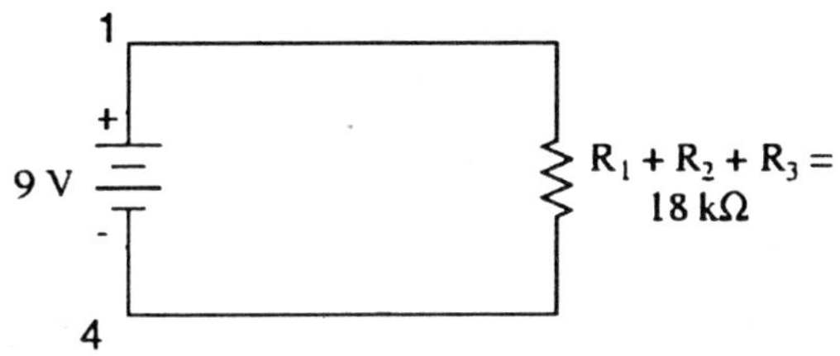

Now we have all the necessary information to calculate circuit current, because we have the voltage between points 1 and 4 (9 volts) and the resistance between points 1 and 4 (18 kΩ):

$$I_{total} = \frac{E_{total}}{R_{total}}$$

$$I_{total} = \frac{9volts}{18k\Omega} = 500\mu A$$

Knowing that current is equal through all components of a series circuit (and we just determined the current through the battery), we can go back to our original circuit schematic and note the current through each component:

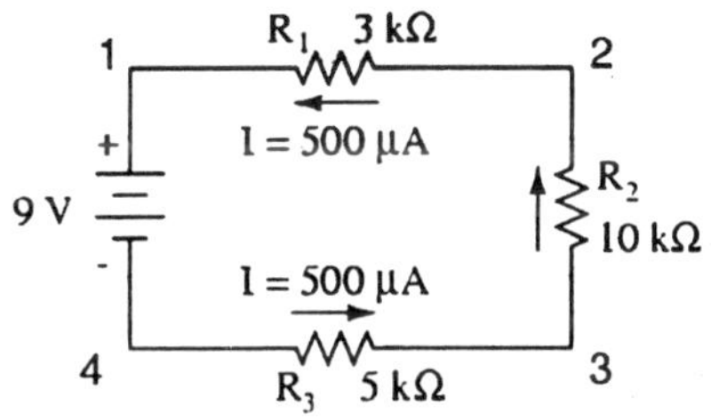

Now that we know the amount of current through each resistor, we can use Ohm's Law to determine the voltage drop across each one (applying Ohm's Law in its proper context):

$$E_{R1} = I_{R1} R_1 \qquad E_{R2} = I_{R2} R_2 \qquad E_{R3} = I_{R3} R_3$$

$$E_{R1} = (500\ \mu A)(3\ k\Omega) = 1.5\ V$$

$$E_{R2} = (500\ \mu A)(10\ k\Omega) = 5\ V$$

$$E_{R3} = (500\ \mu A)(5\ k\Omega) = 2.5\ V$$

Notice the voltage drops across each resistor, and how the sum of the voltage drops (1.5 + 5 + 2.5) is equal to the battery (supply) voltage: 9 volts. This is the third principle of series circuits: that the supply voltage is equal to the sum of the individual voltage drops.

However, the method we just used to analyse this simple series circuit can be streamlined for better understanding. By using a table to list all voltages, currents, and resistances in the circuit, it becomes very easy to see which of those quantities can be properly related in any Ohm's Law equation:

	R_1	R_2	R_3	Total	
E					Volts
I					Amps
R					Ohms

↑ *Ohm's Law* ↑ *Ohm's Law* ↑ *Ohm's Law* ↑ *Ohm's Law*

The rule with such a table is to apply Ohm's Law only to the values within each vertical column. For instance, E_{R1} only with I_{R1} and R_1; E_{R2} only with I_{R2} and R_2; etc. You begin your analysis by filling in those elements of the table that are given to you from the beginning:

	R_1	R_2	R_3	Total	
E				**9**	Volts
I					Amps
R	**3k**	**10k**	**5k**		Ohms

The arrangement of the data, we can't apply the 9 volts of E_T (total voltage) to any of the resistances (R_1, R_2, or R_3) in any Ohm's Law formula because they're in different columns. The 9 volts of battery voltage is *not* applied directly across R_1, R_2, or R_3. However, we can use our "rules" of series circuits to fill in blank spots on a horizontal row. In this case, we can use the series rule of resistances to determine a total resistance from the *sum* of individual resistances:

	R_1	R_2	R_3	Total	
E				9	Volts
I					Amps
R	3k	10k	5k	**18k**	Ohms ←

Rule of series circuits
$R_T = R_1 + R_2 + R_3$

Now, with a value for total resistance inserted into the rightmost ("Total") column, we can apply Ohm's Law of I=E/R to total voltage and total resistance to arrive at a total current of 500 μA:

	R_1	R_2	R_3	Total	
E				9	Volts
I				**500μ**	Amps
R	3k	10k	5k	18k	Ohms

↑ *Ohm's Law*

Then, knowing that the current is shared equally by all components of a series circuit (another "rule" of series circuits), we can fill in the currents for each resistor from the current figure just calculated:

	R_1	R_2	R_3	Total	
E				9	Volts
I	500µ	500µ	500µ	500µ	Amps
R	3k	10k	5k	18k	Ohms

Rule of series circuits
$I_T = I_1 = I_2 = I_3$

Finally, we can use Ohm's Law to determine the voltage drop across each resistor, one column at a time:

	R_1	R_2	R_3	Total	
E	1.5	5	2.5	9	Volts
I	500µ	500µ	500µ	500µ	Amps
R	3k	10k	5k	18k	Ohms
	Ohm's Law	Ohm's Law	Ohm's Law		

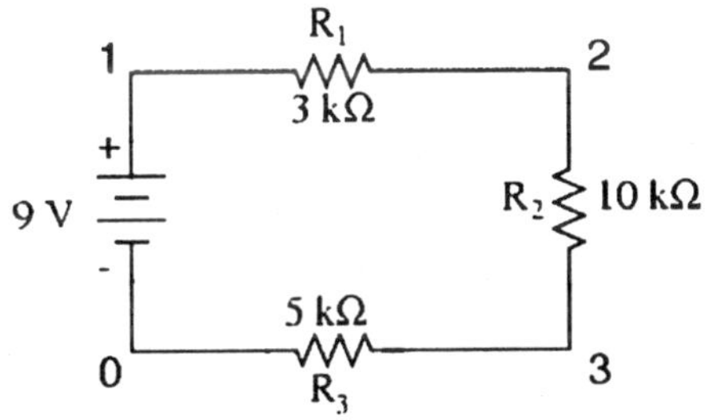

In summary, a series circuit is defined as having only one path for electrons to flow.

From this definition, three rules of series circuits follow: all components share the same current; resistances add to equal a larger, total resistance; and voltage drops add to equal a larger, total voltage. All of these rules find root in the definition of a series circuit. If you understand that definition fully, then the rules are nothing more than footnotes to the definition.

- Components in a series circuit share the same current: $I_{Total} = I_1 = I_2 = \ldots I_n$
- Total resistance in a series circuit is equal to the sum of the individual resistances: $R_{Total} = R_1 + R_2 + \ldots R_n$
- Total voltage in a series circuit is equal to the sum of the individual voltage drops: $E_{Total} = E_1 + E_2 + \ldots E_n$

SIMPLE PARALLEL CIRCUITS

Let's start with a parallel circuit consisting of three resistors and a single battery:

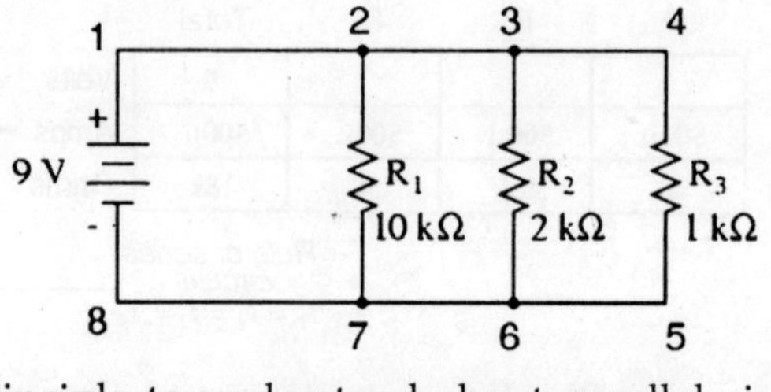

The first principle to understand about parallel circuits is that the voltage is equal across all components in the circuit. This is because there are only two sets of electrically common points in a parallel circuit, and voltage measured between sets of common points must always be the same at any given time. Therefore, in the above circuit, the voltage across R_1 is equal to the voltage across R_2 which is equal to the voltage across R_3 which is equal to the voltage across the battery. This equality of voltages can be represented in another table for our starting values:

	R_1	R_2	R_3	Total	
E	9	9	9	9	Volts
I					Amps
R	10k	2k	1k		Ohms

Just as in the case of series circuits, the same caveat for Ohm's Law applies: values for voltage, current, and resistance must be in the same context in order for the calculations to work correctly. However, in the above example circuit, we can immediately apply Ohm's Law to each resistor to find its current because we know the voltage across each resistor (9 volts) and the resistance of each resistor:

$$I_{R1} = \frac{E_{R1}}{R_1} \qquad I_{R2} = \frac{E_{R2}}{R_2} \qquad I_{R3} = \frac{E_{R3}}{R_3}$$

$$I_{R1} = \frac{9\ \text{V}}{10\ \text{k}\Omega} = 0.9\ \text{mA}$$

$$I_{R2} = \frac{9\ \text{V}}{2\ \text{k}\Omega} = 4.5\ \text{mA}$$

$$I_{R3} = \frac{9\ \text{V}}{1\ \text{k}\Omega} = 9\ \text{mA}$$

At this point we still don't know what the total current or total resistance for this parallel circuit is, so we can't apply Ohm's Law to the rightmost ("Total") column.

	R_1	R_2	R_3	Total	
E	9	9	9	9	Volts
I	**0.9m**	**4.5m**	**9m**		Amps
R	10k	2k	1k		Ohms

↑ *Ohm's Law* ↑ *Ohm's Law* ↑ *Ohm's Law*

However, if we think carefully about what is happening it should become apparent that the total current must equal the sum of all individual resistor ("branch") currents:

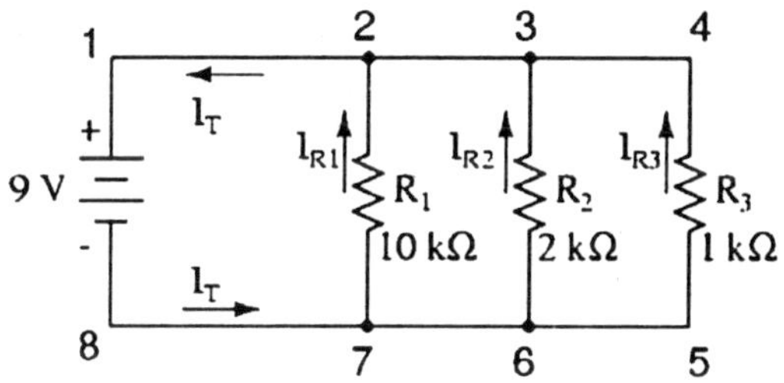

As the total current exits the negative (-) battery terminal at point 8 and travels through the circuit, some of the flow splits off at point 7 to go up through R_1, some more splits off at point 6 to go up through R_2, and the remainder goes up through R_3. Like a river branching into several smaller streams, the combined flow rates of all streams must equal the flow rate of the whole river. The same thing is encountered where the currents through R_1, R_2, and R_3 join to flow back to the positive terminal of the battery (+) toward point 1: the flow of electrons from point 2 to point 1 must equal the sum of the (branch) currents through R_1, R_2, and R_3.

This is the second principle of parallel circuits: the total circuit current is equal to the sum of the individual branch currents. Using this principle, we can fill in the I_T spot on our table with the sum of I_{R1}, I_{R2}, and I_{R3}:

	R_1	R_2	R_3	Total	
E	9	9	9	9	Volts
I	0.9m	4.5m	9m	**14.4m**	Amps ←
R	10k	2k	1k		Ohms

Rule of parallel circuits
$I_{total} = I_1 + I_2 + I_3$

Finally, applying Ohm's Law to the rightmost ("Total") column, we can calculate the total circuit resistance:

	R_1	R_2	R_3	Total	
E	9	9	9	9	Volts
I	0.9m	4.5m	9m	14.4m	Amps
R	10k	2k	1k	625	Ohms

↑ Ohm's Law

$$R_{total} = \frac{E_{total}}{I_{total}} = \frac{9\text{ V}}{14.4\text{ mA}} = 625\ \Omega$$

Please note something very important here. The total circuit resistance is only 625 Ω: *less* than any one of the individual resistors. In the series circuit, where the total resistance was the sum of the individual resistances, the total was bound to be *greater* than any one of the resistors individually. Here in the parallel circuit, however, the opposite is true: we say that the individual resistances *diminish* rather than *add* to make the total.

This principle completes our triad of "rules" for parallel circuits, just as series circuits were found to have three rules for voltage, current, and resistance. Mathematically, the relationship between total resistance and individual resistances in a parallel circuit looks like this:

$$R_{total} = \frac{1}{\frac{1}{R_1} + \frac{1}{R_2} + \frac{1}{R_3}}$$

The same basic form of equation works for *any* number of resistors connected together in parallel, just add as many 1/R terms on the denominator of the fraction as needed to accommodate all parallel resistors in the circuit. In summary, a parallel circuit is defined as one where all components are connected between the same set of electrically common points. Another way of saying this is that all components are connected across each other's terminals.

From this definition, three rules of parallel circuits follow: all components share the same voltage; resistances diminish to equal a smaller, total resistance; and branch currents add to equal a larger, total current. Just as in the case of series circuits, all of these rules find root in the definition of a parallel circuit.

If you understand that definition fully, then the rules are nothing more than footnotes to the definition.

- Components in a parallel circuit share the same voltage:
  $E_{Total} = E_1 = E_2 = \ldots E_n$
- Total resistance in a parallel circuit is *less* than any of the individual resistances: $R_{Total} = 1/\ (1/R_1 + 1/R_2 + \ldots 1/R_n)$
- Total current in a parallel circuit is equal to the sum of the individual branch currents: $I_{Total} = I_1 + I_2 + \ldots I_n$.

CONDUCTANCE

The parallel resistance equation, the natural question to ask is, "Where did *that* thing come from?" It is truly an odd piece of arithmetic, and its origin deserves a good explanation.

Resistance, by definition, is the measure of *friction* a component presents to the flow of electrons through it. Resistance is symbolized by the capital letter "R" and is measured in the unit of "ohm."

However, we can also think of this electrical property in terms of its inverse: how *easy* it is for electrons to flow through a component, rather than how *difficult*.

If *resistance* is the word we use to symbolize the measure of how difficult it is for electrons to flow, then a good word to express how easy it is for electrons to flow would be *conductance.*

Mathematically, conductance is the reciprocal, or inverse, of resistance:

$$\text{Conductance} = \frac{1}{\text{Resistance}}$$

The greater the resistance, the less the conductance, and vice versa. This should make intuitive sense, resistance and conductance being opposite ways to denote the same essential electrical property.

If two components' resistances are compared and it is found that component "A" has one-half the resistance of component "B," then we could alternatively express this relationship by saying that component "A" is *twice* as conductive as component "B."

If component "A" has but one-third the resistance of component "B," then we could say it is *three times* more conductive than component "B," and so on.

Carrying this idea further, a symbol and unit were created to represent conductance.

The symbol is the capital letter "G" and the unit is the *mho,* which is "ohm" spelled backwards (and you didn't think electronics engineers had any sense of humour!).

Despite its appropriateness, the unit of the mho was replaced in later years by the unit of *siemens* (abbreviated by the capital letter "S").

This decision to change unit names is reminiscent of the change from the temperature unit of degrees *Centigrade* to degrees *Celsius*, or the change from the unit of frequency *c.p.s.* (cycles per second) to *Hertz.* If you're looking for a pattern here, Siemens, Celsius, and Hertz are all surnames of famous scientists, the names of which, sadly, tell us less about the nature of the units than the units' original designations.

The unit of siemens is never expressed without the last letter "s." In other words, there is no such thing as a unit of "siemen" as there is in the case of the "ohm" or the "mho."

The reason for this is the proper spelling of the respective scientists' surnames. The unit for electrical resistance was named after someone named "Ohm," whereas the unit for electrical conductance was named after someone named "Siemens," therefore it would be improper to "singularize" the latter unit as its final "s" does not denote plurality.

Back to our parallel circuit example, we should be able to see that multiple paths (branches) for current reduces total resistance for the whole circuit, as electrons are able to flow easier through the whole network of multiple branches than through any one of those branch resistances alone.

In terms of *resistance*, additional branches results in a lesser total (current meets with less opposition). In terms of *conductance*, however, additional branches results in a greater total (electrons flow with greater conductance): Total parallel resistance is *less* than any one of the individual branch resistances because parallel resistors resist less together than they would separately:

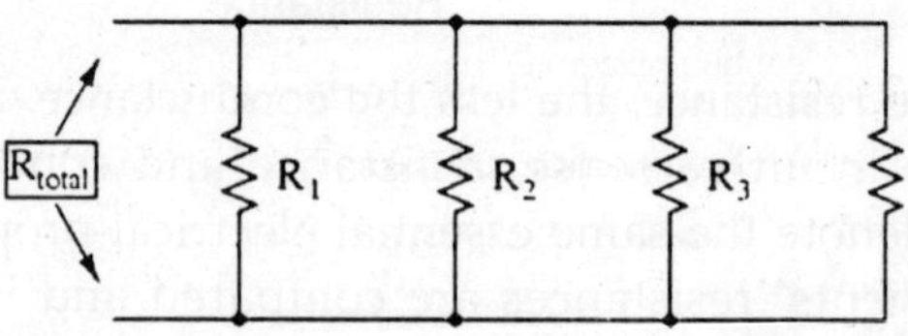

R_{total} is less than R_1, R_2, R_3, or R_4 individually

Total parallel conductance is *greater* than any of the individual branch conductances because parallel resistors conduct better together than they would separately:

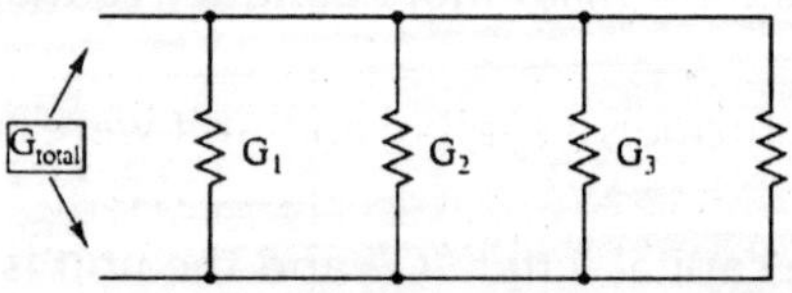

G_{total} is greater than G_1, G_2, G_3, or G_4 individually

To be more precise, the total conductance in a parallel circuit is equal to the sum of the individual conductances:

$$G_{total} = G_1 + G_2 + G_3 + G_4$$

If we know that conductance is nothing more than the mathematical reciprocal (1/x) of resistance, we can translate each term of the above formula into resistance by substituting the reciprocal of each respective conductance:

$$\frac{1}{R_{total}} = \frac{1}{R_1} + \frac{1}{R_2} + \frac{1}{R_3} + \frac{1}{R_4}$$

Solving the above equation for total resistance (instead of the reciprocal of total resistance), we can invert (reciprocate) both sides of the equation:

$$R_{total} = \frac{1}{\frac{1}{R_1} + \frac{1}{R_2} + \frac{1}{R_3} + \frac{1}{R_4}}$$

So, we arrive at our cryptic resistance formula at last! Conductance (G) is seldom used as a practical measurement, and so the above formula is a common one to see in the analysis of parallel circuits.

- Conductance is the opposite of resistance: the measure of how *easy* is it for electrons to flow through something.
- Conductance is symbolized with the letter "G" and is measured in units of *mhos* or *Siemens*.
- Mathematically, conductance equals the reciprocal of resistance: $G = 1/R$

POWER CALCULATIONS

When calculating the power dissipation of resistive components, use any one of the three power equations to derive and answer from values of voltage, current, and/or resistance pertaining to each component:

Power equations

$$P = IE \qquad P = \frac{E^2}{R} \qquad P = I^2R$$

This is easily managed by adding another row to our familiar table of voltages, currents, and resistances:

	R_1	R_2	R_3	Total	
E					Volts
I					Amps
R					Ohms
P					Watts

Power for any particular table column can be found by the appropriate Ohm's Law equation (*appropriate* based on what figures are present for E, I, and R in that column).

An interesting rule for total power versus individual power is that it is additive for *any* configuration of circuit: series, parallel, series/parallel, or otherwise.

Power is a measure of rate of work, and since power dissipated *must* equal the total power applied by the source(s) (as per the Law of Conservation of Energy in physics), circuit configuration has no effect on the mathematics.

Power is additive in *any* configuration of resistive circuit: $P_{Total} = P_1 + P_2 + \ldots P_n$

CORRECT USE OF OHM'S LAW

One of the most common mistakes made by beginning electronics students in their application of Ohm's Laws is mixing the contexts of voltage, current, and resistance. In other words, a student might mistakenly use a value for one resistor and the value for E across a set of interconnected resistors, thinking that they'll arrive at the resistance of that one resistor. The variables used in Ohm's Law equations must be *common* to the same two points in the circuit under consideration. We cannot overemphasize this rule. This is especially important in series-parallel combination circuits where nearby components may have different values for both voltage drop *and* current.

When using Ohm's Law to calculate a variable pertaining to a single component, be sure the voltage you're referencing is solely across that single component and the current you're referencing is solely through that single component and the resistance you're referencing is solely for that single component.

Likewise, when calculating a variable pertaining to a set of components in a circuit, be sure that the voltage, current, and resistance values are specific to that complete set of components only!

A good way to remember this is to pay close attention to the *two points* terminating the component or set of components being analyzed, making sure that the voltage in question is across those two points, that the current in question is the electron flow from one of those points all the way to the other point, that the resistance in question is the equivalent of a single resistor between those two points, and that the power in question is the total power dissipated by all components between those two points.

The "table" method presented for both series and parallel circuits is a good way to keep the context of Ohm's Law correct for any kind of circuit configuration.

You are only allowed to apply an Ohm's Law equation for the values of a single *vertical* column at a time:

	R_1	R_2	R_3	Total	
E					Volts
I					Amps
R					Ohms
P					Watts
	↑ *Ohm's Law*	↑ *Ohm's Law*	↑ *Ohm's Law*	↑ *Ohm's Law*	

Deriving values *horizontally* across columns is allowable as per the principles of series and parallel circuits:

For series circuits:

	R_1	R_2	R_3	Total	
E	→	→	→	Add	Volts
I	→	→	→	Equal	Amps
R	→	→	→	Add	Ohms
P	→	→	→	Add	Watts

$$E_{total} = E_1 + E_2 + E_3$$

$$I_{total} = I_1 = I_2 = I_3$$

$$R_{total} = R_1 + R_2 + R_3$$

$$P_{total} = P_1 + P_2 + P_3$$

For parallel circuits:

	R_1	R_2	R_3	Total	
E	→	→	→	Equal	Volts
I	→	→	→	Add	Amps
R	→	→	→	Diminish	Ohms
P	→	→	→	Add	Watts

$$E_{total} = E_1 = E_2 = E_3$$

$$I_{total} = I_1 + I_2 + I_3$$

$$R_{total} = \frac{1}{\frac{1}{R_1} + \frac{1}{R_2} + \frac{1}{R_3}}$$

$$P_{total} = P_1 + P_2 + P_3$$

Not only does the "table" method simplify the management of all relevant quantities, it also facilitates cross-checking of answers by making it easy to solve for the original unknown variables through other methods, or by working backwards to solve for the initially given values from your solutions.

For example, if you have just solved for all unknown voltages, currents, and resistances in a circuit, you can check your work by adding a row at the bottom for power calculations on each resistor, seeing whether or not all the individual power values add up to the total power.

If not, then you must have made a mistake somewhere! While this technique of "cross-checking" your work is nothing new, using the table to arrange all the data for the cross-check(s) results in a minimum of confusion.

- Apply Ohm's Law to vertical columns in the table.
- Apply rules of series/parallel to horizontal rows in the table.
- Check your calculations by working "backwards" to try to arrive at originally given values (from your first calculated answers), or by solving for a quantity using more than one method (from different given values).

COMPONENT FAILURE ANALYSIS

The job of a technician frequently entails "troubleshooting" (locating and correcting a problem) in malfunctioning circuits. Good troubleshooting is a demanding and rewarding effort, requiring a thorough understanding of the basic concepts, the ability to formulate hypotheses (proposed explanations of an effect), the ability to judge the value of different hypotheses based on their probability (how likely one particular cause may be over another), and a sense of creativity in applying a solution to rectify the problem.

While it is possible to distill these skills into a scientific methodology, most practiced troubleshooters would agree that troubleshooting involves a touch of art, and that it can take years of experience to fully develop this art. An essential skill to have is a ready and intuitive understanding of how component faults affect circuits in different configurations.

Let's start with a simple series circuit:

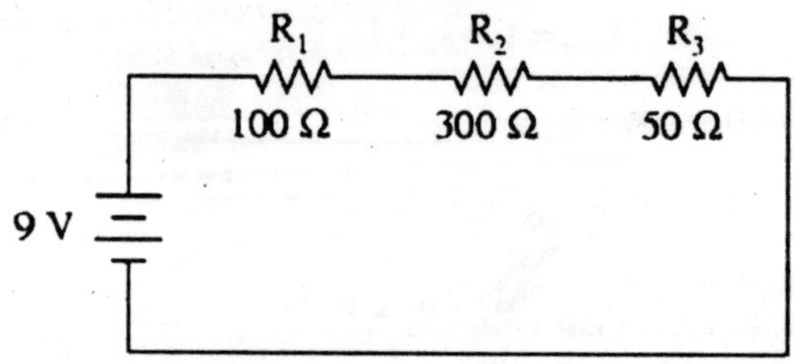

With all components in this circuit functioning at their proper values, we can mathematically determine all currents and voltage drops:

	R_1	R_2	R_3	Total	
E	2	6	1	9	Volts
I	20m	20m	20m	20m	Amps
R	100	300	50	450	Ohms

Now let us suppose that R_2 fails shorted. *Shorted* means that the resistor now acts like a straight piece of wire, with little or no resistance. The circuit will behave as though a "jumper" wire were connected across

R_2 (in case you were wondering, "jumper wire" is a common term for a temporary wire connection in a circuit). What causes the shorted condition of R_2 is no matter to us in this example; we only care about its effect upon the circuit:

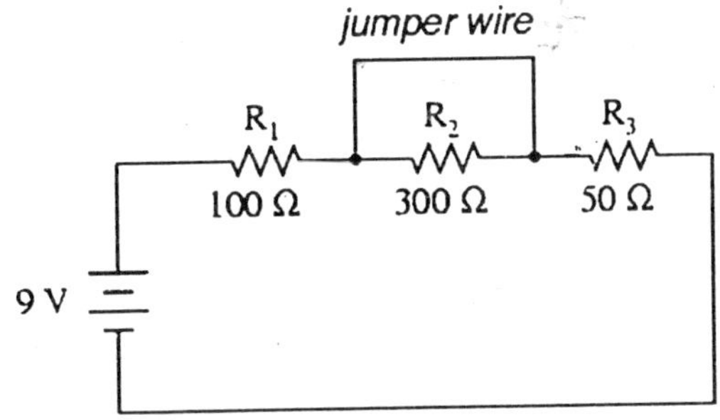

With R_2 shorted, either by a jumper wire or by an internal resistor failure, the total circuit resistance will *decrease*. Since the voltage output by the battery is a constant (at least in our ideal simulation here), a decrease in total circuit resistance means that total circuit current *must increase*:

	R_1	R_2	R_3	Total	
E	6	0	3	9	Volts
I	60m	60m	60m	60m	Amps
R	100	0	50	150	Ohms

↑ *Shorted resistor* (under R_2)

As the circuit current increases from 20 milliamps to 60 milliamps, the voltage drops across R_1 and R_3 (which haven't changed resistances) increase as well, so that the two resistors are dropping the whole 9 volts. R_2, being bypassed by the very low resistance of the jumper wire, is effectively eliminated from the circuit, the resistance from one lead to the other having been reduced to zero. Thus, the voltage drop across R_2, even with the increased total current, is zero volts.

On the other hand, if R_2 were to fail "open" — resistance increasing to nearly infinite levels — it would also create wide-reaching effects in the rest of the circuit:

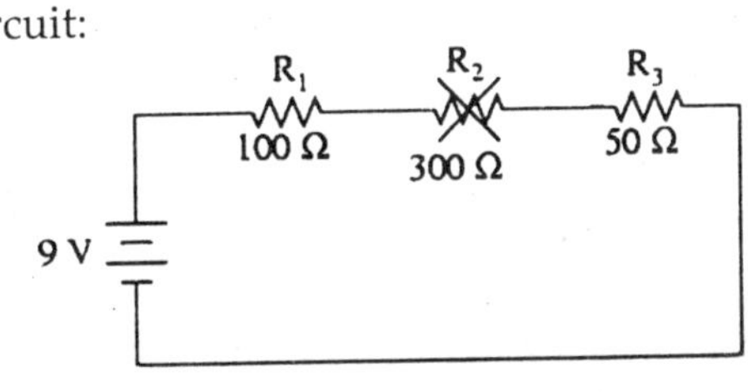

	R_1	R_2	R_3	Total	
E	0	9	0	9	Volts
I	0	0	0	0	Amps
R	100	∞	50	∞	Ohms

↑ *Open resistor* (R_2)

With R_2 at infinite resistance and total resistance being the sum of all individual resistances in a series circuit, the total current decreases to zero. With zero circuit current, there is no electron flow to produce voltage drops across R_1 or R_3. R_2, on the other hand, will manifest the full supply voltage across its terminals. We can apply the same before/after analysis technique to parallel circuits as well. First, we determine what a "healthy" parallel circuit should behave like.

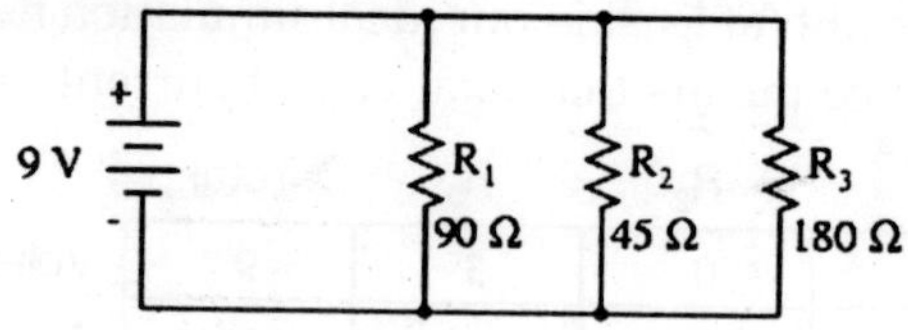

	R_1	R_2	R_3	Total	
E	9	9	9	9	Volts
I	100m	200m	50m	350m	Amps
R	90	45	180	25.714	Ohms

Supposing that R_2 opens in this parallel circuit, here's what the effects will be:

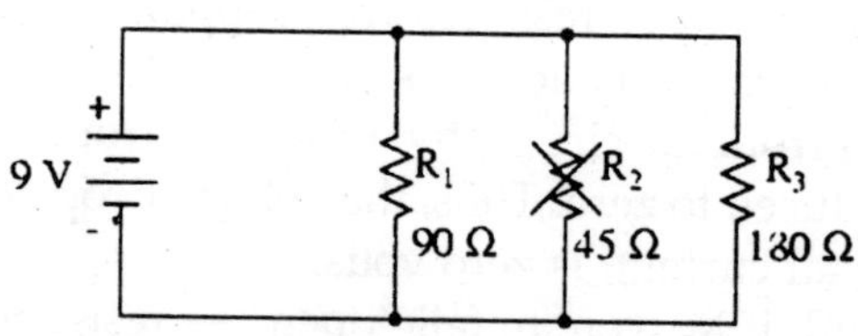

	R_1	R_2	R_3	Total	
E	9	9	9	9	Volts
I	100m	0	50m	150m	Amps
R	90	∞	180	60	Ohms

↑ *Open resistor* (R_2)

Notice that in this parallel circuit, an open branch only affects the current through that branch and the circuit's total current. Total voltage – being shared equally across all components in a parallel circuit, will be the same for all resistors. Due to the fact that the voltage source's tendency is to hold voltage *constant*, its voltage will not change, and being in parallel with all the resistors, it will hold all the resistors' voltages the same as they were before: 9 volts. Being that voltage is the only common parameter in a parallel circuit, and the other resistors haven't changed resistance value, their respective branch currents remain unchanged.

This is what happens in a household lamp circuit: all lamps get their operating voltage from power wiring arranged in a parallel fashion. Turning one lamp on and off (one branch in that parallel circuit closing and opening) doesn't affect the operation of other lamps in the room, only the current in that one lamp (branch circuit) and the total current powering all the lamps in the room:

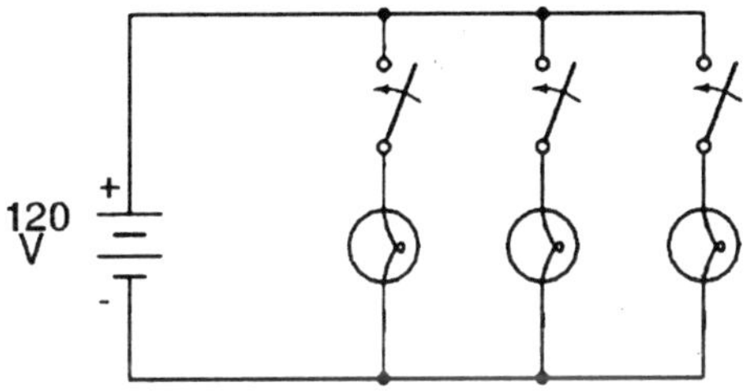

In an ideal case (with perfect voltage sources and zero-resistance connecting wire), shorted resistors in a simple parallel circuit will also have no effect on what's happening in other branches of the circuit. In real life, the effect is not quite the same, and we'll see why in the following example:

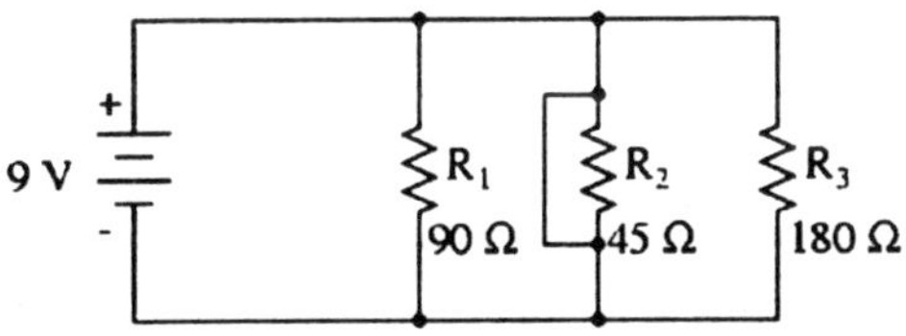

R_2 "shorted" with a jumper wire

	R_1	R_2	R_3	Total	
E	9	9	9	9	Volts
I	100m	∞	50m	∞	Amps
R	90	0	180	0	Ohms

↑ *Shorted resistor* (under R_2)

A shorted resistor (resistance of 0 Ω) would theoretically draw infinite current from any finite source of voltage (I=E/0). In this case, the zero resistance of R_2 decreases the circuit total resistance to zero Ω as well, increasing total current to a value of infinity. As long as the voltage source holds steady at 9 volts, however, the other branch currents (I_{R1} and I_{R3}) will remain unchanged.

The critical assumption in this "perfect" scheme, however, is that the voltage supply will hold steady at its rated voltage while supplying an infinite amount of current to a short-circuit load. This is simply not realistic. Even if the short has a small amount of resistance (as opposed to absolutely zero resistance), no *real* voltage source could arbitrarily supply a huge overload current and maintain steady voltage at the same time. This is primarily due to the internal resistance intrinsic to all electrical power sources, stemming from the inescapable physical properties of the materials they're constructed of:

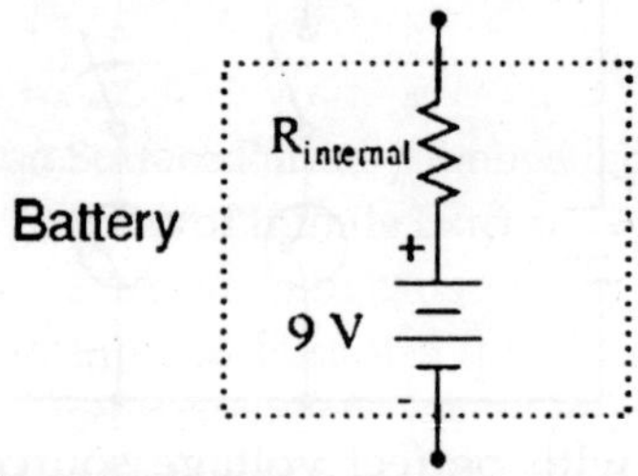

These internal resistances, small as they may be, turn our simple parallel circuit into a series-parallel combination circuit. Usually, the internal resistances of voltage sources are low enough that they can be safely ignored, but when high currents resulting from shorted components are encountered, their effects become very noticeable. In this case, a shorted R_2 would result in almost all the voltage being dropped across the internal resistance of the battery, with almost no voltage left over for resistors R_1, R_2, and R_3:

Suffice it to say, intentional direct short-circuits across the terminals of any voltage source is a bad idea. Even if the resulting high current (heat, flashes, sparks) causes no harm to people nearby, the voltage source will likely sustain damage, unless it has been specifically designed to handle short-circuits, which most voltage sources are not.

The analysis of circuits *without the use of any numbers*, that is, analyzing the effects of component failure in a circuit without knowing exactly how many volts the battery produces, how many ohms of resistance is in each resistor, etc. This section serves as an introductory step to that kind of analysis.

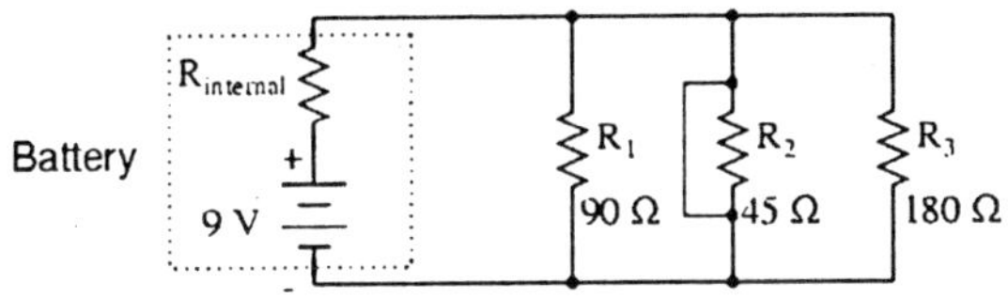

R_2 "shorted" with a jumper wire

	R_1	R_2	R_3	Total	
E	low	low	low	low	Volts
I	low	high	low	high	Amps
R	90	0	180	0	Ohms

↑ *Shorted resistor* (under R_2)

↑ *Supply voltage decrease due to voltage drop across internal resistance* (under Total)

Whereas the normal application of Ohm's Law and the rules of series and parallel circuits is performed with numerical quantities (*"quantitative"*), this new kind of analysis without precise numerical figures something I like to call *qualitative* analysis. In other words, we will be analyzing the *qualities* of the effects in a circuit rather than the precise *quantities*.

The result, for you, will be a much deeper intuitive understanding of electric circuit operation.

- To determine what would happen in a circuit if a component fails, re-draw that circuit with the equivalent resistance of the failed component in place and re-calculate all values.
- The ability to intuitively determine what will happen to a circuit with any given component fault is a *crucial* skill for any electronics troubleshooter to develop. The best way to learn is to experiment with circuit calculations and real-life circuits, paying close attention to what changes with a fault, what remains the same, and *why*!
- A *shorted* component is one whose resistance has dramatically decreased.
- An *open* component is one whose resistance has dramatically increased. For the record, resistors tend to fail open more often than fail shorted, and they almost never fail unless physically or electrically overstressed.

SERIES-PARALLEL CIRCUIT

With simple series circuits, all components are connected end-to-end to form only one path for electrons to flow through the circuit:

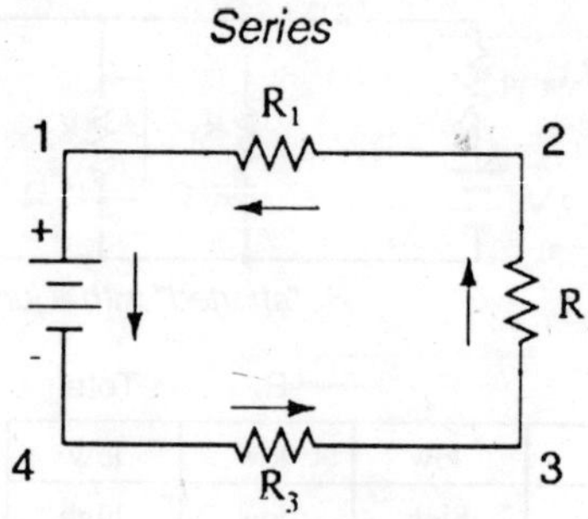

With simple parallel circuits, all components are connected between the same two sets of electrically common points, creating multiple paths for electrons to flow from one end of the battery to the other:

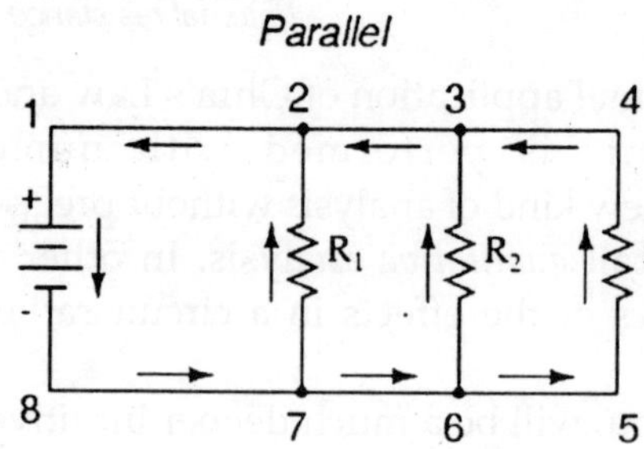

With each of these two basic circuit configurations, we have specific sets of rules describing voltage, current, and resistance relationships.

- Series Circuits:
- Voltage drops add to equal total voltage.
- All components share the same (equal) current.
- Resistances add to equal total resistance.
- Parallel Circuits:
- All components share the same (equal) voltage.
- Branch currents add to equal total current.
- Resistances diminish to equal total resistance.

However, if circuit components are series-connected in some parts and parallel in others, we won't be able to apply a *single* set of rules to every part of that circuit.

Instead, we will have to identify which parts of that circuit are series and which parts are parallel, then selectively apply series and parallel rules as necessary to determine what is happening. Take the following circuit, for instance:

This circuit is neither simple series nor simple parallel. Rather, it contains elements of both. The current exits the bottom of the battery, splits up to travel through R_3 and R_4, rejoins, then splits up again to travel through R_1 and R_2, then rejoins again to return to the top of the battery. There exists more than one path for current to travel (not series), yet there are more than two sets of electrically common points in the circuit (not parallel).

A series-parallel combination circuit

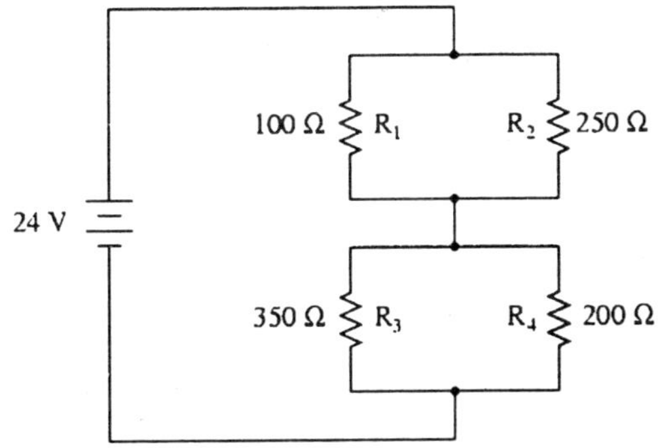

	R_1	R_2	R_3	R_4	Total	
E					24	Volts
I						Amps
R	100	250	350	200		Ohms

Because the circuit is a combination of both series and parallel, we cannot apply the rules for voltage, current, and resistance "across the table" to begin analysis like we could when the circuits were one way or the other.

For instance, if the above circuit were simple series, we could just add up R_1 through R_4 to arrive at a total resistance, solve for total current, and then solve for all voltage drops. Likewise, if the above circuit were simple parallel, we could just solve for branch currents, add up branch currents to figure the total current, and then calculate total resistance from total voltage and total current. However, this circuit's solution will be more complex. The table will still help us manage the different values for series-parallel combination circuits, but we'll have to be careful how and where we apply the different rules for series and parallel. Ohm's Law, of course, still works just the same for determining values within a vertical column in the table. If we are able to identify which parts of the circuit are series and which parts are parallel, we can analyse it in stages, approaching each part one at a time, using the appropriate rules to determine the relationships of voltage, current, and resistance.

The rules of series and parallel circuits must be applied selectively to circuits containing both types of interconnections.

Analysis Technique

The goal of series-parallel resistor circuit analysis is to be able to determine all voltage drops, currents, and power dissipations in a circuit. The general strategy to accomplish this goal is as follows:

- Step 1: Assess which resistors in a circuit are connected together in simple series or simple parallel.
- Step 2: Re-draw the circuit, replacing each of those series or parallel resistor combinations identified in step 1 with a single, equivalent-value resistor. If using a table to manage variables, make a new table column for each resistance equivalent.
- Step 3: Repeat steps 1 and 2 until the entire circuit is reduced to one equivalent resistor.
- Step 4: Calculate total current from total voltage and total resistance (I=E/R).
- Step 5: Taking total voltage and total current values, go back to last step in the circuit reduction process and insert those values where applicable.
- Step 6: From known resistances and total voltage/ total current values from step 5, use Ohm's Law to calculate unknown values (voltage or current) (E=IR or I=E/R).
- Step 7: Repeat steps 5 and 6 until all values for voltage and current are known in the original circuit configuration. Essentially, you will proceed step-by-step from the simplified version of the circuit back into its original, complex form, plugging in values of voltage and current where appropriate until all values of voltage and current are known.
- Step 8: Calculate power dissipations from known voltage, current, and/or resistance values.

A series-parallel combination circuit

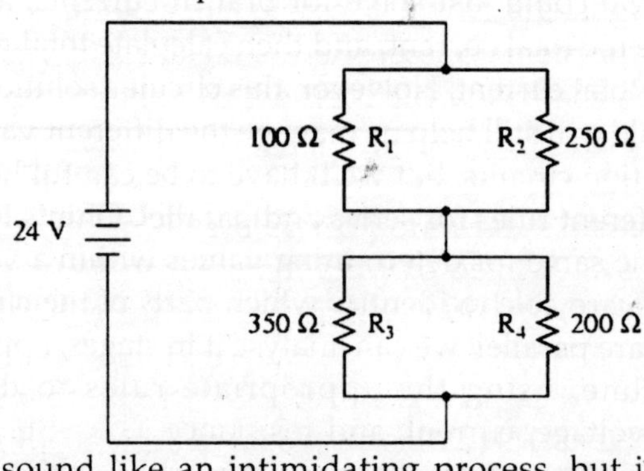

This may sound like an intimidating process, but it's much easier understood through example than through description.

	R_1	R_2	R_3	R_4	Total	
E					24	Volts
I						Amps
R	100	250	350	200		Ohms

In the example circuit above, R_1 and R_2 are connected in a simple parallel arrangement, as are R_3 and R_4. Having been identified, these sections need to be converted into equivalent single resistors, and the circuit re-drawn:

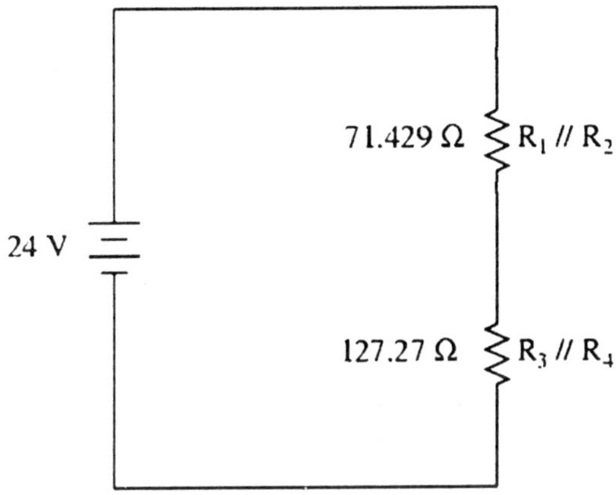

The double slash (//) symbols represent "parallel" to show that the equivalent resistor values were calculated using the 1/(1/R) formula. The 71.429 Ω resistor at the top of the circuit is the equivalent of R_1 and R_2 in parallel with each other.

The 127.27 Ω resistor at the bottom is the equivalent of R_3 and R_4 in parallel with each other.

Our table can be expanded to include these resistor equivalents in their own columns:

	R_1	R_2	R_3	R_4	$R_1 // R_2$	$R_3 // R_4$	Total	
E							24	Volts
I								Amps
R	100	250	350	200	**71.429**	**127.27**		Ohms

It should be apparent now that the circuit has been reduced to a simple series configuration with only two (equivalent) resistances.

The final step in reduction is to add these two resistances to come up with a total circuit resistance.

When we add those two equivalent resistances, we get a resistance of 198.70 Ω. Now, we can re-draw the circuit as a single equivalent resistance and add the total resistance figure to the rightmost column of our table.

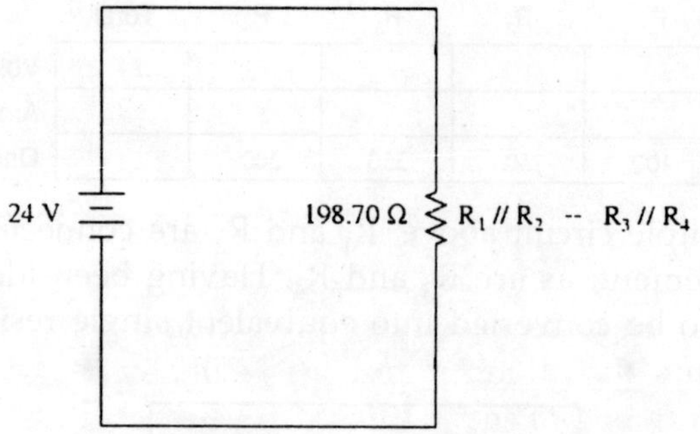

Note that the "Total" column has been relabeled (R_1//R_2—R_3//R_4) to indicate how it relates electrically to the other columns of figures. The "—" symbol is used here to represent "series," just as the "//" symbol is used to represent "parallel."

	R_1	R_2	R_3	R_4	R_1 // R_2	R_3 // R_4	R_1 // R_2 — R_3 // R_4 Total	
E							24	Volts
I								Amps
R	100	250	350	200	71.429	127.27	198.70	Ohms

Now, total circuit current can be determined by applying Ohm's Law (I=E/R) to the "Total" column in the table:

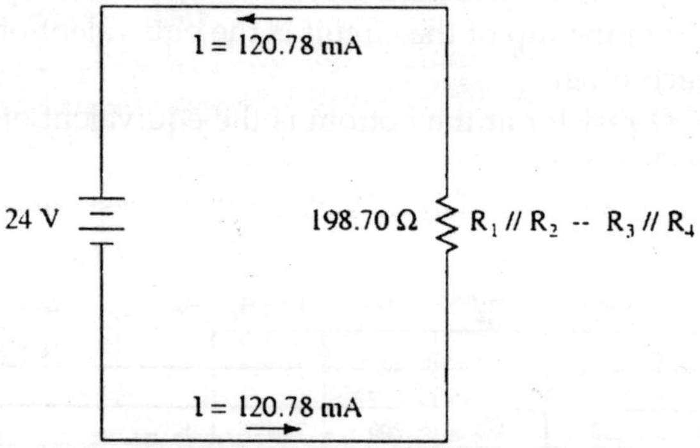

Back to our equivalent circuit drawing, our total current value of 120.78 milliamps is shown as the only current here:

Now we start to work backwards in our progression of circuit re-drawings to the original configuration. The next step is to go to the circuit where R_1//R_2 and R_3//R_4 are in series:

Since R_1//R_2 and R_3//R_4 are in series with each other, the current through those two sets of equivalent resistances must be the same. Furthermore, the current through them must be the same as the total current, so we can fill in our table with the appropriate current values, simply copying the current figure from the Total column to the $R_1//R_2$ and $R_3//R_4$ columns:

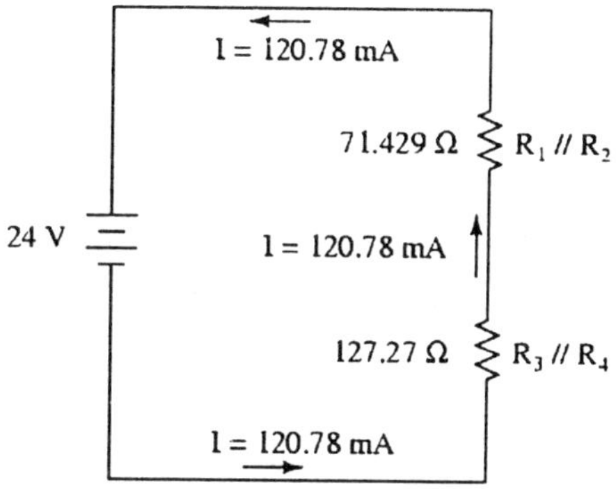

	R_1	R_2	R_3	R_4	$R_1 // R_2$	$R_3 // R_4$	$R_1 // R_2$ — $R_3 // R_4$ Total	
E							24	Volts
I					**120.78m**	**120.78m**	120.78m	Amps
R	100	250	350	200	71.429	127.27	198.70	Ohms

Now, knowing the current through the equivalent resistors $R_1//R_2$ and $R_3//R_4$, we can apply Ohm's Law (E=IR) to the two right vertical columns to find voltage drops across them:

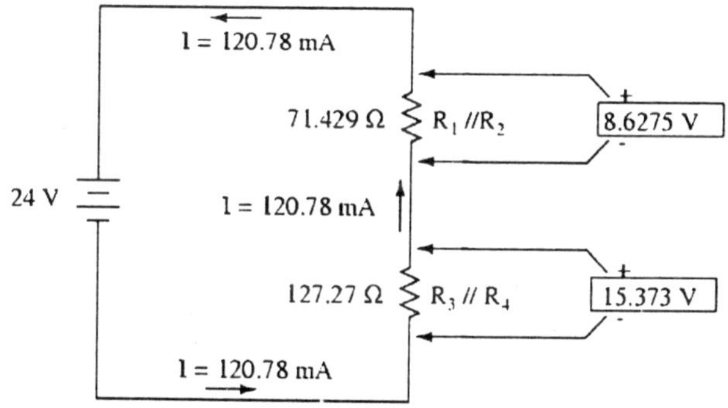

	R_1	R_2	R_3	R_4	$R_1 // R_2$	$R_3 // R_4$	$R_1 // R_2$ — $R_3 // R_4$ Total	
E					**8.6275**	**15.373**	24	Volts
I					120.78m	120.78m	120.78m	Amps
R	100	250	350	200	71.429	127.27	198.70	Ohms

Because we know $R_1//R_2$ and $R_3//R_4$ are parallel resistor equivalents, and we know that voltage drops in parallel circuits are the same, we can transfer the respective voltage drops to the appropriate columns on the table for those individual resistors.

In other words, we take another step backwards in our drawing sequence to the original configuration, and complete the table accordingly:

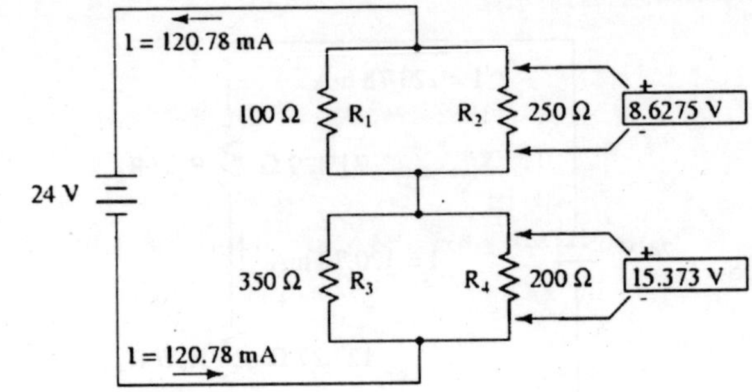

	R_1	R_2	R_3	R_4	$R_1 // R_2$	$R_3 // R_4$	$R_1 // R_2$ — $R_3 // R_4$ Total	
E	8.6275	8.6275	15.373	15.373	8.6275	15.373	24	Volts
I					120.78m	120.78m	120.78m	Amps
R	100	250	350	200	71.429	127.27	198.70	Ohms

Finally, the original section of the table (columns R_1 through R_4) is complete with enough values to finish. Applying Ohm's Law to the remaining vertical columns (I=E/R), we can determine the currents through R_1, R_2, R_3, and R_4 individually:

	R_1	R_2	R_3	R_4	$R_1 // R_2$	$R_3 // R_4$	$R_1 // R_2$ — $R_3 // R_4$ Total	
E	8.6275	8.6275	15.373	15.373	8.6275	15.373	24	Volts
I	**86.275m**	**34.510m**	**43.922m**	**76.863m**	120.78m	120.78m	120.78m	Amps
R	100	250	350	200	71.429	127.27	198.70	Ohms

Having found all voltage and current values for this circuit, we can show those values in the schematic diagram as such:

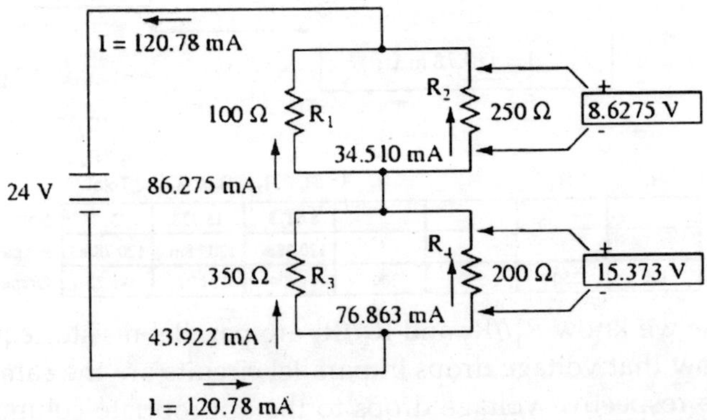

As a final check of our work, we can see if the calculated current values add up as they should to the total. Since R_1 and R_2 are in parallel, their combined currents should add up to the total of 120.78 mA. Likewise, since R_3 and R_4 are in parallel, their combined currents should also add up to the total of 120.78 mA.

All the figures do agree with the our calculated values.

- To analyse a series-parallel combination circuit, follow these steps:
- Reduce the original circuit to a single equivalent resistor, re-drawing the circuit in each step of reduction as simple series and simple parallel parts are reduced to single, equivalent resistors.
- Solve for total resistance.
- Solve for total current (I=E/R).
- Determine equivalent resistor voltage drops and branch currents one stage at a time, working backwards to the original circuit configuration again.

Re-drawing Complex Schematics

Typically, complex circuits are not arranged in nice, neat, clean schematic diagrams for us to follow. They are often drawn in such a way that makes it difficult to follow which components are in series and which are in parallel with each other.

The purpose of this section is to show you a method useful for re-drawing circuit schematics in a neat and orderly fashion. Like the stage-reduction strategy for solving series-parallel combination circuits, it is a method easier demonstrated than described.

Let's start with the following (convoluted) circuit diagram. Perhaps this diagram was originally drawn this way by a technician or engineer. Perhaps it was sketched as someone traced the wires and connections of a real circuit. In any case, here it is in all its ugliness:

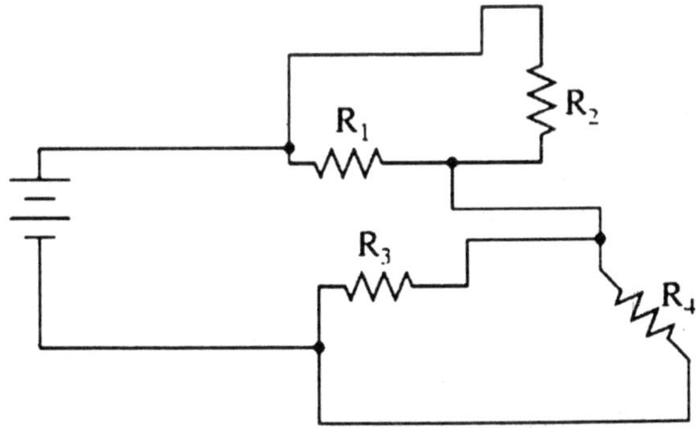

With electric circuits and circuit diagrams, the length and routing of wire connecting components in a circuit matters little. (Actually, in some AC circuits it becomes critical, and very long wire lengths can contribute unwanted resistance to both AC and DC circuits, but in most cases wire length is irrelevant.) What this means for us is that we can lengthen, shrink, and/or bend connecting wires without affecting the operation of our circuit.

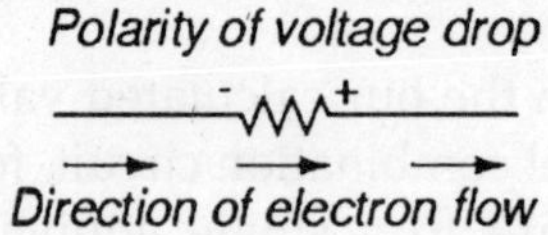

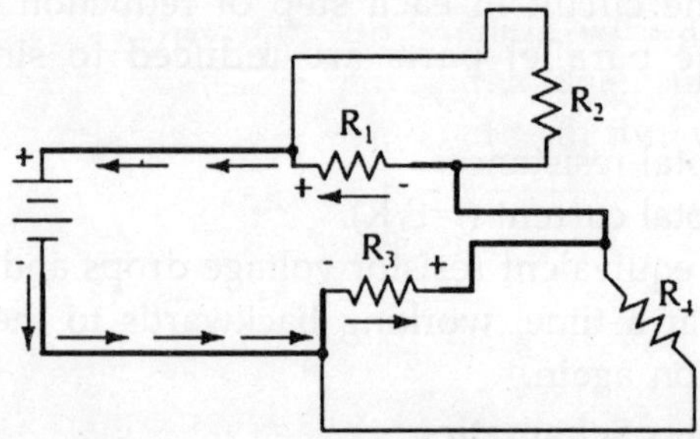

The strategy found easiest to apply is to start by tracing the current from one terminal of the battery around to the other terminal, following the loop of components closest to the battery and ignoring all other wires and components for the time being. While tracing the path of the loop, mark each resistor with the appropriate polarity for voltage drop. At the negative terminal of the battery and finish at the positive terminal, in the same general direction as the electrons would flow.

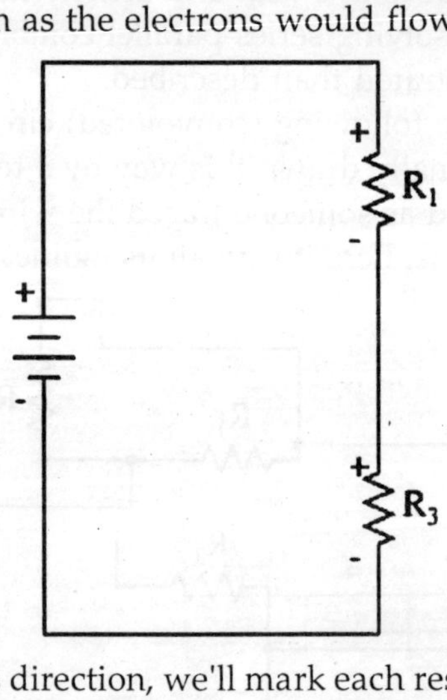

When tracing this direction, we'll mark each resistor with the polarity of negative on the entering side and positive on the exiting side, for that is how the actual polarity will be as electrons (negative in charge) enter and exit a resistor: Any components encountered along this short loop are drawn vertically in order: Now, proceed to trace any loops of components connected around components that were just traced. In this case, there's a loop around R_1 formed by R_2, and another loop around R_3 formed by R_4:

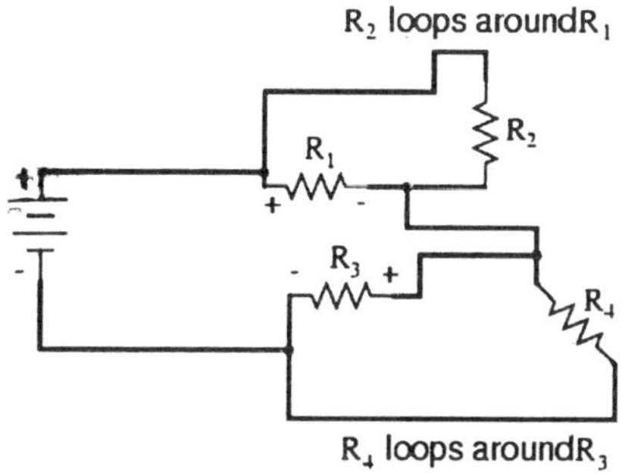

Tracing those loops, I draw R_2 and R_4 iPn parallel with R_1 and R_3 (respectively) on the vertical diagram. Noting the polarity of voltage drops across R_3 and R_1, I mark R_4 and R_2 likewise:

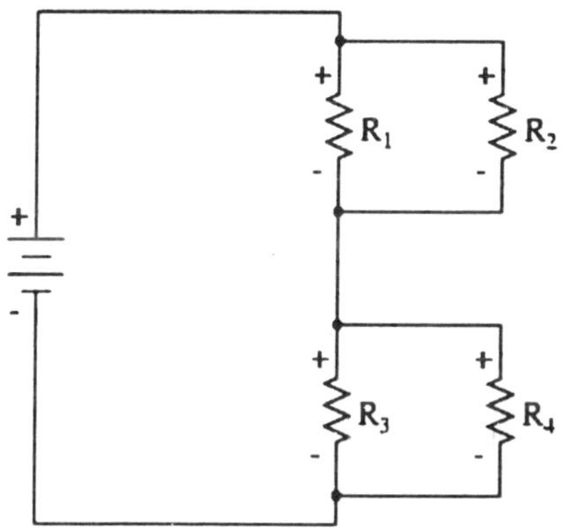

Now we have a circuit that is very easily understood and analyzed. In this case, it is identical to the four-resistor series-parallel configuration.

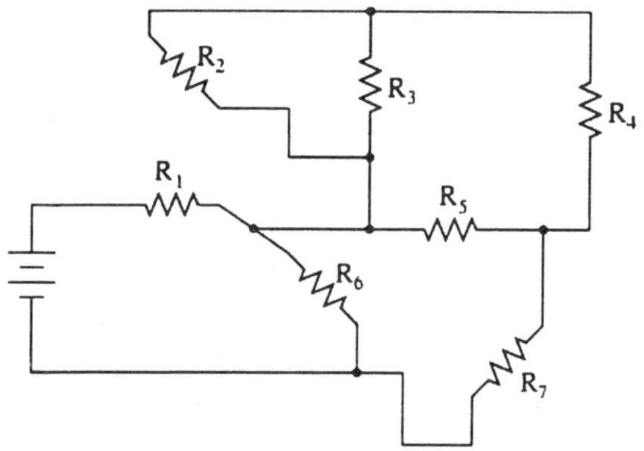

Let's look at another example, even uglier than the one before: The first loop I'll trace is from the negative (-) side of the battery, through R_6, through R_1, and back to the positive (+) end of the battery:

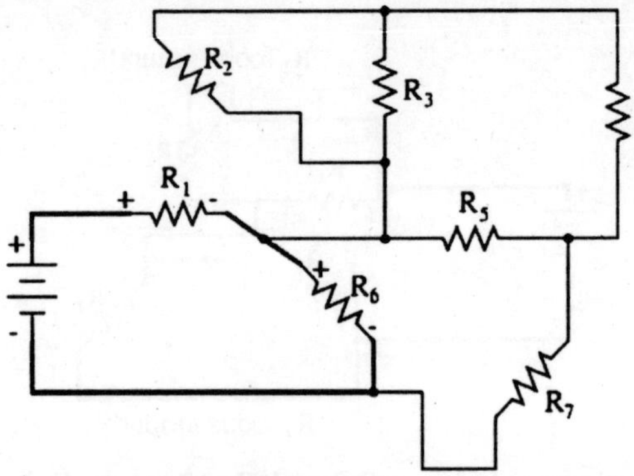

Re-drawing vertically and keeping track of voltage drop polarities along the way, our equivalent circuit starts out looking like this:

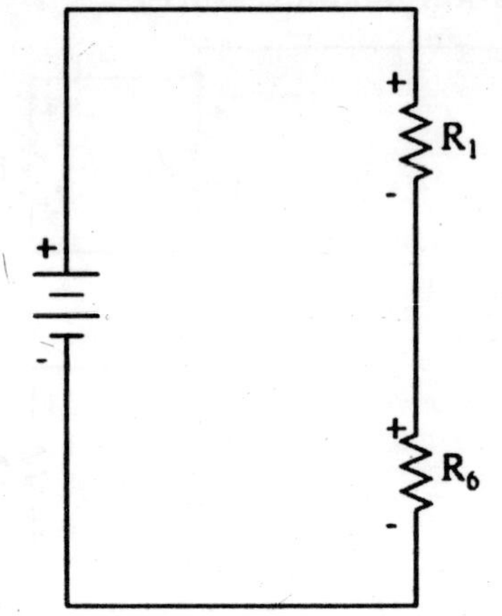

Next, we can proceed to follow the next loop around one of the traced resistors (R_6), in this case, the loop formed by R_5 and R_7. We start at the negative end of R_6 and proceed to the positive end of R_6, marking voltage drop polarities across R_7 and R_5 as we go:

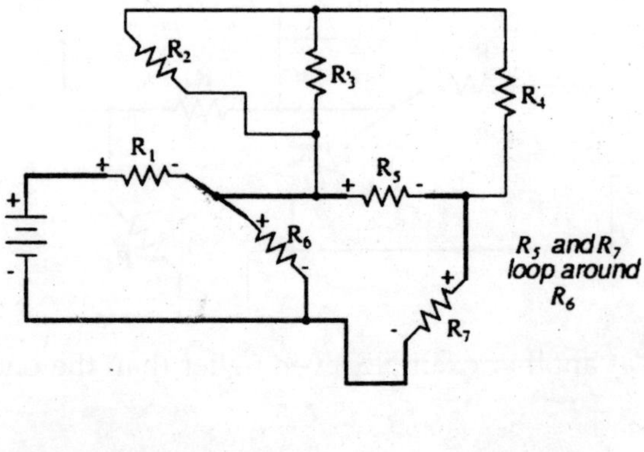

Now we add the R_5—R_7 loop to the vertical drawing. Notice how the voltage drop polarities across R_7 and R_5 correspond with that of R_6, and how this is the same as what we found tracing R_7 and R_5 in the original circuit:

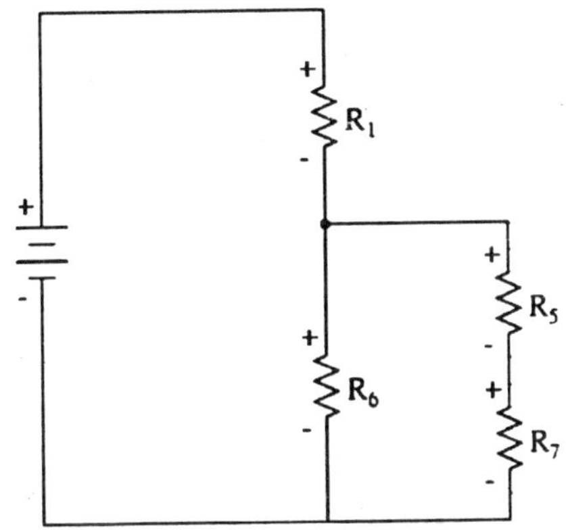

We repeat the process again, identifying and tracing another loop around an already-traced resistor. In this case, the R_3—R_4 loop around R_5 looks like a good loop to trace next:

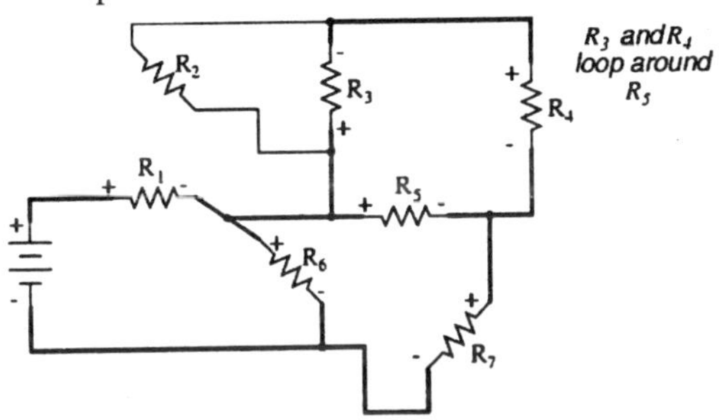

Adding the R_3—R_4 loop to the vertical drawing, marking the correct polarities as well:

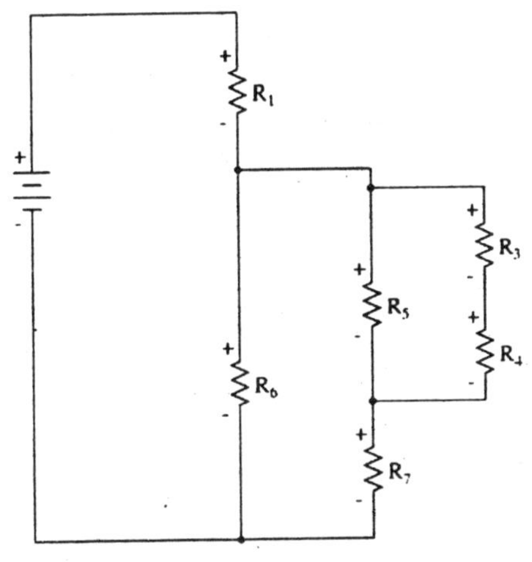

With only one remaining resistor left to trace, then next step is obvious: trace the loop formed by R_2 around R_3:

R_2 loops aroundR_3

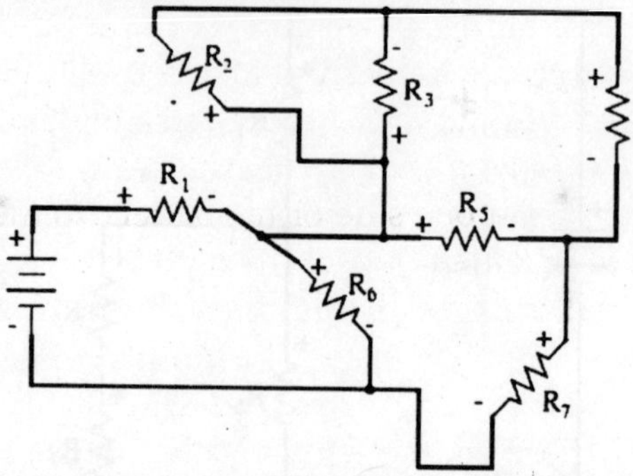

Adding R_2 to the vertical drawing, and we're finished! The result is a diagram that's very easy to understand compared to the original:

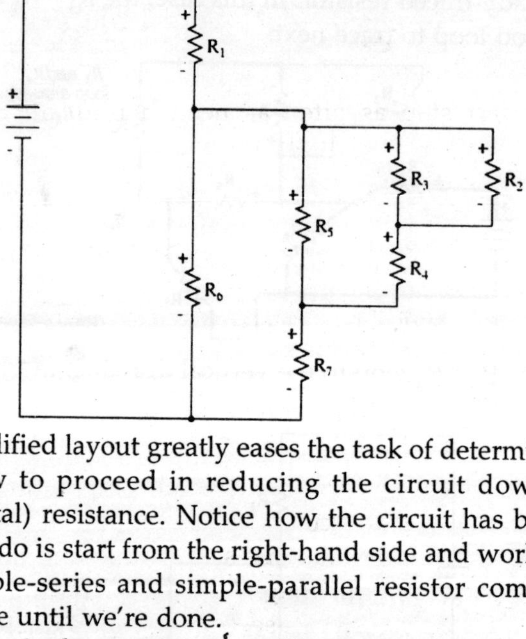

This simplified layout greatly eases the task of determining where to start and how to proceed in reducing the circuit down to a single equivalent (total) resistance. Notice how the circuit has been re-drawn, all we have to do is start from the right-hand side and work our way left, reducing simple-series and simple-parallel resistor combinations one group at a time until we're done.

In this particular case, we would start with the simple parallel combination of R_2 and R_3, reducing it to a single resistance. Then, we would take that equivalent resistance ($R_2//R_3$) and the one in series with it (R_4), reducing them to another equivalent resistance ($R_2//R_3—R_4$). Next, we would proceed to calculate the parallel equivalent of that resistance ($R_2//R_3—R_4$) with R_5, then in series with R_7, then in parallel with R_6, then in series with R_1 to give us a grand total resistance for the circuit as a whole. From there we could calculate total current from total voltage and total resistance (I=E/R), then "expand" the circuit back into its original form one stage at a time, distributing the appropriate values of voltage and current to the resistances as we go.

- Wires in diagrams and in real circuits can be lengthened, shortened, and/or moved without affecting circuit operation.
- To simplify a convoluted circuit schematic, follow these steps:
- Trace current from one side of the battery to the other, following any single path ("loop") to the battery. Sometimes it works better to start with the loop containing the most components, but regardless of the path taken the result will be accurate. Mark polarity of voltage drops across each resistor as you trace the loop. Draw those components you encounter along this loop in a vertical schematic.
- Mark traced components in the original diagram and trace remaining loops of components in the circuit. Use polarity marks across traced components as guides for what connects where. Document new components in loops on the vertical re-draw schematic as well.
- Repeat last step as often as needed until all components in original diagram have been traced.

Component Failure Analysis

There is a lot of truth to that quote from Dirac. With a little modification, I can extend his wisdom to electric circuits.

We briefly considered how circuits could be analyzed in a *qualitative* rather than *quantitative* manner.

Building this skill is an important step towards becoming a proficient troubleshooter of electric circuits. Once you have a thorough understanding of how any particular failure will affect a circuit, it will be much easier to work the other way around: pinpointing the source of trouble by assessing how a circuit is behaving.

We may take this technique one step further and adapt it for total qualitative analysis. By *"qualitative"* I mean working with symbols representing "increase," "decrease," and "same" instead of precise numerical figures. We can still use the principles of series and parallel circuits, and the concepts of Ohm's Law, we'll just use symbolic *qualities* instead of numerical *quantities*. By doing this, we can gain more of an intuitive "feel" for how circuits work rather than leaning on abstract equations, attaining Dirac's definition of "understanding."

Enough talk. Let's try this technique on a real circuit example and see how it works:

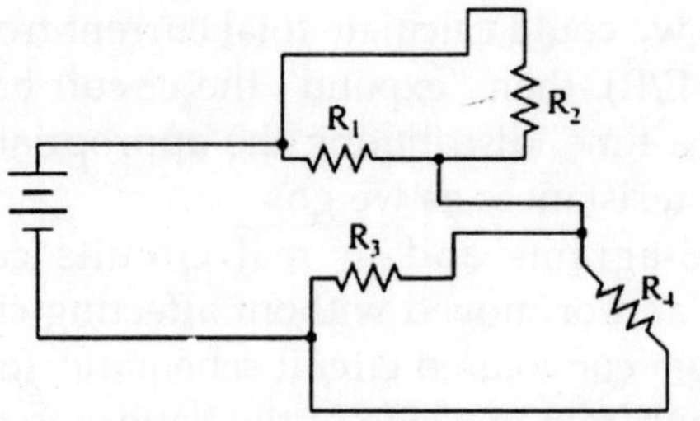

This is the first "convoluted" circuit we straightened out for analysis in the last section. Since you already know how this particular circuit reduces to series and parallel sections.

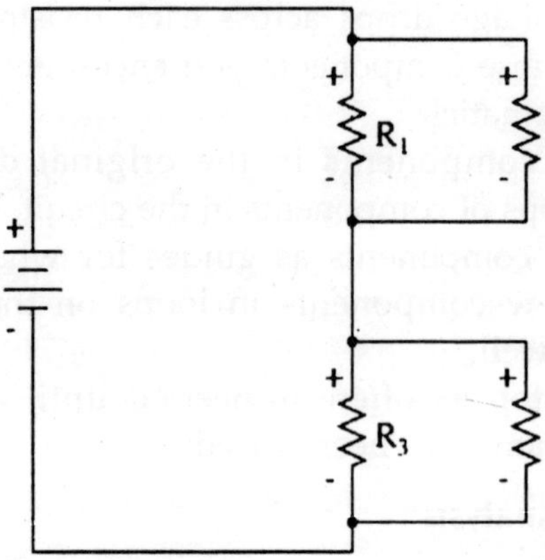

R_3 and R_4 are in parallel with each other; so are R_1 and R_2. The parallel equivalents of $R_3//R_4$ and $R_1//R_2$ are in series with each other. Expressed in symbolic form, the total resistance for this circuit is as follows:

$R_{Total} = (R_1//R_2)—(R_3//R_4)$

First, we need to formulate a table with all the necessary rows and columns for this circuit:

	R_1	R_2	R_3	R_4	$R_1 // R_2$	$R_3 // R_4$	Total	
E								Volts
I								Amps
R								Ohms

Next, we need a failure scenario. Let's suppose that resistor R_2 were to fail shorted. We will assume that all other components maintain their original values. Because we'll be analyzing this circuit qualitatively rather than quantitatively, we won't be inserting any real numbers into the table. For any quantity unchanged after the component failure, we'll use the word "same" to represent "no change from before." For any quantity that has changed as a result of the failure, we'll use a down arrow for "decrease" and an up arrow for "increase." As usual, we start by filling in the spaces of the table for individual resistances and total voltage, our "given" values:

	R_1	R_2	R_3	R_4	$R_1//R_2$	$R_3//R_4$	Total	
E							same	Volts
I								Amps
R	same	↓	same	same				Ohms

The only "given" value different from the normal state of the circuit is R_2, which we said was failed shorted (abnormally low resistance). All other initial values are the same as they were before, as represented by the "same" entries. All we have to do now is work through the familiar Ohm's Law and series-parallel principles to determine what will happen to all the other circuit values.

First, we need to determine what happens to the resistances of parallel subsections $R_1//R_2$ and $R_3//R_4$. If neither R_3 nor R_4 have changed in resistance value, then neither will their parallel combination. However, since the resistance of R_2 has decreased while R_1 has stayed the same, their parallel combination must decrease in resistance as well:

	R_1	R_2	R_3	R_4	$R_1//R_2$	$R_3//R_4$	Total	
E							same	Volts
I								Amps
R	same	↓	same	same	↓	same		Ohms

Now, we need to figure out what happens to the total resistance. This part is easy: when we're dealing with only one component change in the circuit, the change in total resistance will be in the same direction as the change of the failed component. This is not to say that the *magnitude* of change between individual component and total circuit will be the same, merely the *direction* of change. In other words, if any single resistor decreases in value, then the total circuit resistance must also decrease, and vice versa. In this case, since R_2 is the only failed component, and its resistance has decreased, the total resistance *must* decrease:

	R_1	R_2	R_3	R_4	$R_1//R_2$	$R_3//R_4$	Total	
E							same	Volts
I								Amps
R	same	↓	same	same	↓	same	↓	Ohms

Now we can apply Ohm's Law (qualitatively) to the Total column in the table. Given the fact that total voltage has remained the same and total resistance has decreased, we can conclude that total current must increase ($I=E/R$).

In case you're not familiar with the qualitative assessment of an equation, it works like this. First, we write the equation as solved for the unknown quantity. In this case, we're trying to solve for current, given voltage and resistance:

$$I = \frac{E}{R}$$

Now that our equation is in the proper form, change given the change(s) to "E" and "R":

$$I = \frac{E \text{ (same)}}{R \downarrow}$$

If the denominator of a fraction decreases in value while the numerator stays the same, then the overall value of the fraction must increase:

$$\uparrow I = \frac{E \text{ (same)}}{R \downarrow}$$

Therefore, Ohm's Law (I=E/R) tells us that the current (I) will increase. We'll mark this conclusion in our table with an "up" arrow:

	R_1	R_2	R_3	R_4	$R_1//R_2$	$R_3//R_4$	Total	
E							same	Volts
I							↑	Amps
R	same	↓	same	same	↓	same	↓	Ohms

With all resistance places filled in the table and all quantities determined in the Total column, we can proceed to determine the other voltages and currents. Knowing that the total resistance in this table was the result of $R_1//R_2$ and $R_3//R_4$ in *series,* we know that the value of total current will be the same as that in $R_1//R_2$ and $R_3//R_4$ (because series components share the same current). Therefore, if total current increased, then current through $R_1//R_2$ and $R_3//R_4$ must also have increased with the failure of R_2:

	R_1	R_2	R_3	R_4	$R_1//R_2$	$R_3//R_4$	Total	
E							same	Volts
I					↑	↑	↑	Amps
R	same	↓	same	same	↓	same	↓	Ohms

Fundamentally, what we're doing here with a qualitative usage of Ohm's Law and the rules of series and parallel circuits is no different from what we've done before with numerical figures. In fact, it's a lot easier because you don't have to worry about making an arithmetic or calculator keystroke error in a calculation. Instead, you're just focusing on the *principles* behind the equations.

From our table above, we can see that Ohm's Law should be applicable to the $R_1//R_2$ and $R_3//R_4$ columns. For $R_3//R_4$, we figure what happens to the voltage, given an increase in current and no change in resistance. Intuitively, we can see that this must result in an increase in voltage across the parallel combination of $R_3//R_4$:

	R_1	R_2	R_3	R_4	$R_1//R_2$	$R_3//R_4$	Total	
E						↑	same	Volts
I					↑	↑	↑	Amps
R	same	↓	same	same	↓	same	↓	Ohms

But how do we apply the same Ohm's Law formula (E = IR) to the $R_1//R_2$ column, where we have resistance decreasing *and* current increasing? It's easy to determine if only one variable is changing, as it was with $R_3//R_4$, but with two variables moving around and no definite numbers to work with, Ohm's Law isn't going to be much help.

However, there is another rule we can apply *horizontally* to determine what happens to the voltage across $R_1//R_2$: the rule for voltage in series circuits.

If the voltages across $R_1//R_2$ and $R_3//R_4$ add up to equal the total (battery) voltage and we know that the $R_3//R_4$ voltage has increased while total voltage has stayed the same, then the voltage across $R_1//R_2$ *must* have decreased with the change of R_2's resistance value:

	R_1	R_2	R_3	R_4	$R_1//R_2$	$R_3//R_4$	Total	
E					↓	↑	same	Volts
I					↑	↑	↑	Amps
R	same	↓	same	same	↓	same	↓	Ohms

Now we're ready to proceed to some new columns in the table. Knowing that R_3 and R_4 comprise the parallel subsection $R_3//R_4$, and knowing that voltage is shared equally between parallel components, the increase in voltage seen across the parallel combination $R_3//R_4$ must also be seen across R_3 and R_4 individually:

	R_1	R_2	R_3	R_4	$R_1//R_2$	$R_3//R_4$	Total	
E			↑	↑	↓	↑	same	Volts
I					↑	↑	↑	Amps
R	same	↓	same	same	↓	same	↓	Ohms

The same goes for R_1 and R_2. The voltage decrease seen across the parallel combination of R_1 and R_2 will be seen across R_1 and R_2 individually:

	R_1	R_2	R_3	R_4	$R_1//R_2$	$R_3//R_4$	Total	
E	↓	↓	↑	↑	↓	↑	same	Volts
I					↑	↑	↑	Amps
R	same	↓	same	same	↓	same	↓	Ohms

Applying Ohm's Law vertically to those columns with unchanged ("same") resistance values, we can tell what the current will do through those components. Increased voltage across an unchanged resistance leads to increased current. Conversely, decreased voltage across an unchanged resistance leads to decreased current:

	R_1	R_2	R_3	R_4	$R_1//R_2$	$R_3//R_4$	Total	
E	↓	↓	↑	↑	↓	↑	same	Volts
I	↓		↑	↑	↑	↑	↑	Amps
R	same	↓	same	same	↓	same	↓	Ohms

Once again we find ourselves in a position where Ohm's Law can't help us: for R_2, both voltage and resistance have decreased, but without knowing *how much* each one has changed, we can't use the I=E/R formula to qualitatively determine the resulting change in current.

However, we can still apply the rules of series and parallel circuits *horizontally*. We know that the current through the $R_1//R_2$ parallel combination has increased, and we also know that the current through R_1 has decreased.

One of the rules of parallel circuits is that total current is equal to the sum of the individual branch currents. In this case, the current through $R_1//R_2$ is equal to the current through R_1 added to the current through R_2. If current through $R_1//R_2$ has increased while current through R_1 has decreased, current through R_2 *must* have increased:

	R_1	R_2	R_3	R_4	$R_1//R_2$	$R_3//R_4$	Total	
E	↓	↓	↑	↑	↓	↑	same	Volts
I	↓	↑	↑	↑	↑	↑	↑	Amps
R	same	↓	same	same	↓	same	↓	Ohms

And with that, our table of qualitative values stands completed. This particular exercise may look laborious due to all the detailed commentary, but the actual process can be performed very quickly with some practice. An important thing to realise here is that the general procedure is little different from quantitative analysis: start with the known values, then proceed to determining total resistance, then total current, then transfer figures of voltage and current as allowed by the rules of series and parallel circuits to the appropriate columns.

A few general rules can be memorized to assist and/or to check your progress when proceeding with such an analysis:

- For any *single* component failure (open or shorted), the total resistance will always change in the same direction (either increase or decrease) as the resistance change of the failed component.
- When a component fails shorted, its resistance always decreases. Also, the current through it will increase, and the voltage across it *may* drop. In some cases it will remain the same.
- When a component fails open, its resistance always increases. The current through that component will decrease to zero,

because it is an incomplete electrical path. This *may* result in an increase of voltage across it. The same exception stated above applies here as well: in a simple parallel circuit with an ideal voltage source, the voltage across an open-failed component will remain unchanged.

Building Series-parallel Resistor Circuits

Once again, when building battery/resistor circuits, the student or hobbyist is faced with several different modes of construction. Perhaps the most popular is the *solderless breadboard*: a platform for constructing temporary circuits by plugging components and wires into a grid of interconnected points.

A breadboard appears to be nothing but a plastic frame with hundreds of small holes in it. Underneath each hole, though, is a spring clip which connects to other spring clips beneath other holes. The connection pattern between holes is simple and uniform:

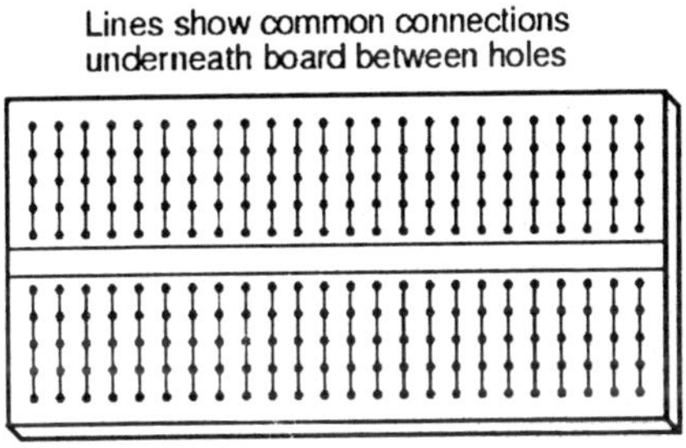

Suppose we wanted to construct the following series-parallel combination circuit on a breadboard:

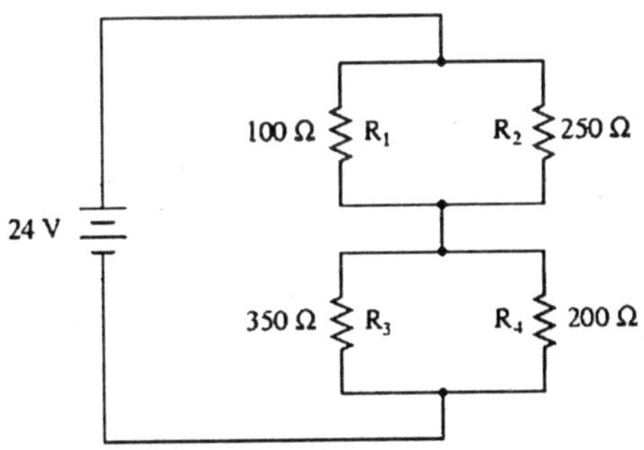

The recommended way to do so on a breadboard would be to arrange the resistors in approximately the same pattern as seen in the schematic, for ease of relation to the schematic. If 24 volts is required and we only have 6-volt batteries available, four may be connected in series to achieve the same effect:

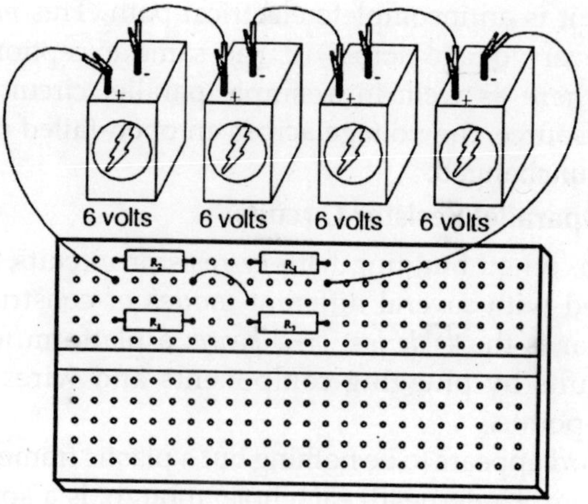

This is by no means the only way to connect these four resistors together to form the circuit shown in the schematic. Consider this alternative layout:

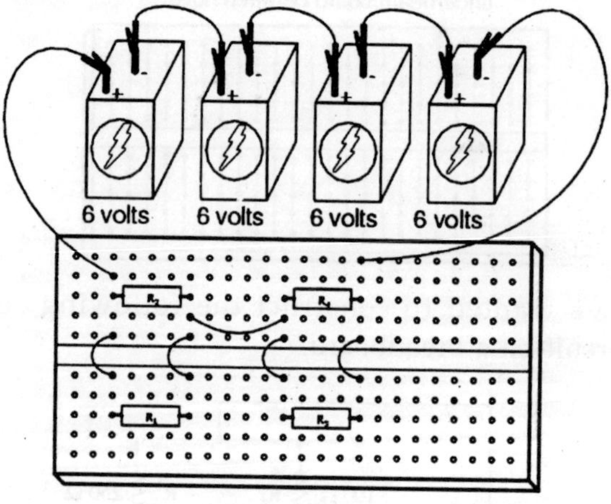

If greater permanence is desired without resorting to soldering or wire-wrapping, one could choose to construct this circuit on a *terminal strip* (also called a *barrier strip*, or *terminal block*).

In this method, components and wires are secured by mechanical tension underneath screws or heavy clips attached to small metal bars. The metal bars, in turn, are mounted on a nonconducting body to keep them electrically isolated from each other.

Building a circuit with components secured to a terminal strip isn't as easy as plugging components into a breadboard, principally because the components cannot be physically arranged to resemble the schematic layout.

Instead, the builder must understand how to "bend" the schematic's representation into the real-world layout of the strip. Consider one example of how the same four-resistor circuit could be built on a terminal strip:

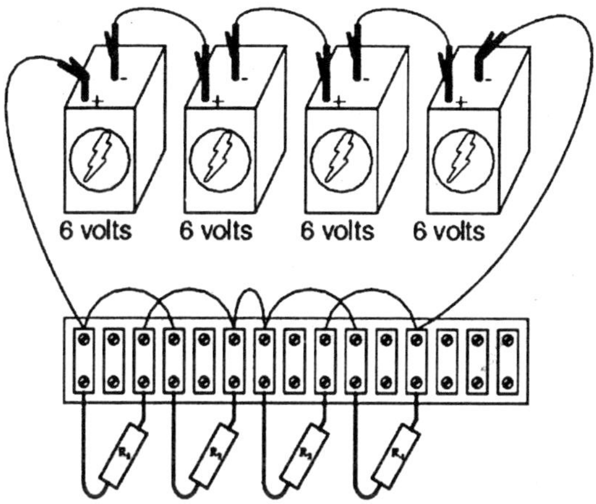

Another terminal strip layout, simpler to understand and relate to the schematic, involves anchoring parallel resistors ($R_1//R_2$ and $R_3//R_4$) to the same two terminal points on the strip like this:

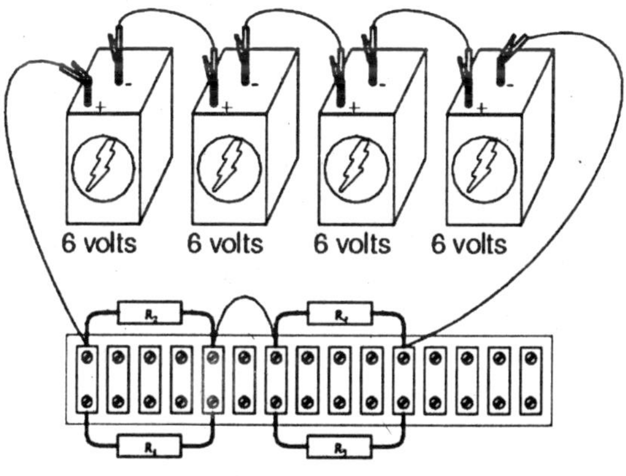

Building more complex circuits on a terminal strip involves the same spatial-reasoning skills, but of course requires greater care and planning. Take for instance this complex circuit, represented in schematic form:

The terminal strip used in the prior example barely has enough terminals to mount all seven resistors required for this circuit! It will be a challenge to determine all the necessary wire connections between resistors, but with patience it can be done.

First, begin by installing and labeling all resistors on the strip. The original schematic diagram will be shown next to the terminal strip circuit for reference:

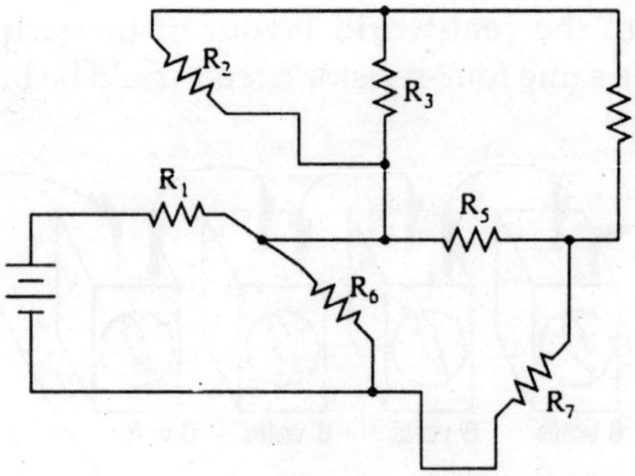

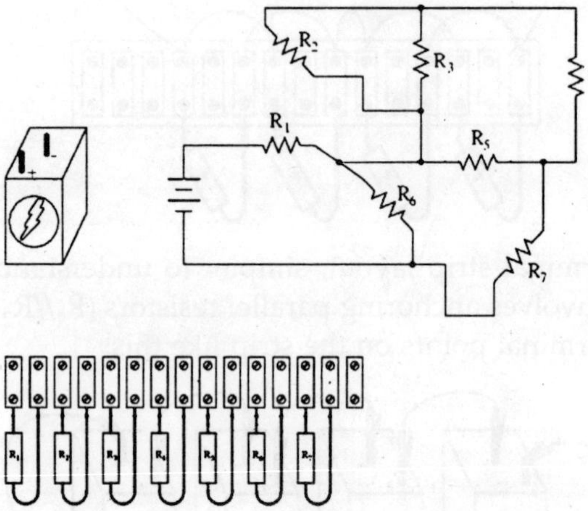

Next, begin connecting components together wire by wire as shown in the schematic. Over-draw connecting lines in the schematic to indicate completion in the real circuit.

Watch this sequence of illustrations as each individual wire is identified in the schematic, then added to the real circuit:

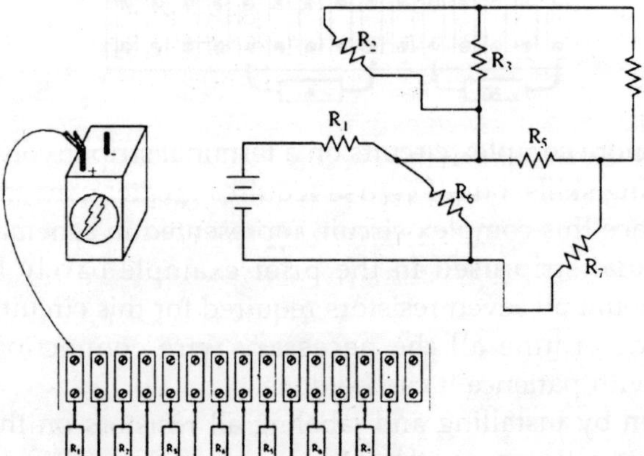

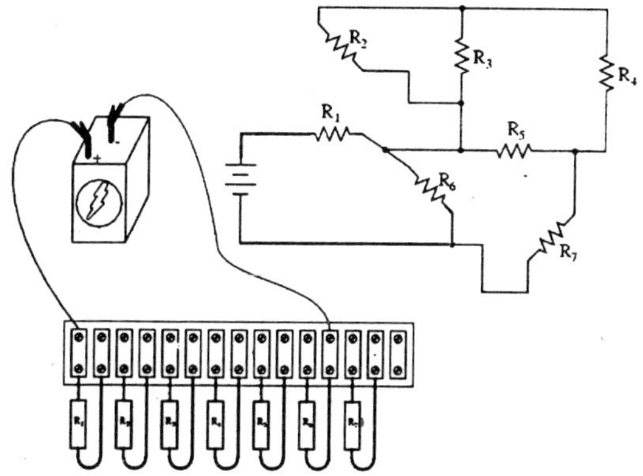

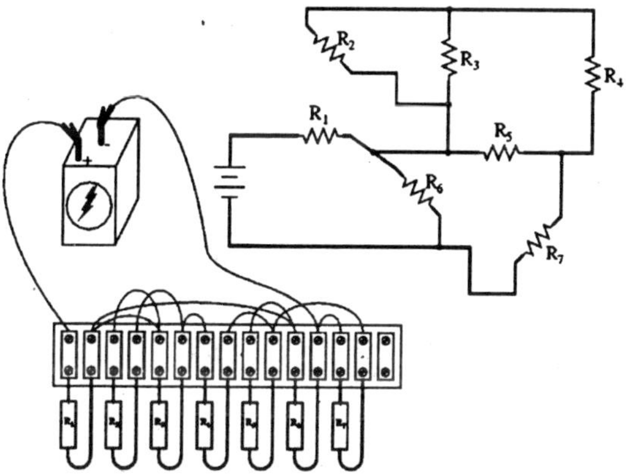

Although there are minor variations possible with this terminal strip circuit, the choice of connections shown in this example sequence is both electrically accurate (electrically identical to the schematic diagram) and carries the additional benefit of not burdening any one screw terminal on the strip with more than two wire ends, a good practice in any terminal strip circuit. An example of a "variant" wire connection might be the very last wire added, which we placed between the left terminal of R_2 and the left terminal of R_3. This last wire completed the parallel connection between R_2 and R_3 in the circuit.

However, I could have placed this wire instead between the left terminal of R_2 and the right terminal of R_1, since the right terminal of R_1 is already connected to the left terminal of R_3 and so is electrically common with that one point.

Doing this, though, would have resulted in *three* wires secured to the right terminal of R_1 instead of two, which is a *faux pax* in terminal strip etiquette. Would the circuit have worked this way? Certainly! It's just that more than two wires secured at a single terminal makes for a "messy" connection: one that is aesthetically unpleasing and may place undue stress on the screw terminal. Another variation would be to reverse the terminal connections for resistor R_7. The voltage polarity across R_7 is negative on the left and positive on the right (–, +), whereas all the other resistor polarities are positive on the left and negative on the right (+, –):

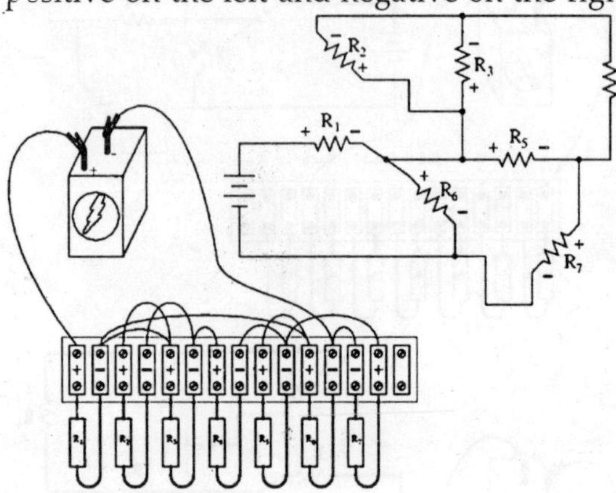

While this poses no electrical problem, it might cause confusion for anyone measuring resistor voltage drops with a voltmeter, especially an analog voltmeter which will "peg" downscale when subjected to a voltage of the wrong polarity. For the sake of consistency, it might be wise to arrange all wire connections so that all resistor voltage drop polarities are the same, like this:

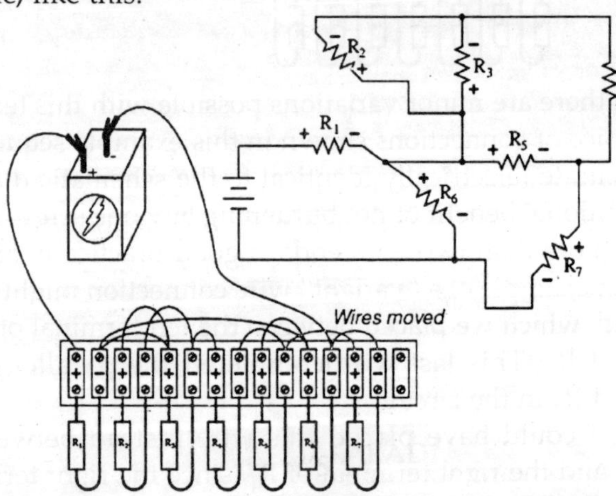

Though electrons do not care about such consistency in component layout, people do. This illustrates an important aspect of any engineering endeavor: the human factor. Whenever a design may be modified for easier comprehension and/or easier maintenance with no sacrifice of functional performance it should be done so.

Chapter 4

Magnetism and Electromagnetism

PERMANENT MAGNETS

Centuries ago, it was discovered that certain types of mineral rock possessed unusual properties of attraction to the metal iron. One particular mineral, called *lodestone,* or *magnetite,* is found mentioned in very old historical records (about 2500 years ago in Europe, and much earlier in the Far East) as a subject of curiosity.

Later, it was employed in the aid of navigation, as it was found that a piece of this unusual rock would tend to orient itself in a north-south direction if left free to rotate (suspended on a string or on a float in water). A scientific study undertaken in 1269 by Peter Peregrinus revealed that steel could be similarly "charged" with this unusual property after being rubbed against one of the "poles" of a piece of lodestone.

Unlike electric charges (such as those observed when amber is rubbed against cloth), magnetic objects possessed two poles of opposite effect, denoted "north" and "south" after their self-orientation to the earth. As Peregrinus found, it was impossible to isolate one of these poles by itself by cutting a piece of lodestone in half: each resulting piece possessed its own pair of poles:

Like electric charges, there were only two types of poles to be found: north and south (by analogy, positive and negative). Just as with electric charges, same poles repel one another, while opposite poles attract. This force, like that caused by static electricity, extended itself invisibly over space, and could even pass through objects such as paper and wood with little effect upon strength.

The philosopher-scientist Rene Descartes noted that this invisible "field" could be mapped by placing a magnet underneath a flat piece of cloth or wood and sprinkling iron filings on top. The filings will align themselves with the magnetic field, "mapping" its shape. The result shows how the field continues unbroken from one pole of a magnet to the other:

As with any kind of field (electric, magnetic, gravitational), the total quantity, or effect, of the field is referred to as a *flux,* while the "push" causing the flux to form in space is called a *force*. Michael Faraday coined the term "tube" to refer to a string of magnetic flux in space (the term "line" is more commonly used now). Indeed, the measurement of magnetic field flux is often defined in terms of the number of flux lines, although it is doubtful that such fields exist in individual, discrete lines of constant value.

Modern theories of magnetism maintain that a magnetic field is produced by an electric charge in motion, and thus it is theorized that the magnetic field of a so-called "permanent" magnets such as lodestone is the result of electrons within the atoms of iron spinning uniformly in the same direction. Whether or not the electrons in a material's atoms are subject to this kind of uniform spinning is dictated by the atomic structure of the material (not unlike how electrical conductivity is dictated by the electron binding in a material's atoms). Thus, only certain types of substances react with magnetic fields, and even fewer have the ability to permanently sustain a magnetic field.

Iron is one of those types of substances that readily magnetizes. If a piece of iron is brought near a permanent magnet, the electrons within the atoms in the iron orient their spins to match the magnetic field force produced by the permanent magnet, and the iron becomes "magnetized." The iron will magnetize in such a way as to incorporate the magnetic flux lines into its shape, which attracts it toward the permanent magnet, no matter which pole of the permanent magnet is offered to the iron:

The unmagnetized iron becomes magnetized as it is brought closer to the permanent magnet. No matter what pole of the permanent magnet is extended toward the iron, the iron will magnetize in such a way as to be attracted toward the magnet:

Referencing the natural magnetic properties of iron (Latin = "ferrum"), a *ferromagnetic* material is one that readily magnetizes (its constituent atoms easily orient their electron spins to conform to an external magnetic field force). All materials are magnetic to some degree, and those that are not considered ferromagnetic (easily magnetized) are classified as either *paramagnetic* (slightly magnetic) or *diamagnetic* (tend to exclude magnetic fields). Of the two, diamagnetic materials are the strangest. In the presence of an external magnetic field, they actually become slightly magnetized in the opposite direction, so as to repel the external field!

If a ferromagnetic material tends to retain its magnetization after an external field is removed, it is said to have good *retentivity*. This, of course, is a necessary quality for a permanent magnet.

ELECTROMAGNETISM

The discovery of the relationship between magnetism and electricity was, like so many other scientific discoveries, stumbled upon almost by accident. The Danish physicist Hans Christian Oersted was lecturing one day in 1820 on the *possibility* of electricity and magnetism being related to one another, and in the process demonstrated it conclusively by experiment in front of his whole class! By passing an electric current through a metal wire suspended above a magnetic compass, Oersted was able to produce a definite motion of the compass needle in response to the current.

What began as conjecture at the start of the class session was confirmed as fact at the end. Needless to say, Oersted had to revise his lecture notes for future classes! His serendipitous discovery paved the way for a whole new branch of science: electromagnetics.

Detailed experiments showed that the magnetic field produced by an electric current is always oriented perpendicular to the direction of flow. A simple method of showing this relationship is called the *left-hand rule*. Simply stated, the left-hand rule says that the magnetic flux lines produced by a current-carrying wire will be oriented the same direction as the curled fingers of a person's left hand (in the "hitchhiking" position), with the thumb pointing in the direction of electron flow:

The magnetic field encircles this straight piece of current-carrying wire, the magnetic flux lines having no definite "north" or "south' poles.

While the magnetic field surrounding a current-carrying wire is indeed interesting, it is quite weak for common amounts of current, able to deflect a compass needle and not much more.

To create a stronger magnetic field force (and consequently, more field flux) with the same amount of electric current, we can wrap the wire into a coil shape, where the circling magnetic fields around the wire will join to create a larger field with a definite magnetic (north and south) polarity:

The amount of magnetic field force generated by a coiled wire is proportional to the current through the wire multiplied by the number of "turns" or "wraps" of wire in the coil. This field force is called *magnetomotive force* (mmf), and is very much analogous to electromotive force (E) in an electric circuit.

An *electromagnet* is a piece of wire intended to generate a magnetic field with the passage of electric current through it. Though all current-carrying conductors produce magnetic fields, an electromagnet is usually constructed in such a way as to maximize the strength of the magnetic field it produces for a special purpose. Electromagnets find frequent application in research, industry, medical, and consumer products. As an electrically-controllable magnet, electromagnets find application in a wide variety of "electromechanical" devices: machines that effect mechanical force or motion through electrical power. Perhaps the most obvious example of such a machine is the *electric motor*. Another example is the *relay*, an electrically-controlled switch. If a switch contact mechanism is built so that it can be actuated (opened and closed) by the application of a magnetic field, and an electromagnet coil is placed in the near vicinity to produce that requisite field, it will be possible to open and close the switch by the application of a current through the coil. In effect, this gives us a device that enables elelctricity to control electricity:

Relays can be constructed to actuate multiple switch contacts, or operate them in "reverse" (energizing the coil will *open* the switch contact, and unpowering the coil will allow it to spring closed again).

MAGNETIC UNITS OF MEASUREMENT

If the burden of two systems of measurement for common quantities (English vs. metric) throws your mind into confusion, this is not the place for you! Due to an early lack of standardization in the science of magnetism, we have been plagued with no less than three complete systems of measurement for magnetic quantities.

First, we need to become acquainted with the various quantities associated with magnetism. There are quite a few more quantities to be dealt with in magnetic systems than for electrical systems. With electricity, the basic quantities are Voltage (E), Current (I), Resistance (R), and Power (P). The first three are related to one another by Ohm's Law (E=IR; I=E/R; R=E/I), while Power is related to voltage, current, and resistance by Joule's Law (P=IE; $P=I^2R$; $P=E^2/R$).

With magnetism, we have the following quantities to deal with:

Magnetomotive Force: The quantity of magnetic field force, or "push." Analogous to electric voltage (electromotive force).

Field Flux: The quantity of total field effect, or "substance" of the field. Analogous to electric current.

Field Intensity: The amount of field force (mmf) distributed over the length of the electromagnet. Sometimes referred to as *Magnetizing Force.*

Flux Density: The amount of magnetic field flux concentrated in a given area.

Reluctance: The opposition to magnetic field flux through a given volume of space or material. Analogous to electrical resistance.

Permeability: The specific measure of a material's acceptance of magnetic flux, analogous to the specific resistance of a conductive material (ñ), except inverse (greater permeability means easier passage of magnetic flux, whereas greater specific resistance means more difficult passage of electric current).

But wait... the fun is just beginning! Not only do we have more quantities to keep track of with magnetism than with electricity, but we have several different systems of unit measurement for each of these quantities. As with common quantities of length, weight, volume, and temperature, we have both English and metric systems. However, there is actually more than one metric system of units, and multiple metric systems are used in magnetic field measurements! One is called the *cgs*, which stands for Centimeter-Gram-Second, denoting the root measures upon which the whole system is based.

The other was originally known as the *mks* system, which stood for Meter-Kilogram-Second, which was later revised into another system, called *rmks*, standing for Rationalized Meter-Kilogram-Second. This ended up being adopted as an international standard and renamed *SI* (Systeme International).

And yes, the μ symbol is really the same as the metric prefix "micro." I find this especially confusing, using the exact same alphabetical character to symbolize both a specific quantity and a general metric prefix.

The relationship between field force, field flux, and reluctance is much the same as that between the electrical quantities of electromotive force (E), current (I), and resistance (R). This provides something akin to an Ohm's Law for magnetic circuits: And, given that permeability is inversely analogous to specific resistance, the equation for finding the reluctance of a magnetic material is very similar to that for finding the resistance of a conductor:

In either case, a longer piece of material provides a greater opposition, all other factors being equal. Also, a larger cross-sectional area makes for less opposition, all other factors being equal.

The major caveat here is that the reluctance of a material to magnetic flux actually *changes* with the concentration of flux going through it. This makes the "Ohm's Law" for magnetic circuits nonlinear and far more difficult to work with than the electrical version of Ohm's Law. It would be analogous to having a resistor that changed resistance as the current through it varied (a circuit composed of *var*istors instead of *res*istors).

PERMEABILITY AND SATURATION

The nonlinearity of material permeability may be graphed for better understanding. We'll place the quantity of field intensity (H), equal to field force (mmf) divided by the length of the material, on the horizontal axis of the graph. On the vertical axis, we'll place the quantity of flux density (B), equal to total flux divided by the cross-sectional area of the material.

We will use the quantities of field intensity (H) and flux density (B) instead of field force (mmf) and total flux (ϕ) so that the shape of our graph remains independent of the physical dimensions of our test material. What we're trying to do here is show a mathematical relationship between field force and flux for *any* chunk of a particular substance, in the same spirit as describing a material's *specific resistance* in ohm-cmil/ft instead of its actual *resistance* in ohms.

This is called the *normal magnetization curve,* or *B-H curve,* for any particular material. Notice how the flux density for any of the above materials (cast iron, cast steel, and sheet steel) levels off with increasing amounts of field intensity. This effect is known as *saturation.* When there is little applied magnetic force (low H), only a few atoms are in alignment, and the rest are easily aligned with additional force.

However, as more flux gets crammed into the same cross-sectional area of a ferromagnetic material, fewer atoms are available within that material to align their electrons with additional force, and so it takes more and more force (H) to get less and less "help" from the material in creating more flux density (B). To put this in economic terms, we're seeing a case of diminishing returns (B) on our investment (H). Saturation is a phenomenon limited to iron-core electromagnets. Air-core electromagnets don't saturate, but on the other hand they don't produce nearly as much magnetic flux as a ferromagnetic core for the same number of wire turns and current.

Another quirk to confound our analysis of magnetic flux versus force is the phenomenon of magnetic *hysteresis.* Hysteresis means a lag between input and output in a system upon a change in direction. Anyone who's ever driven an old automobile with "loose" steering knows what hysteresis is: to change from turning left to turning right (or vice versa), you have to rotate the steering wheel an additional amount to overcome the built-in "lag" in the mechanical linkage system between the steering wheel and the front wheels of the car. In a magnetic system, hysteresis is seen in a ferromagnetic material that tends to stay magnetized after an applied field force has been removed, if the force is reversed in polarity.

Let's use the same graph again, only extending the axes to indicate both positive and negative quantities. First we'll apply an increasing field force (current through the coils of our electromagnet). We should see the flux density increase (go up and to the right) according to the normal magnetization curve:

Next, we'll stop the current going through the coil of the electromagnet and see what happens to the flux, leaving the first curve still on the graph:

Due to the retentivity of the material, we still have a magnetic flux with no applied force (no current through the coil). Our electromagnet core is acting as a permanent magnet at this point. Now we will slowly apply the same amount of magnetic field force in the *opposite* direction to our sample: The flux density has now reached a point equivalent to what it was with a full positive value of field intensity (H), except in the negative, or opposite, direction. Let's stop the current going through the coil again and see how much flux remains:

Once again, due to the natural retentivity of the material, it will hold a magnetic flux with no power applied to the coil, except this time its in a direction opposite to that of the last time we stopped current through the coil. If we re-apply power in a positive direction again, we should see the flux density reach its prior peak in the upper-right corner of the graph again: The "S"-shaped curve traced by these steps form what is called the *hysteresis curve* of a ferromagnetic material for a given set of field intensity extremes (-H and +H). If this doesn't quite make sense, consider a hysteresis graph for the automobile steering scenario described earlier, one graph depicting a "tight" steering system and one depicting a "loose" system:

Just as in the case of automobile steering systems, hysteresis can be a problem. If you're designing a system to produce precise amounts of magnetic field flux for given amounts of current, hysteresis may hinder this design goal (due to the fact that the amount of flux density would depend on the current *and* how strongly it was magnetized before!). Similarly, a loose steering system is unacceptable in a race car, where precise, repeatable steering response is a necessity.

Also, having to overcome prior magnetization in an electromagnet can be a waste of energy if the current used to energize the coil is alternating back and forth (AC). The area within the hysteresis curve gives a rough estimate of the amount of this wasted energy.

Other times, magnetic hysteresis is a desirable thing. Such is the case when magnetic materials are used as a means of storing information (computer disks, audio and video tapes). In these applications, it is desirable to be able to magnetize a speck of iron oxide (ferrite) and rely on that material's retentivity to "remember" its last magnetized state.

Another productive application for magnetic hysteresis is in filtering high-frequency electromagnetic "noise" (rapidly alternating surges of voltage) from signal wiring by running those wires through the middle of a ferrite ring. The energy consumed in overcoming the hysteresis of ferrite attenuates the strength of the "noise" signal. Interestingly enough, the hysteresis curve of ferrite is quite extreme:

ELECTROMAGNETIC INDUCTION

While Oersted's surprising discovery of electromagnetism paved the way for more practical *applications* of electricity, it was Michael Faraday who gave us the key to the practical *generation* of electricity: electromagnetic induction. Faraday discovered that a voltage would be generated across a length of wire if that wire was exposed to a perpendicular magnetic field flux of changing intensity.

An easy way to create a magnetic field of changing intensity is to move a permanent magnet next to a wire or coil of wire. Remember: the magnetic field must increase or decrease in intensity *perpendicular* to the wire (so that the lines of flux "cut across" the conductor), or else no voltage will be induced:

Faraday was able to mathematically relate the rate of change of the magnetic field flux with induced voltage (note the use of a lower-case letter "e" for voltage. This refers to *instantaneous* voltage, or voltage at a specific point in time, rather than a steady, stable voltage.):

The "d" terms are standard calculus notation, representing rate-of-change of flux over time. "N" stands for the number of turns, or wraps, in the wire coil (assuming that the wire is formed in the shape of a coil for maximum electromagnetic efficiency).

This phenomenon is put into obvious practical use in the construction of electrical generators, which use mechanical power to move a magnetic field past coils of wire to generate voltage. However, this is by no means the only practical use for this principle.

If we recall that the magnetic field produced by a current-carrying wire was always perpendicular to that wire, and that the flux intensity of that magnetic field varied with the amount of current through it, we can see that a wire is capable of inducing a voltage *along its own length* simply due to a change in current through it.

This effect is called *self-induction:* a changing magnetic field produced by changes in current through a wire inducing voltage along the length of that same wire. If the magnetic field flux is enhanced by bending the wire into the shape of a coil, and/or wrapping that coil around a material of high permeability, this effect of self-induced voltage will be more intense. A device constructed to take advantage of this effect is called an *inductor*.

MUTUAL INDUCTANCE

If two coils of wire are brought into close proximity with each other so the magnetic field from one links with the other, a voltage will be generated in the second coil as a result. This is called *mutual inductance:* when voltage impressed upon one coil induces a voltage in another.

A device specifically designed to produce the effect of mutual inductance between two or more coils is called a *transformer*. The device shown in the above photograph is a kind of transformer, with two concentric wire coils. It is actually intended as a precision standard unit for mutual inductance, but for the purposes of illustrating what the essence of a transformer is, it will suffice.

The two wire coils can be distinguished from each other by colour: the bulk of the tube's length is wrapped in green-insulated wire (the first coil) while the second coil (wire with bronze-colored insulation) stands in the middle of the tube's length. The wire ends run down to connection terminals at the bottom of the unit. Most transformer units are not built with their wire coils exposed like this.

Because magnetically-induced voltage only happens when the magnetic field flux is *changing* in strength relative to the wire, mutual inductance between two coils can only happen with alternating (changing — AC) voltage, and not with direct (steady — DC) voltage. The only applications for mutual inductance in a DC system is where some means is available to switch power on and off to the coil (thus creating a *pulsing* DC voltage), the induced voltage peaking at every pulse.

A very useful property of transformers is the ability to transform voltage and current levels according to a simple ratio, determined by the ratio of input and output coil turns. If the energized coil of a transformer is energized by an AC voltage, the amount of AC voltage induced in the unpowered coil will be equal to the input voltage multiplied by the ratio of output to input wire turns in the coils.

Conversely, the current through the windings of the output coil compared to the input coil will follow the opposite ratio: if the voltage is increased from input coil to output coil, the current will be decreased by the same proportion. This action of the transformer is analogous to that of mechanical gear, belt sheave, or chain sprocket ratios:

A transformer designed to output more voltage than it takes in across the input coil is called a "step-up" transformer, while one designed to do the opposite is called a "step-down," in reference to the transformation of voltage that takes place. The current through each respective coil, of course, follows the exact opposite proportion.

MAGNETIC INDUCTION

FARADAY'S LAW

The phenomenon of magnetic induction plays a crucial role in three very useful electrical devices: the *electric generator*, the *electric motor*, and the *transformer*. Without these devices, modern life would be impossible in its present form. Magnetic induction was discovered in 1830 by the English physicist Michael Faraday. The American physicist Joseph Henry independently made the same discovery at about the same time. Both physicists were intrigued by the fact that an electric current flowing around a circuit can generate a magnetic field.

Surely, they reasoned, if an electric current can generate a magnetic field then a magnetic field must somehow be able to generate an electric current. However, it took many years of fruitless experimentation before they were able to find the essential ingredient which allows a magnetic field to generate an electric current. This ingredient is *time variation.*

Consider a planar loop C of conducting wire of cross-sectional area *A*. Let us place this loop in a magnetic field whose strength B is approximately uniform over the extent of the loop. Suppose that the direction of the magnetic field subtends an angle θ with the normal direction to the loop. The *magnetic flux* Φ_B through the loop is defined as the product of the area of the loop and the component of the magnetic field perpendicular to the loop. Thus,

$$\Phi_B = AB_\perp = AB\cos\theta.$$

If the loop is wrapped around itself *N* times (*i.e.*, if the loop has *N turns*) then the magnetic flux through the loop is simply *N* times the magnetic flux through a single turn:

$$\Phi_B = NAB_\perp.$$

Finally, if the magnetic field is not uniform over the loop, or the loop does not lie in one plane, then we must evaluate the magnetic flux as a surface integral

$$\Phi_B = \int_S \mathrm{B} \cdot dS.$$

Here, *S* is some surface attached to C. If the loop has *N* turns then the flux is *N* times the above value. The SI unit of magnetic flux is the weber (Wb). One tesla is equivalent to one weber per meter squared:

$$1\mathrm{T} \equiv 1\mathrm{Wbm}^{-2}$$

Faraday discovered that if the magnetic field through a loop of wire *varies in time* then an emf is induced around the loop. Faraday was able to observe this effect because the emf gives rise to a current circulating in the loop. Faraday found that the magnitude of the emf is directly proportional to the time rate of change of the magnetic field. He also discovered that an emf is generated when a loop of wire *moves* from a region of low magnetic field-strength to one of high magnetic field-strength, and *vice versa.*

The emf is directly proportional to the velocity with which the loop moves between the two regions. Finally, Faraday discovered that an emf is generated around a loop which *rotates* in a uniform magnetic field of constant strength. In this case, the emf is directly proportional to the rate at which the loop rotates. Faraday was eventually able to propose a single law which could account for all of his many and varied observations. This law, which is known as *Faraday's law of magnetic induction*, is as follows:

The emf induced in a circuit is proportional to the time rate of change of the magnetic flux linking that circuit.

SI units have been fixed so that the constant of proportionality in this law is *unity*. Thus, if the magnetic flux through a circuit changes by an amount $d\Phi_B$ in a time interval then the emf ε generated in the circuit is

$$\varepsilon = \frac{d\Phi_B}{dt}.$$

There are many different ways in which the magnetic flux linking an electric circuit can change. Either the magnetic field-strength can change, or the direction of the magnetic field can change, or the position of the circuit can change, or the shape of the circuit can change, or the orientation of the circuit can change. Faraday's law states that all of these ways are completely *equivalent* as far as the generation of an emf around the circuit is concerned.

LENZ'S LAW

We still have not specified in which direction the emf generated by a time-varying magnetic flux linking an electric circuit acts. In order to help specify this direction, we need to make use of a right-hand rule. Suppose that a current *I* circulates around a planar loop of conducting wire, and, thereby, generates a magnetic field B. What is the direction of this magnetic field as it passes through the middle of the loop? Well, if the fingers of a right-hand circulate in the same direction as the current, then the thumb indicates the direction of the magnetic field as it passes through the centre of the loop.

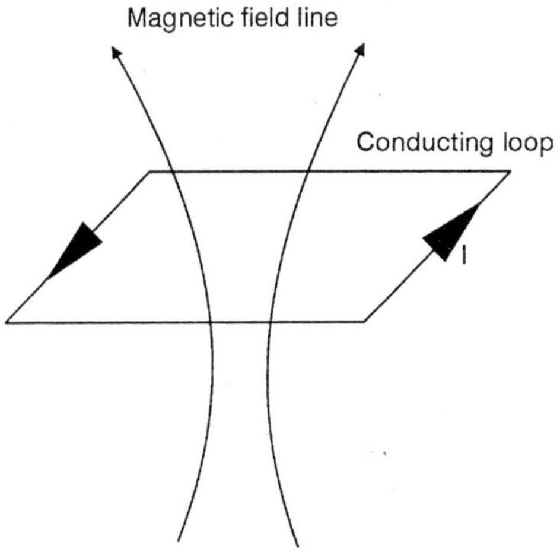

Fig. Magnetic Field Generated by a Planar Current-carrying Loop.

Consider a plane loop of conducting wire which is linked by magnetic flux. By convention, the direction in which current would have to flow around the loop in order to *increase* the magnetic flux linking the loop is termed the *positive* direction. Likewise, the direction in which current would have to flow around the loop in order to *decrease* the magnetic flux linking the loop is termed the *negative* direction. Suppose that the magnetic flux linking the loop is increased. In accordance with Faraday's law, an emf is generated around the loop.

Does this emf act in the positive direction, so as to drive a current around the loop which further increases the magnetic flux, or does it act in the negative direction, so as to drive a current around the loop which decreases the magnetic flux ϕ. It is easily demonstrated experimentally that the emf acts in the negative direction. Thus:

The emf induced in an electric circuit always acts in such a direction that the current it drives around the circuit opposes the change in magnetic flux which produces the emf.

This result is known as *Lenz's law,* after the nineteenth century Russian scientist Heinrich Lenz, who first formulated it. Faraday's law, combined with Lenz's law, is usually written

$$\varepsilon = \frac{d\Phi_B}{dt}.$$

The minus sign is to remind us that the emf always acts to oppose the change in magnetic flux which generates the emf.

MAGNETIC INDUCTION

Consider a one-turn loop of conducting wire C which is placed in a magnetic field B. The magnetic flux Φ_B linking loop C can be written

$$\Phi_B = \int_S B \cdot dS$$

where is any surface attached to the loop.

Suppose that the magnetic field changes in time, causing the magnetic flux Φ_B linking circuit C to vary. Let the flux change by an amount $d\Phi_B$ in the time interval dt. According to Faraday's law, the emf ε induced around loop C is given by

$$\varepsilon = \frac{d\Phi_B}{dt}.$$

If ε is *positive* then the emf acts around the loop in the *same* sense as that indicated by the fingers of a right-hand, when the thumb points in the direction of the mean magnetic field passing through the loop. Likewise, if ε is *negative* then the emf acts around the loop in the *opposite* sense to that indicated by the fingers of a right-hand, when the thumb points in the direction of the mean magnetic field passing through the loop. In the former case, we say that the emf acts in the *positive* direction, whereas in the latter case we say it acts in the *negative* direction.

Suppose that $\varepsilon > 0$, so that the emf acts in the positive direction. How, exactly, is this emf produced? In order to answer this question, we need to remind ourselves what an emf actually is. When we say that an emf ε acts around the loop C in the positive direction, what we really mean is that a charge q which circulates once around the loop in the positive direction acquires the energy $q\varepsilon$. How does the charge acquire this energy? Clearly, either an electric field or a magnetic field, or some combination of the two, must perform the work $q\varepsilon$ on the charge as it circulates around the loop. However, we have already seen, that a magnetic field is unable to do work on a charged particle. Thus, the charge must acquire the energy $q\varepsilon$ from an *electric* field as it circulates once around the loop in the positive direction.

According to Sect. 5, the work that the electric field does on the charge as it goes around the loop is

$$W = q\oint_C E \cdot dr,$$

where is a line element of the loop. Hence, by energy conservation, we can write $W = q\varepsilon$, or

$$\varepsilon = \oint_C E \cdot dr.$$

The term on the right-hand side of the above expression can be recognized as the *line integral* of the electric field around loop C in the positive direction. Thus, the emf generated around the circuit in the positive direction is equal to the line integral of the electric field around the circuit in the same direction.

Equations can be combined to give

$$\oint_C E \cdot dr = -\frac{d\Phi_B}{dt}.$$

Thus, Faraday's law implies that the line integral of the electric field around circuit (in the positive direction) is equal to minus the time rate of change of the magnetic flux linking this circuit. Does this law just apply to conducting circuits, or can we apply it to an arbitrary closed loop in space? Well, the difference between a conducting circuit and an arbitrary closed loop is that electric current can flow around a circuit, whereas current cannot, in general, flow around an arbitrary loop. In fact, the emf induced around a conducting circuit drives a current $I = \varepsilon/R$ around that circuit, where is the resistance of the circuit.

However, we can make this resistance arbitrarily large without invalidating Eq. In the limit in which tends to infinity, no current flows around the circuit, so the circuit becomes indistinguishable from an arbitrary loop. Since we can place such a circuit anywhere in space, and Eq. still holds, we are forced to the conclusion that Eq. is valid for *any* closed loop in space, and not just for conducting circuits.

Equation describes how a time-varying magnetic field *generates* an electric field which fills space. The strength of the electric field is directly proportional to the rate of change of the magnetic field. The field-lines associated with this electric field form loops in the plane perpendicular to the magnetic field.

If the magnetic field is increasing then the electric field-lines circulate in the opposite sense to the fingers of a right-hand, when the thumb points in the direction of the field. If the magnetic field is decreasing then the electric field-lines circulate in the same sense as the fingers of a right-hand, when the thumb points in the direction of the field.

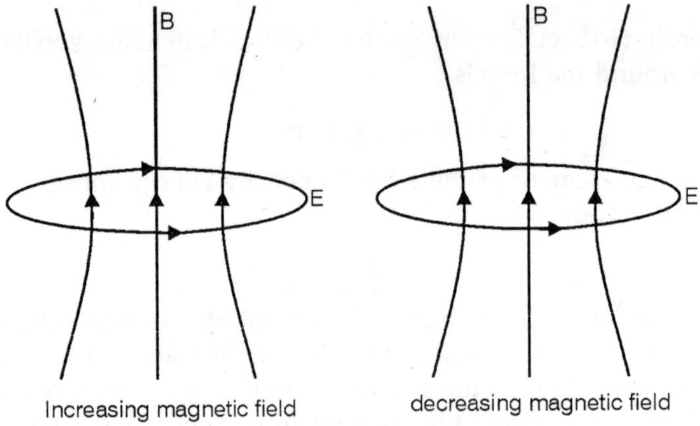

Fig. Inductively generated electric fields

We can now appreciate that when a conducting circuit is placed in a time-varying magnetic field, it is the electric field induced by the changing magnetic field which gives rise to the emf around the circuit. If the loop has a finite resistance then this electric field also drives a current around the circuit. Note, however, that the electric field is generated irrespective of the presence of a conducting circuit.

The electric field generated by a time-varying magnetic field is quite different in nature to that generated by a set of stationary electric charges. In the latter case, the electric field-lines begin on positive charges, end on negative charges, and *never* form closed loops in free space. In the former case, the electric field-lines *never* begin or end, and *always* form closed loops in free space.

In fact, the electric field-lines generated by magnetic induction behave in much the same manner as magnetic field-lines. An electric field generated by fixed charges is unable to do net work on a charge which circulates in a closed loop. On the other hand, an electric field generated by magnetic induction certainly can do work on a charge which circulates in a closed loop. This is basically how a current is induced in a conducting loop placed in a time-varying magnetic field.

One consequence of this fact is that the work done in slowly moving a charge between two points in an inductive electric field *does* depend on the path taken between the two points. It follows that we cannot calculate a *unique* potential difference between two points in an inductive electric field. In fact, the whole idea of electric potential breaks down in a such a field (fortunately, there is a way of salvaging the idea of electric potential in an inductive field, but this topic lies beyond the scope of this course).

Note, however, that it is still possible to calculate a *unique* value for the emf generated around a conducting circuit by an inductive electric field, because, in this case, the path taken by electric charges is uniquely specified: *i.e.*, the charges have to follow the circuit.

MOTIONAL EMF

We now understand how an emf is generated around a *fixed* circuit placed in a time-varying magnetic field. But, according to Faraday's law, an emf is also generated around a *moving* circuit placed in a magnetic field which does not vary in time. According to Equation. no space-filling inductive electric field is generated in the latter case, since the magnetic field is steady. So, how do we account for the emf in the latter case?

In order to help answer this question, let us consider a simple circuit in which a conducting rod of length slides along a U-shaped conducting frame in the presence of a uniform magnetic field. This circuit. Suppose, for the sake of simplicity, that the magnetic field is directed perpendicular to the plane of the circuit. Suppose, further, that we move the rod to the right with the constant velocity.

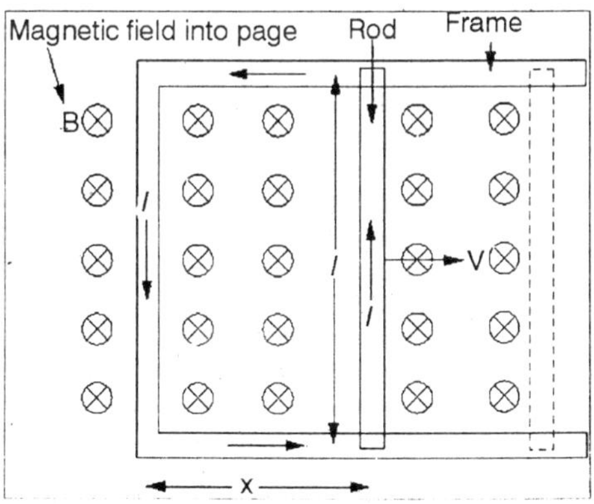

Fig. Motional emf.

The magnetic flux linked by the circuit is simply the product of the perpendicular magnetic field-strength, and the area of the circuit, where determines the position of the sliding rod. Thus,

$$\Phi_B = Blx.$$

Now, the rod moves a distance $dx = v$ in a time interval, so in the same time interval the magnetic flux linking the circuit *increases* by

$$d\Phi_B = Bl\,dx = Bl\,v\,dt.$$

It follows, from Faraday's law, that the magnitude of the emf generated around the circuit is given by

$$\varepsilon = \frac{d\Phi_B}{dt} = Blv.$$

Thus, the emf generated in the circuit by the moving rod is simply the product of the magnetic field-strength, the length of the rod, and the velocity of the rod. If the magnetic field is not perpendicular to the circuit, but instead subtends an angle with respect to the normal direction to the plane of the circuit, then it is easily demonstrated that the *motional emf* generated in the circuit by the moving rod is

$$\varepsilon = B_{\perp} lv,$$

where B^ = $B \cos\theta$ is the component of the magnetic field which is perpendicular to the plane of the circuit.

Since the magnetic flux linking the circuit *increases* in time, the emf acts in the *negative* direction (*i.e.*, in the opposite sense to the fingers of a right-hand, if the thumb points along the direction of the magnetic field). The emf, ε, therefore, acts in the *anti-clockwise* direction in the figure. If is the total resistance of the circuit, then this emf drives an anti-clockwise electric current of magnitude $I = \varepsilon/R$ around the circuit.

But, where does the emf come from? Let us again remind ourselves what an emf is. When we say that an emf ε acts around the circuit in the anti-clockwise direction, what we really mean is that a charge q which circulates once around the circuit in the anti-clockwise direction acquires the energy $q\varepsilon$.

The only way in which the charge can acquire this energy is if something does *work* on it as it circulates. Let us assume that the charge circulates very *slowly*. The magnetic field exerts a negligibly small force on the charge when it is traversing the non-moving part of the circuit (since the charge is moving very slowly). However, when the charge is traversing the moving rod it experiences an *upward* (in the figure) magnetic force of magnitude $f = qvB$ (assuming that $q > 0$). The net work done on the charge by this force as it traverses the rod is

$$W' = qv\,Bl = q\varepsilon,$$

since $\varepsilon = Blv$. Thus, it would appear that the motional emf generated around the circuit can be accounted for in terms of the magnetic force exerted on charges traversing the moving rod. But, if we think carefully, we can see that there is something seriously wrong with the above explanation. We seem to be saying that the charge acquires the energy $q\varepsilon$ from the *magnetic field* as it moves around the circuit once in the anti-clockwise direction. But, this is impossible, because a magnetic field *cannot* do work on an electric charge.

Let us look at the problem from the point of view of a charge q traversing the moving rod. In the frame of reference of the rod, the charge only moves very slowly, so the magnetic force on it is negligible. In fact, only an electric field can exert a significant force on a slowly moving charge.

In order to account for the motional emf generated around the circuit, we need the charge to experience an upward force of magnitude $qv\, B$. The only way in which this is possible is if the charge sees an upward pointing *electric field* of magnitude

$$E_0 = vB.$$

In other words, although there is no electric field in the laboratory frame, there is an electric field in the frame of reference of the moving rod, and it is this field which does the necessary amount of work on charges moving around the circuit to account for the existence of the motional emf, $\varepsilon = Blv$.

More generally, if a conductor moves v in the laboratory frame with velocity in the presence of a magnetic field B then a charge q inside the conductor experiences a magnetic force $f = qv \times B$. In the frame of the conductor, in which the charge is essentially stationary, the same force takes the form of an electric force $f = qE_0$, where E_0 is the electric field in the frame of reference of the conductor. Thus, if a conductor moves with velocity v through a magnetic field B then the electric field E_0 which appears in the rest frame of the conductor is given by

$$E_0 = v \times B.$$

This electric field is the ultimate origin of the motional emfs which are generated whenever circuits move with respect to magnetic fields.

We can now appreciate that Faraday's law is due to a combination of two apparently distinct effects. The first is the space-filling electric field generated by a changing magnetic field. The second is the electric field generated inside a conductor when it moves through a magnetic field. In reality, these effects are two aspects of the same basic phenomenon, which explains why no real distinction is made between them in Faraday's law.

EDDY CURRENTS

We have seen, in the above example, that when a conductor is moved in a magnetic field a motional emf is generated. Moreover, this emf drives a current which heats the conductor, and, when combined with the magnetic field, also gives rise to a magnetic force acting on the conductor which opposes its motion. In turns out that these results are quite general. Incidentally, the induced currents which circulate inside a moving conductor in a static magnetic field, or a stationary conductor in a time-varying magnetic field, are usually called *eddy currents*.

Consider a metal disk which rotates in a perpendicular magnetic field which only extends over a small rectangular portion of the disk. Such a field could be produced by the pole of a horseshoe magnetic. The motional emf induced in the disk, as it moves through the field-containing region, acts in the direction v × B, where v is the velocity of the disk, and B the magnetic field. It follows from figure that the emf acts downward.

The emf drives currents which are also directed downward. However, these currents must form closed loops, and, hence, they are directed upward in those regions of the disk immediately adjacent to the field-containing region.

It can be seen that the induced currents flow in little eddies. Hence, the name "eddy currents." According to the right-hand rule, the downward currents in the field-containing region give rise to a magnetic force on the disk which acts to the right. In other words, the magnetic force acts to prevent the rotation of the disk. Clearly, external work must be done on the disk in order to keep it rotating at a constant angular velocity. This external work is ultimately dissipated as heat by the eddy currents circulating inside the disk.

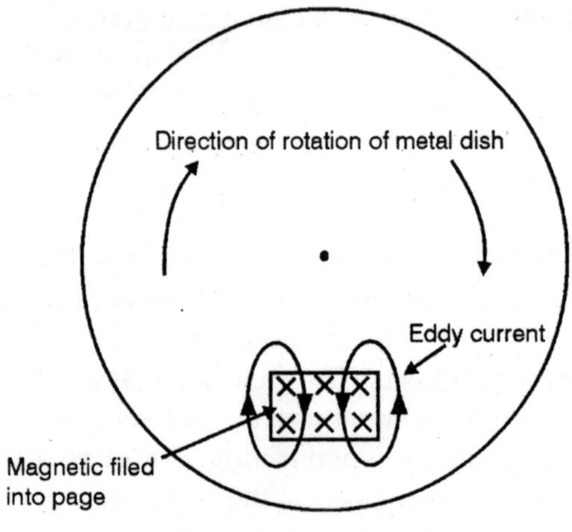

Fig. Eddy currents

Eddy currents can be very useful. For instance, some cookers work by using eddy currents. The cooking pots, which are usually made out of aluminium, are placed on plates which generate oscillating magnetic fields.

These fields induce eddy currents in the pots which heat them up. The heat is then transmitted to the food inside the pots.

This type of cooker is particularly useful for food which needs to be cooked gradually over a long period of time: *i.e.*, over many hours, or even days. Eddy currents can also be used to heat small pieces of metal until they become white-hot by placing them in a very rapidly oscillating magnetic field. This technique is sometimes used in brazing. Heating conductors by means of eddy currents is called *inductive heating*. Eddy currents can also be used to damp motion. This technique, which is called *eddy current damping*, is often employed in galvanometers.

THE ALTERNATING CURRENT GENERATOR

An electric generator, or dynamo, is a device which converts mechanical energy into electrical energy. The simplest practical generator consists of a rectangular coil rotating in a uniform magnetic field. The magnetic field is usually supplied by a permanent magnet.

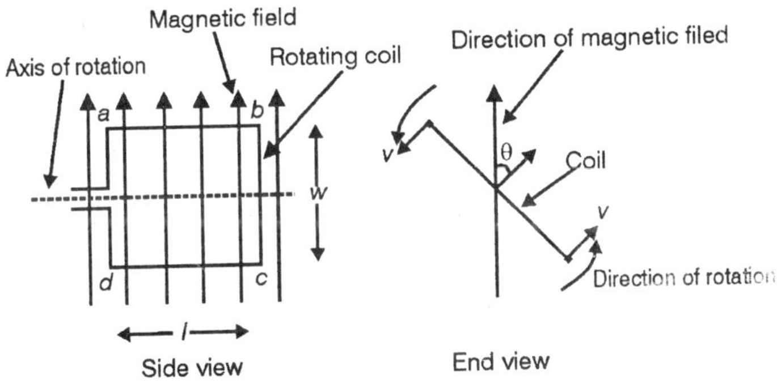

Fig. An Alternating Current Generator.

Let l be the length of the coil along its axis of rotation, and the width of the coil perpendicular to this axis. Suppose that the coil rotates with constant angular velocity in a uniform magnetic field of strength B.

The velocity v with which the the two long sides of the coil (*i.e.*, sides *ab* and *cd*) move through the magnetic field is simply the product of the angular velocity of rotation w and the distance $w/2$ of each side from the axis of rotation, so $v = w\, w/2$. The motional emf induced in each side is given by $e = B_{\perp} lv$, where $B_{\perp}$ is the component of the magnetic field perpendicular to instantaneous direction of motion of the side in question. If the direction of the magnetic field subtends an angle θ with the normal direction to the coil, then $B_{\perp} = B \sin\theta$. Thus, the magnitude of the motional emf generated in sides *ab* and *cd* is

$$\varepsilon_{ab} = \frac{Bwlw \sin\theta}{2} = \frac{BAw \sin\theta}{2},$$

where $A = w$ is the area of the coil. The emf is zero when $\theta = 0°$ or $180°$, since the direction of motion of sides *ab* and *cd* is *parallel* to the direction of the magnetic field in these cases. The emf attains its maximum value when $\theta = 90°$ or $270°$, since the direction of motion of sides *ab* and *cd* is *perpendicular* to the direction of the magnetic field in these cases. Incidentally, it is clear, from symmetry, that no net motional emf is generated in sides and of the coil.

Suppose that the direction of rotation of the coil is such that side is moving, whereas side is moving out of the page. The motional emf induced in side *ab* acts from *a* to *b*. Likewise, the motional emf induce in side *cd* acts from *c* to ε. It can be seen that both emfs act in the clockwise direction around the coil. Thus, the net emf ε acting around the coil is $2\varepsilon_{ab}$. If the coil has *N* turns then the net emf becomes $2N\varepsilon_{ab}$. Thus, the general expression for the emf generated around a steadily rotating, multi-turn coil in a uniform magnetic field is

$$\varepsilon = NBAw \sin(wt),$$

where we have written $\theta = w$ for a steadily rotating coil (assuming that $\theta = 0$ at). This expression can also be written

$$\varepsilon = \varepsilon_{max} \sin(2\pi f t),$$

where

$$\varepsilon_{max} = 2\pi N B A f$$

is the peak emf produced by the generator, and $f = w/2\pi$ is the number of complete rotations the coils executes per second. Thus, the peak emf is directly proportional to the area of the coil, the number of turns in the coil, the rotation frequency of the coil, 'and the magnetic field-strength.

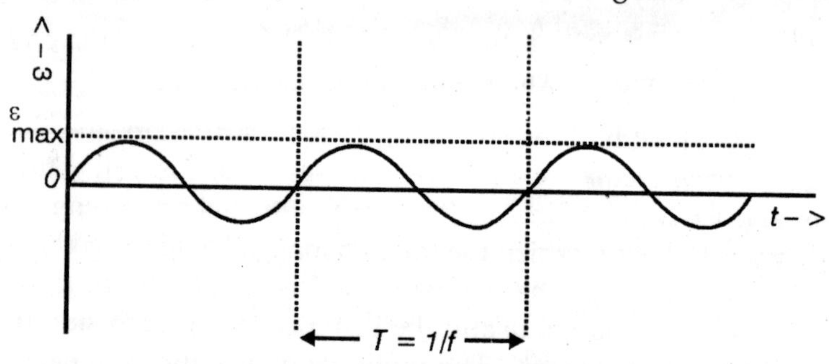

Fig. Emf Generated by a Steadily Rotating AC Generator.

The emf specified in Eq. plotted as a function of time. It can be seen that the variation of the emf with time is *sinusoidal* in nature. The emf attains its peak values when the plane of the coil is parallel to the plane of the magnetic field, passes through zero when the plane of the coil is perpendicular to the magnetic field, and reverses sign every half period of revolution of the coil. The emf is periodic (*i.e.*, it continually repeats the same pattern in time), with period $T = 1/f$ (which is, of course, the rotation period of the coil).

Suppose that some load (*e.g.*, a light-bulb, or an electric heating element) of resistance is connected across the terminals of the generator. In practice, this is achieved by connecting the two ends of the coil to rotating rings which are then connected to the external circuit by means of metal brushes. According to Ohm's law, the current which flows in the load is given by

$$I = \frac{\varepsilon}{R} = \frac{\varepsilon_{max}}{R} \sin(2\pi f t).$$

Note that this current is constantly changing direction, just like the emf of the generator. Hence, the type of generator described above is usually termed an *alternating current*, or *AC*, generator.

The current I which flows through the load must also flow around the coil. Since the coil is situated in a magnetic field, this current gives rise to a torque on the coil which, as is easily demonstrated, acts to slow down its rotation. The braking torque τ acting on the coil is given by

$$\tau = N I B_{\parallel} A,$$

where $B_{\parallel} = B \sin\theta$ is the component of the magnetic field which lies in the plane of the coil. It follows from equation that

$$\tau = \frac{\varepsilon I}{w},$$

since $\varepsilon = N B_{\parallel} Aw$. An external torque which is equal and opposite to the breaking torque must be applied to the coil if it is to rotate *uniformly*, as assumed above. The rate P at which this external torque does work is equal to the product of the torque and the angular velocity ω of the coil. Thus,

$$P = \tau\omega = \varepsilon I.$$

Not surprisingly, the rate at which the external torque performs works exactly matches the rate εIat which electrical energy is generated in the circuit comprising the rotating coil and the load.

Equations yield

$$\tau = \tau_{max} \sin^2(2\pi f t),$$

where $\tau_{max} = (\varepsilon_{max})^2 / (2\pi f R)$, . Figure shows the breaking torque plotted as a function of time, according to Eq. It can be seen that the torque is always of the same sign (*i.e.*, it always acts in the same direction, so as to continually oppose the rotation of the coil), but is not constant in time. Instead, it *pulsates* periodically with period T. The breaking torque attains its maximum value whenever the plane of the coil is parallel to the plane of the magnetic field, and is zero whenever the plane of the coil is perpendicular to the magnetic field. It is clear that the external torque needed to keep the coil rotating at a constant angular velocity must also pulsate in time with period T. A constant external torque would give rise to a non-uniformly rotating coil, and, hence, to an alternating emf which varies with time in a more complicated manner than sin $(2\pi f t)$.

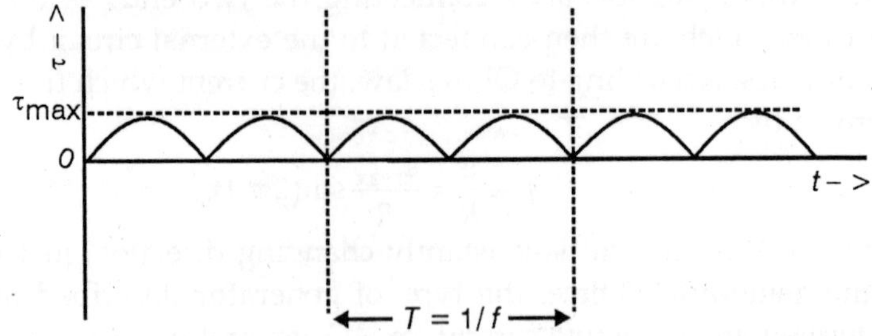

Fig. The Braking Torque in a Steadily Rotating AC Generator.

Virtually all commercial power stations generate electricity using AC generators. The external power needed to turn the generating coil is usually supplied by a steam turbine (steam blasting against fan-like blades which are forced into rotation). Water is vaporized to produce high pressure steam by burning coal, or by using the energy released inside a nuclear reactor.

Of course, in hydroelectric power stations, the power needed to turn the generator coil is supplied by a water turbine (which is similar to a steam turbine, except that falling water plays the role of the steam). Recently, a new type of power station has been developed in which the power needed to rotate the generating coil is supplied by a gas turbine (basically, a large jet engine which burns natural gas).

In the United States and Canada, the alternating emf generated by power stations oscillates at $f = 60$ Hz, which means that the generator coils in power stations rotate exactly sixty times a second. In Europe, and much of the rest of the world, the oscillation frequency of commercially generated electricity is $f = 50$ Hz.

Chapter 5

Transistors

The invention of the bipolar transistor in 1948 ushered in a revolution in electronics. Technical feats previously requiring relatively large, mechanically fragile, power-hungry vacuum tubes were suddenly achievable with tiny, mechanically rugged, power-thrifty specks of crystalline silicon. This revolution made possible the design and manufacture of lightweight, inexpensive electronic devices that we now take for granted. Understanding how transistors function is of paramount importance to anyone interested in understanding modern electronics.

My intent here is to focus as exclusively as possible on the practical function and application of bipolar transistors, rather than to explore the quantum world of semiconductor theory. A bipolar transistor consists of a three-layer "sandwich" of doped (extrinsic) semiconductor materials, either P-N-P or N-P-N. Each layer forming the transistor has a specific name, and each layer is provided with a wire contact for connection to a circuit. Shown here are schematic symbols and physical diagrams of these two transistor types:

***PNP* transistor**

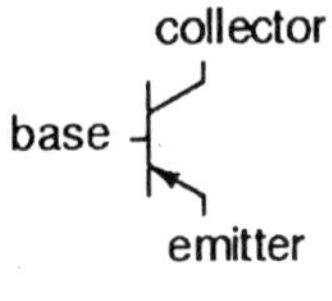

schematic symbol

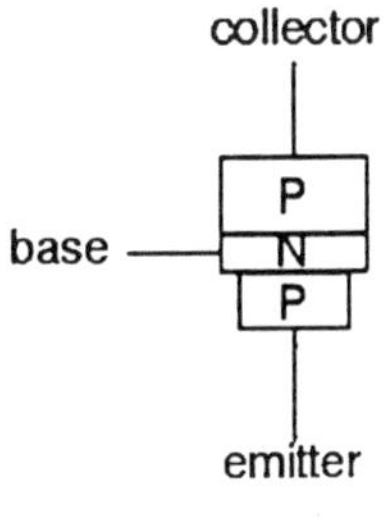

physical diagram

The only functional difference between a PNP transistor and an NPN transistor is the proper biasing (polarity) of the junctions when operating. For any given state of operation, the current directions and voltage polarities for each type of transistor are exactly opposite each other. Bipolar transistors work as current-controlled current *regulators*. In other words, they restrict the amount of current that can go through them according to a smaller, controlling current.

***NPN* transistor**

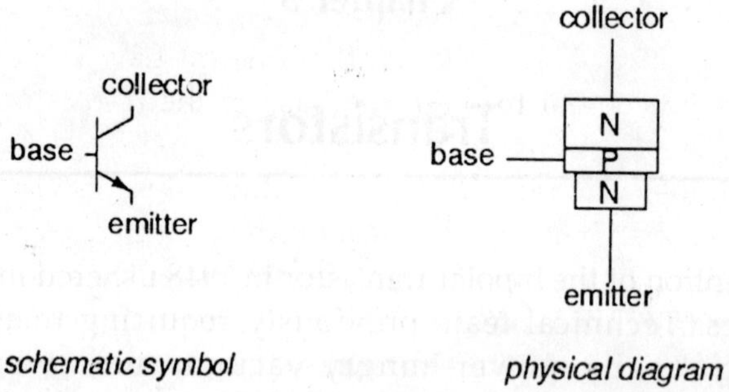

The main current that is *controlled* goes from collector to emitter, or from emitter to collector, depending on the type of transistor it is (PNP or NPN, respectively).

The small current that *controls* the main current goes from base to emitter, or from emitter to base, once again depending on the type of transistor it is (PNP or NPN, respectively). According to the confusing standards of semiconductor symbology, the arrow always points *against* the direction of electron flow:

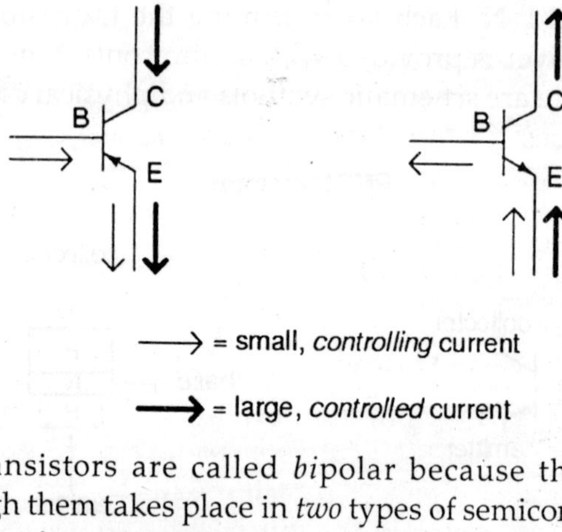

Bipolar transistors are called *bi*polar because the main flow of electrons through them takes place in *two* types of semiconductor material: P and N, as the main current goes from emitter to collector (or vice versa). In other words, two types of charge carriers — electrons and holes — comprise this main current through the transistor.

The *controlling* current and the *controlled* current always mesh together through the emitter wire, and their electrons always flow *against* the direction of the transistor's arrow. This is the first and foremost rule in the use of transistors: all currents must be going in the proper directions for the device to work as a current regulator.

The small, controlling current is usually referred to simply as the *base current* because it is the only current that goes through the base wire of the transistor. Conversely, the large, controlled current is referred to as the *collector current* because it is the only current that goes through the collector wire. The emitter current is the sum of the base and collector currents, in compliance with Kirchhoff's Current Law.

If there is no current through the base of the transistor, it shuts off like an open switch and prevents current through the collector. If there is a base current, then the transistor turns on like a closed switch and allows a proportional amount of current through the collector. Collector current is primarily limited by the base current, regardless of the amount of voltage available to push it. The next section will explore in more detail the use of bipolar transistors as switching elements.

- Bipolar transistors are so named because the controlled current must go through *two* types of semiconductor material: P and N. The current consists of both electron and hole flow, in different parts of the transistor.
- Bipolar transistors consist of either a P-N-P or an N-P-N semiconductor "sandwich" structure.
- The three leads of a bipolar transistor are called the *Emitter, Base,* and *Collector.*
- Transistors function as current regulators by allowing a small current to *control* a larger current. The amount of current allowed between collector and emitter is primarily determined by the amount of current moving between base and emitter.
- In order for a transistor to properly function as a current regulator, the controlling (base) current and the controlled (collector) currents must be going in the proper directions: meshing additively at the emitter and going *against* the emitter arrow symbol.

THE TRANSISTOR AS A SWITCH

Because a transistor's collector current is proportionally limited by its base current, it can be used as a sort of current-controlled switch. A relatively small flow of electrons sent through the base of the transistor has the ability to exert control over a much larger flow of electrons through the collector.

Suppose we had a lamp that we wanted to turn on and off by means of a switch. Such a circuit would be extremely simple: For the sake of illustration, let's insert a transistor in place of the switch to show how it can control the flow of electrons through the lamp. Remember that the controlled current through a transistor must go between collector and emitter.

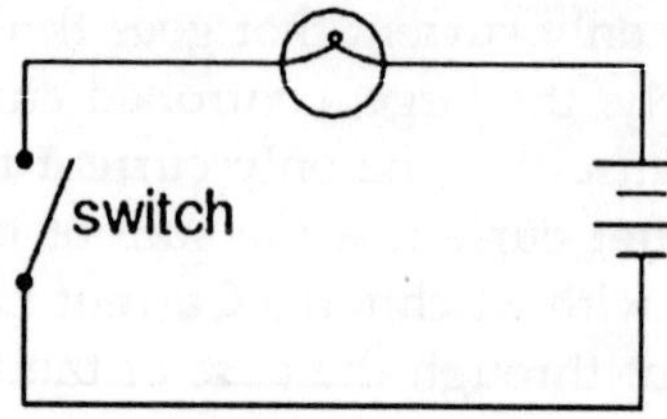

Since it's the current through the lamp that we want to control, we must position the collector and emitter of our transistor where the two contacts of the switch are now. We must also make sure that the lamp's current will move *against* the direction of the emitter arrow symbol to ensure that the transistor's junction bias will be correct:

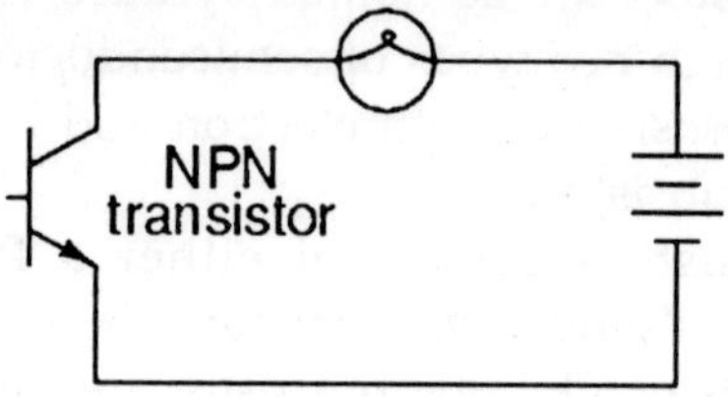

In this example I happened to choose an NPN transistor. A PNP transistor could also have been chosen for the job, and its application would look like this:

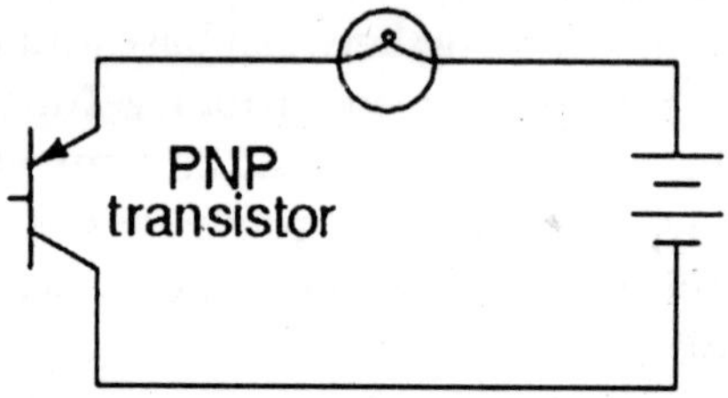

The choice between NPN and PNP is really arbitrary. All that matters is that the proper current directions are maintained for the sake of correct junction biasing (electron flow going *against* the transistor symbol's arrow).

Going back to the NPN transistor in our example circuit, we are faced with the need to add something more so that we can have base current. Without a connection to the base wire of the transistor, base current will be zero, and the transistor cannot turn on, resulting in a lamp that is always off. Remember that for an NPN transistor, base current must consist of electrons flowing from emitter to base (against the emitter arrow symbol, just like the lamp current). Perhaps the simplest thing to do would be to connect a switch between the base and collector wires of the transistor like this:

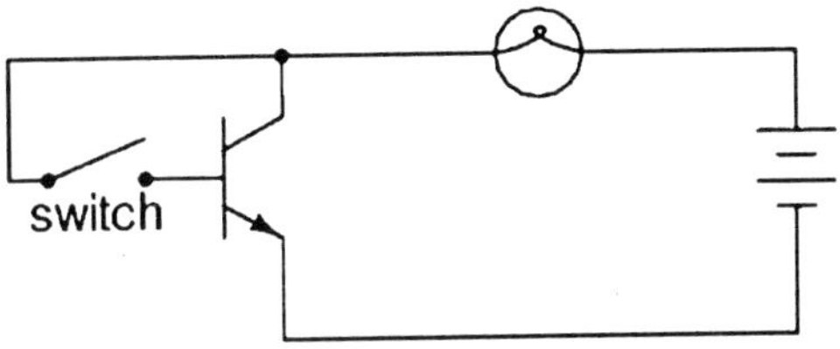

If the switch is open, the base wire of the transistor will be left "floating" (not connected to anything) and there will be no current through it. In this state, the transistor is said to be *cutoff*. If the switch is closed, however, electrons will be able to flow from the emitter through to the base of the transistor, through the switch and up to the left side of the lamp, back to the positive side of the battery. This base current will enable a much larger flow of electrons from the emitter through to the collector, thus lighting up the lamp. In this state of maximum circuit current, the transistor is said to be *saturated*.

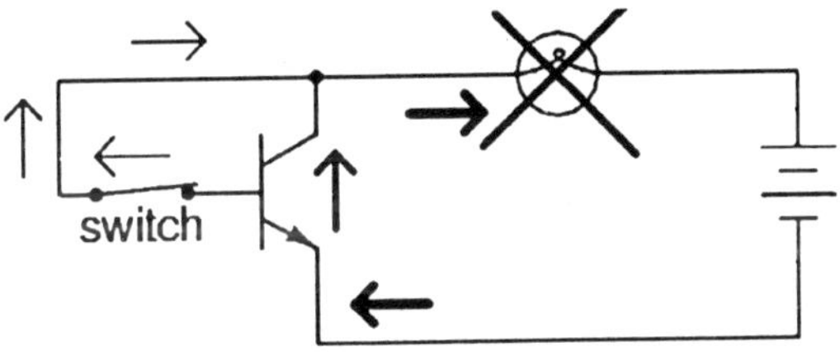

Of course, it may seem pointless to use a transistor in this capacity to control the lamp. After all, we're still using a switch in the circuit, aren't we? If we're still using a switch to control the lamp — if only indirectly — then what's the point of having a transistor to control the current? Why not just go back to our original circuit and use the switch directly to control the lamp current?

There are a couple of points to be made here, actually. First is the fact that when used in this manner, the switch contacts need only handle what little base current is necessary to turn the transistor on, while the transistor itself handles the majority of the lamp's current.

This may be an important advantage if the switch has a low current rating: a small switch may be used to control a relatively high-current load. Perhaps more importantly, though, is the fact that the current-controlling behaviour of the transistor enables us to use something completely different to turn the lamp on or off. Consider this example, where a solar cell is used to control the transistor, which in turn controls the lamp:

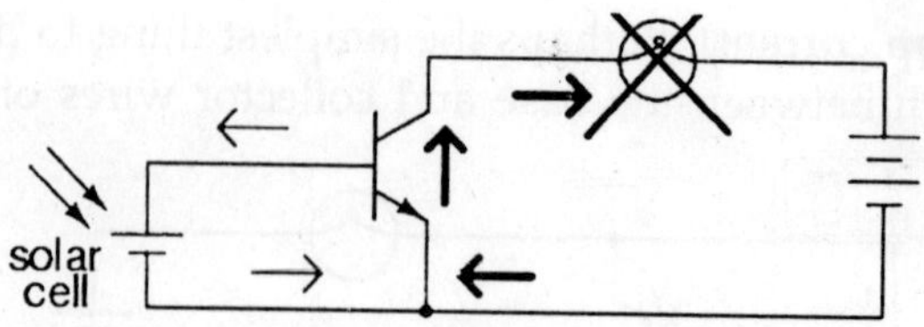

Or, we could use a thermocouple to provide the necessary base current to turn the transistor on:

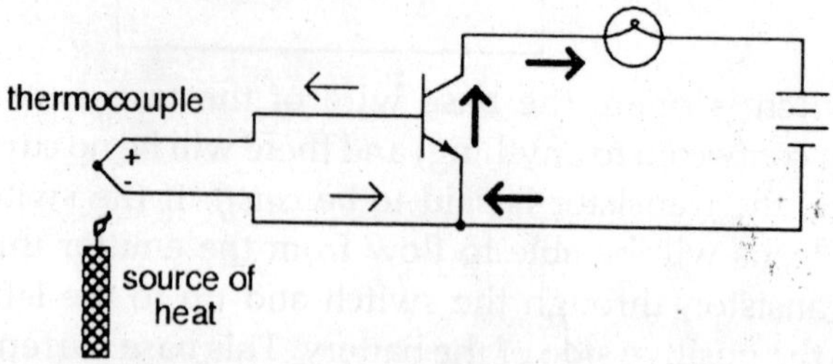

Even a microphone of sufficient voltage and current output could be used to turn the transistor on, provided its output is rectified from AC to DC so that the emitter-base PN junction within the transistor will always be forward-biased:

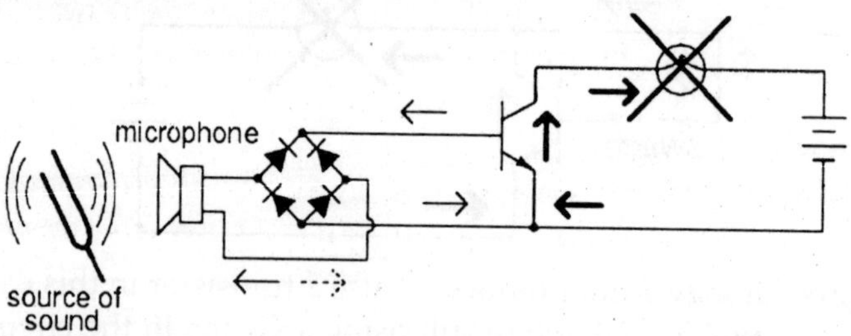

The point should be quite apparent by now: *any* sufficient source of DC current may be used to turn the transistor on, and that source of current need only be a fraction of the amount of current needed to energize the lamp. Here we see the transistor functioning not only as a switch, but as a true *amplifier*: using a relatively low-power signal to *control* a relatively large amount of power. Please note that the actual power for lighting up the lamp comes from the battery to the right of the schematic.

It is not as though the small signal current from the solar cell, thermocouple, or microphone is being magically transformed into a greater amount of power. Rather, those small power sources are simply *controlling* the battery's power to light up the lamp.

METER CHECK OF A TRANSISTOR

Bipolar transistors are constructed of a three-layer semiconductor "sandwich," either PNP or NPN. As such, they register as two diodes connected back-to-back when tested with a multimeter's "resistance" or "diode check" functions:

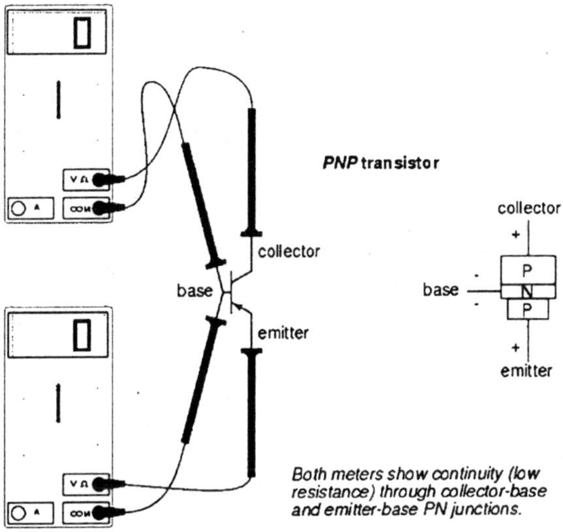

Here I'm assuming the use of a multimeter with only a single continuity range (resistance) function to check the PN junctions. Some multimeters are equipped with two separate continuity check functions: resistance and "diode check," each with its own purpose. If your meter has a designated "diode check" function, use that rather than the "resistance" range, and the meter will display the actual forward voltage of the PN junction and not just whether or not it conducts current.

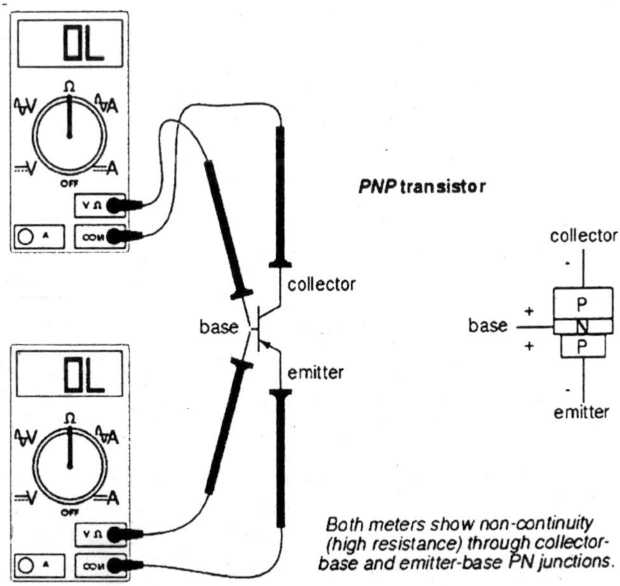

Meter readings will be exactly opposite, of course, for an NPN transistor, with both PN junctions facing the other way. If a multimeter with a "diode check" function is used in this test, it will be found that the emitter-base junction possesses a slightly greater forward voltage drop than the collector-base junction.

This forward voltage difference is due to the disparity in doping concentration between the emitter and collector regions of the transistor: the emitter is a much more heavily doped piece of semiconductor material than the collector, causing its junction with the base to produce a higher forward voltage drop. Knowing this, it becomes possible to determine which wire is which on an unmarked transistor. This is important because transistor packaging, unfortunately, is not standardized.

All bipolar transistors have three wires, of course, but the positions of the three wires on the actual physical package are not arranged in any universal, standardized order.

Suppose a technician finds a bipolar transistor and proceeds to measure continuity with a multimeter set in the "diode check" mode. Measuring between pairs of wires and recording the values displayed by the meter, the technician obtains the following data:

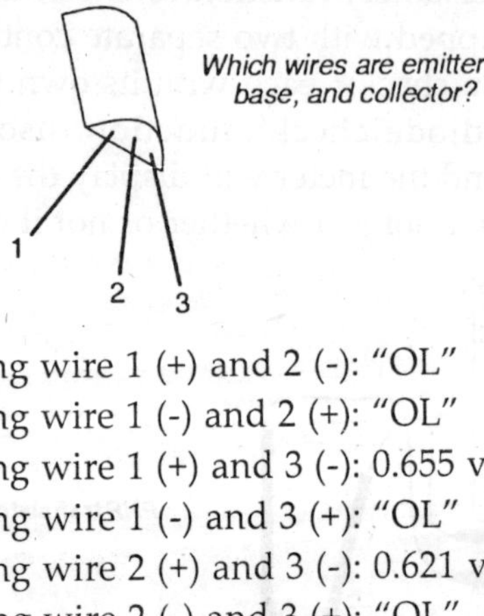

- Meter touching wire 1 (+) and 2 (-): "OL"
- Meter touching wire 1 (-) and 2 (+): "OL"
- Meter touching wire 1 (+) and 3 (-): 0.655 volts
- Meter touching wire 1 (-) and 3 (+): "OL"
- Meter touching wire 2 (+) and 3 (-): 0.621 volts
- Meter touching wire 2 (-) and 3 (+): "OL"

The only combinations of test points giving conducting meter readings are wires 1 and 3 (red test lead on 1 and black test lead on 3), and wires 2 and 3 (red test lead on 2 and black test lead on 3). These two readings *must* indicate forward biasing of the emitter-to-base junction (0.655 volts) and the collector-to-base junction (0.621 volts).

Now we look for the one wire common to both sets of conductive readings. It must be the base connection of the transistor, because the base is the only layer of the three-layer device common to both sets of PN junctions (emitter-base and collector-base).

In this example, that wire is number 3, being common to both the 1-3 and the 2-3 test point combinations. In both those sets of meter readings, the *black* (-) meter test lead was touching wire 3, which tells us that the base of this transistor is made of N-type semiconductor material (black = negative). Thus, the transistor is an PNP type with base on wire 3, emitter on wire 1 and collector on wire 2:

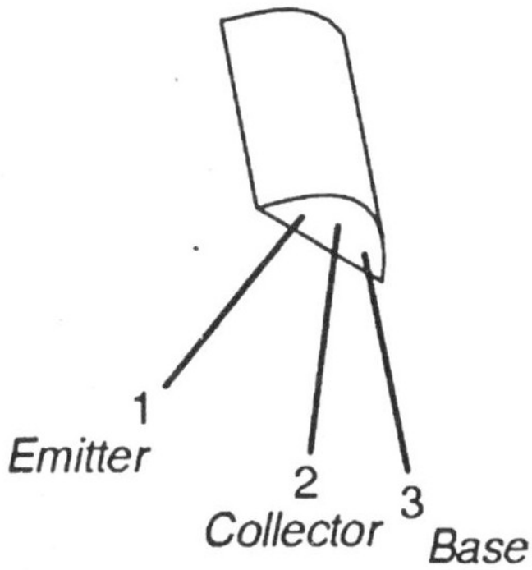

Please note that the base wire in this example is *not* the middle lead of the transistor, as one might expect from the three-layer "sandwich" model of a bipolar transistor. This is quite often the case, and tends to confuse new students of electronics. The only way to be sure which lead is which is by a meter check, or by referencing the manufacturer's "data sheet" documentation on that particular part number of transistor.

Knowing that a bipolar transistor behaves as two back-to-back diodes when tested with a conductivity meter is helpful for identifying an unknown transistor purely by meter readings. It is also helpful for a quick functional check of the transistor.

If the technician were to measure continuity in any more than two or any less than two of the six test lead combinations, he or she would immediately know that the transistor was defective (or else that it *wasn't* a bipolar transistor but rather something else — a distinct possibility if no part numbers can be referenced for sure identification!). However, the "two diode" model of the transistor fails to explain how or why it acts as an amplifying device.

To better illustrate this paradox, let's examine one of the transistor switch circuits using the physical diagram rather than the schematic symbol to represent the transistor. This way the two PN junctions will be easier to see: A grey-colored diagonal arrow shows the direction of electron flow through the emitter-base junction. This part makes sense, since the electrons are flowing from the N-type emitter to the P-type base: the junction is obviously forward-biased. However, the base-collector junction is another matter entirely.

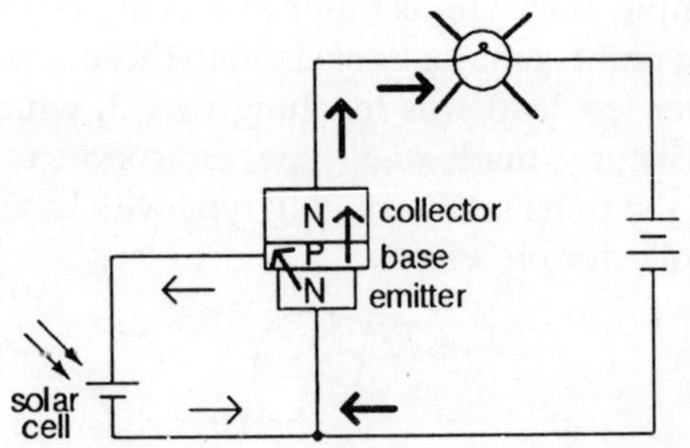

Notice how the grey-colored thick arrow is pointing in the direction of electron flow (upwards) from base to collector. With the base made of P-type material and the collector of N-type material, this direction of electron flow is clearly backwards to the direction normally associated with a PN junction! A normal PN junction wouldn't permit this "backward" direction of flow, at least not without offering significant opposition. However, when the transistor is saturated, there is very little opposition to electrons all the way from emitter to collector, as evidenced by the lamp's illumination! Clearly then, something is going on here that defies the simple "two-diode" explanatory model of the bipolar transistor. When we were first learning about transistor operation, we tried to construct my own transistor from two back-to-back diodes, like this:

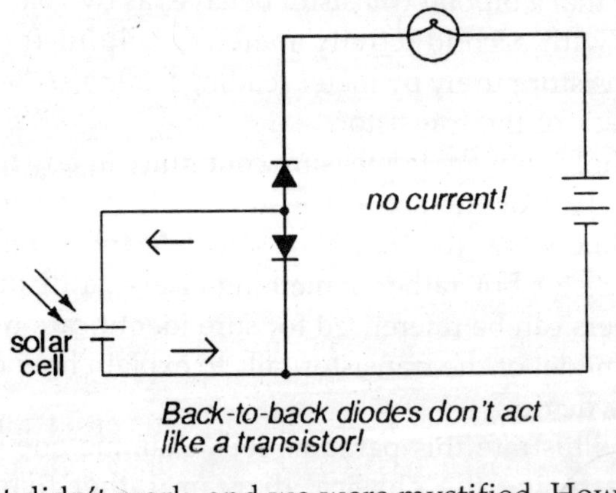

Back-to-back diodes don't act like a transistor!

Our circuit didn't work, and we were mystified. However useful the "two diode" description of a transistor might be for testing purposes, it doesn't explain how a transistor can behave as a controlled switch.

What happens in a transistor is this, the reverse bias of the base-collector junction prevents collector current when the transistor is in cutoff mode (that is, when there is no base current). However, when the base-emitter junction is forward biased by the controlling signal, the normally-blocking action of the base-collector junction is overridden and current is permitted through the collector, despite the fact that electrons are going the "wrong way" through that PN junction.

This action is dependent on the quantum physics of semiconductor junctions, and can only take place when the two junctions are properly spaced and the doping concentrations of the three layers are properly proportioned. Two diodes wired in series fail to meet these criteria, and so the top diode can never "turn on" when it is reversed biased, no matter how much current goes through the bottom diode in the base wire loop.

That doping concentrations play a crucial part in the special abilities of the transistor is further evidenced by the fact that collector and emitter are not interchangeable. If the transistor is merely viewed as two back-to-back PN junctions, or merely as a plain N-P-N or P-N-P sandwich of materials, it may seem as though either end of the transistor could serve as collector or emitter.

This, however, is not true. If connected "backwards" in a circuit, a base-collector current will fail to control current between collector and emitter. Despite the fact that both the emitter and collector layers of a bipolar transistor are of the same doping *type* (either N or P), they are definitely not identical! So, current through the emitter-base junction allows current through the reverse-biased base-collector junction. The action of base current can be thought of as "opening a gate" for current through the collector. More specifically, any given amount of emitter-to-base current *permits a limited amount* of base-to-collector current.

For every electron that passes through the emitter-base junction and on through the base wire, there is allowed a certain, restricted number of electrons to pass through the base-collector junction and no more.

- Tested with a multimeter in the "resistance" or "diode check" modes, a transistor behaves like two back-to-back PN (diode) junctions.
- The emitter-base PN junction has a slightly greater forward voltage drop than the collector-base PN junction, due to more concentrated doping of the emitter semiconductor layer.
- The reverse-biased base-collector junction normally blocks any current from going through the transistor between emitter and collector. However, that junction begins to conduct if current is drawn through the base wire. Base current can be thought of as "opening a gate" for a certain, limited amount of current through the collector.

ACTIVE MODE OPERATION

When a transistor is in the fully-off state (like an open switch), it is said to be *cutoff*. Conversely, when it is fully conductive between emitter and collector (passing as much current through the collector as the collector power supply and load will allow), it is said to be *saturated*. These are the two modes of operation explored thus far in using the transistor as a switch. However, bipolar transistors don't have to be restricted to these two extreme modes of operation. Base current "opens a gate" for a limited amount of current through the collector.

If this limit for the controlled current is greater than zero but less than the maximum allowed by the power supply and load circuit, the transistor will "throttle" the collector current in a mode somewhere between cutoff and saturation.

This mode of operation is called the *active* mode. An automotive analogy for transistor operation is as follows: *cutoff* is the condition where there is no motive force generated by the mechanical parts of the car to make it move. In cutoff mode, the brake is engaged (zero base current), preventing motion (collector current).

Active mode is when the automobile is cruising at a constant, controlled speed (constant, controlled collector current) as dictated by the driver. *Saturation* is when the automobile is driving up a steep hill that prevents it from going as fast as the driver would wish. In other words, a "saturated" automobile is one where the accelerator pedal is pushed all the way down (base current calling for more collector current than can be provided by the power supply/load circuit).

This voltage/current relationship is entirely different from what we're used to seeing across a resistor. With a resistor, current increases linearly as the voltage across it increases. Here, with a transistor, current from emitter to collector stays limited at a fixed, maximum value no matter how high the voltage across emitter and collector increases.

Often it is useful to superimpose several collector current/voltage graphs for different base currents on the same graph. A collection of curves like this — one curve plotted for each distinct level of base current — for a particular transistor is called the transistor's *characteristic curves*:

Each curve on the graph reflects the collector current of the transistor, plotted over a range of collector-to-emitter voltages, for a given amount of base current.

Since a transistor tends to act as a current regulator, limiting collector current to a proportion set by the base current, it is useful to express this proportion as a standard transistor performance measure. Specifically, the ratio of collector current to base current is known as the *Beta* ratio (symbolized by the Greek letter β):

$$\beta = \frac{I_{collector}}{I_{base}} \quad \beta \text{ is also known as } h_{fe}$$

Sometimes the β ratio is designated as "h_{fe}," a label used in a branch of mathematical semiconductor analysis known as "hybrid parameters" which strives to achieve very precise predictions of transistor performance with detailed equations. Hybrid parameter variables are many, but they are all labeled with the general letter "h" and a specific subscript. The variable "h_{fe}" is just another (standardized) way of expressing the ratio of collector current to base current, and is interchangeable with "β." Like all ratios, β is unitless.

β for any transistor is determined by its design: it cannot be altered after manufacture. However, there are so many physical variables impacting β that it is rare to have two transistors of the same design exactly match. If a circuit design relies on equal β ratios between multiple transistors, "matched sets" of transistors may be purchased at extra cost. However, it is generally considered bad design practice to engineer circuits with such dependencies.

It would be nice if the β of a transistor remained stable for all operating conditions, but this is not true in real life.

For an actual transistor, the β ratio may vary by a factor of over 3 within its operating current limits.

Sometimes it is helpful for comprehension to "model" complex electronic components with a collection of simpler, better-understood components. The following is a popular model shown in many introductory electronics texts:

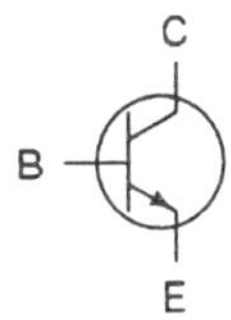

NPN diode-rheostat model

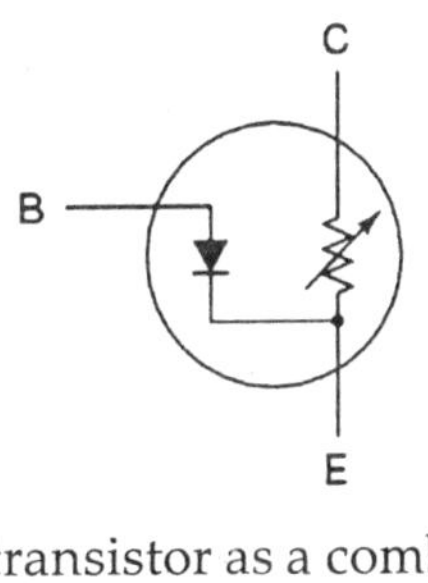

This model casts the transistor as a combination of diode and rheostat (variable resistor). Current through the base-emitter diode controls the resistance of the collector-emitter rheostat (as implied by the dashed line connecting the two components), thus controlling collector current. An NPN transistor is modeled in the figure shown, but a PNP transistor would be only slightly different (only the base-emitter diode would be reversed). This model succeeds in illustrating the basic concept of transistor amplification: how the base current signal can exert control over the collector current.

However, we don't like this model because it tends to miscommunicate the notion of a set amount of collector-emitter resistance for a given amount of base current. If this were true, the transistor wouldn't *regulate* collector current at all like the characteristic curves show. Instead of the collector current curves flattening out after their brief rise as the collector-emitter voltage increases, the collector current would be directly proportional to collector-emitter voltage, rising steadily in a straight line on the graph.

A Better Transistor Model:

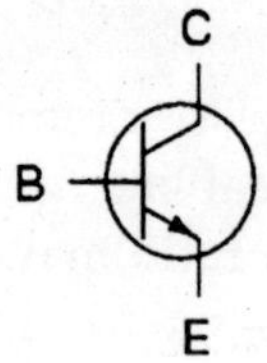

NPN diode-current source model

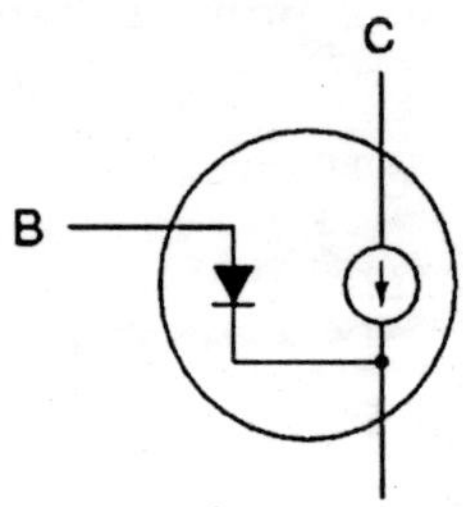

It casts the transistor as a combination of diode and current source, the output of the current source being set at a multiple (β ratio) of the base current. This model is far more accurate in depicting the true input/output characteristics of a transistor: base current establishes a certain amount of collector *current,* rather than a certain amount of collector-emitter *resistance* as the first model implies.

Also, this model is favored when performing network analysis on transistor circuits, the current source being a well-understood theoretical component. Unfortunately, using a current source to model the transistor's current-controlling behaviour can be misleading: in no way will the transistor ever act as a *source* of electrical energy, which the current source symbol implies is a possibility.

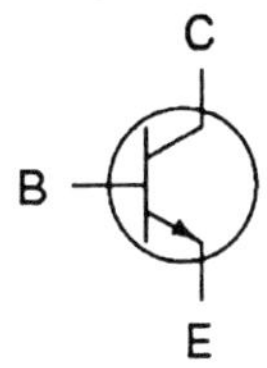

NPN diode-regulating diode model

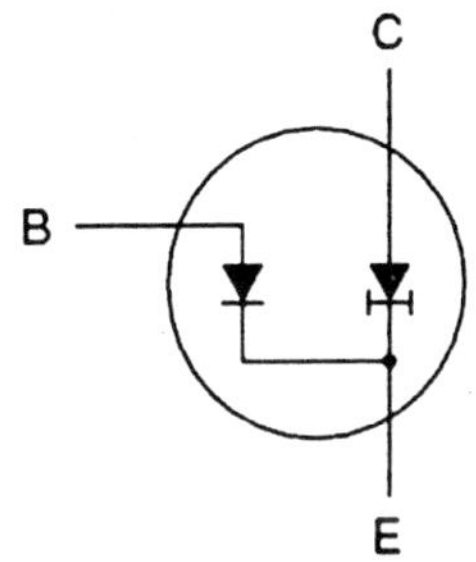

My own personal suggestion for a transistor model substitutes a constant-current diode for the current source:

Since no diode ever acts as a *source* of electrical energy, this analogy escapes the false implication of the current source model as a source of power, while depicting the transistor's constant-current behaviour better than the rheostat model.

Another way to describe the constant-current diode's action would be to refer to it as a *current regulator,* so this transistor illustration of mine might also be described as a *diode-current regulator* model. The greatest disadvantage we see to this model is the relative obscurity of constant-current diodes. Many people may be unfamiliar with their symbology or even of their existence, unlike either rheostats or current sources, which are commonly known.

- A transistor is said to be in its *active* mode if it is operating somewhere between fully on (saturated) and fully off (cutoff).
- Base current tends to regulate collector current. By *regulate,* we mean that no more collector current may exist than what is allowed by the base current.
- The ratio between collector current and base current is called "Beta" (β) or "h_{fe}".
- β ratios are different for every transistor, and they tend to change for different operating conditions.

THE COMMON-EMITTER AMPLIFIER

We saw how transistors could be used as switches, operating in either their "saturation" or "cutoff" modes. In the last section we saw how transistors behave within their "active" modes, between the far limits of saturation and cutoff. Because transistors are able to control current in an analog (infinitely divisible) fashion, they find use as amplifiers for analog signals. One of the simpler transistor amplifier circuits to study is the one used previously for illustrating the transistor's switching ability:

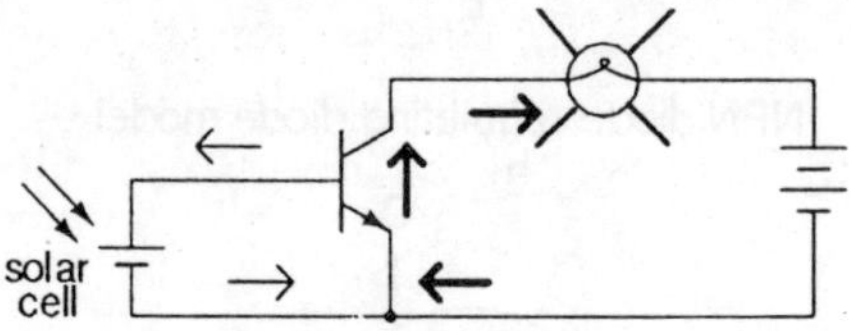

It is called the *common-emitter* configuration because (ignoring the power supply battery) both the signal source and the load share the emitter lead as a common connection point.

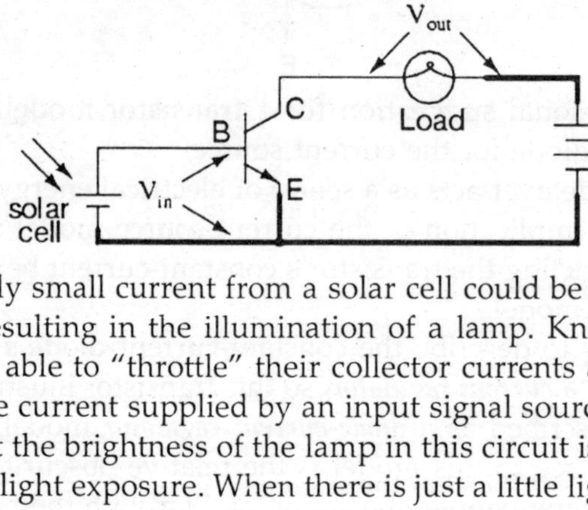

A relatively small current from a solar cell could be used to saturate a transistor, resulting in the illumination of a lamp. Knowing now that transistors are able to "throttle" their collector currents according to the amount of base current supplied by an input signal source, we should be able to see that the brightness of the lamp in this circuit is controllable by the solar cell's light exposure. When there is just a little light shone on the solar cell, the lamp will glow dimly. The lamp's brightness will steadily increase as more light falls on the solar cell.

Suppose that we were interested in using the solar cell as a light intensity instrument. We want to be able to measure the intensity of incident light with the solar cell by using its output current to drive a meter movement. It is possible to directly connect a meter movement to a solar cell for this purpose. In fact, the simplest light-exposure meters for photography work are designed like this:

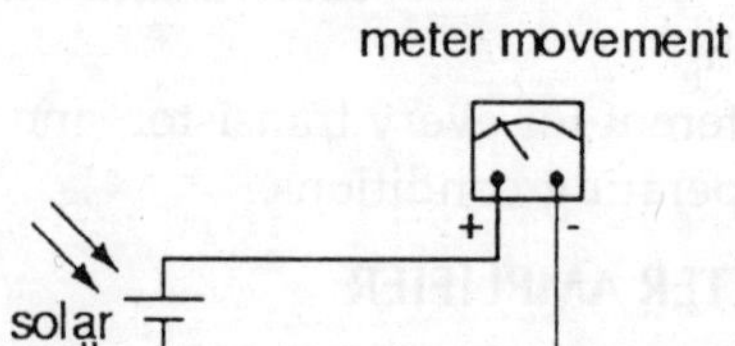

While this approach might work for moderate light intensity

measurements, it would not work as well for low light intensity measurements. Because the solar cell has to supply the meter movement's power needs, the system is necessarily limited in its sensitivity. Supposing that our need here is to measure very low-level light intensities, we are pressed to find another solution. Perhaps the most direct solution to this measurement problem is to use a transistor to *amplify* the solar cell's current so that more meter movement needle deflection may be obtained for less incident light. Consider this approach:

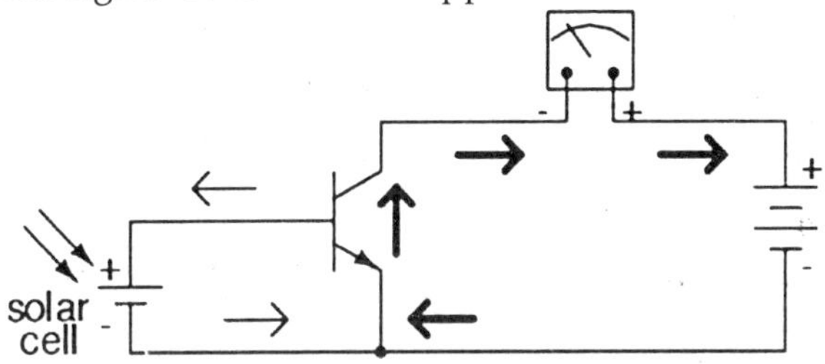

Current through the meter movement in this circuit will be β times the solar cell current. With a transistor β of 100, this represents a substantial increase in measurement sensitivity. It is prudent to point out that the additional power to move the meter needle comes from the battery on the far right of the circuit, not the solar cell itself. All the solar cell's current does is *control* battery current to the meter to provide a greater meter reading than the solar cell could provide unaided.

Because the transistor is a current-regulating device, and because meter movement indications are based on the amount of current through their movement coils, meter indication in this circuit should depend only on the amount of current from the solar cell, not on the amount of voltage provided by the battery.

This means the accuracy of the circuit will be independent of battery condition, a significant feature! All that is required of the battery is a certain minimum voltage and current output ability to be able to drive the meter full-scale if needed.

Another way in which the common-emitter configuration may be used is to produce an output *voltage* derived from the input signal, rather than a specific output *current*. Let's replace the meter movement with a plain resistor and measure voltage between collector and emitter:

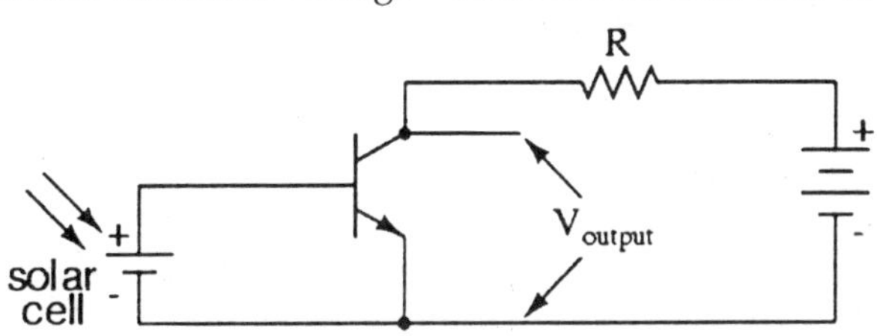

With the solar cell darkened (no current), the transistor will be in cutoff mode and behave as an open switch between collector and emitter. This will produce maximum voltage drop between collector and emitter for maximum V_{output}, equal to the full voltage of the battery.

At full power (maximum light exposure), the solar cell will drive the transistor into saturation mode, making it behave like a closed switch between collector and emitter. The result will be minimum voltage drop between collector and emitter, or almost zero output voltage. In actuality, a saturated transistor can never achieve zero voltage drop between collector and emitter due to the two PN junctions through which collector current must travel. However, this "collector-emitter saturation voltage" will be fairly low, around several tenths of a volt, depending on the specific transistor used.

For light exposure levels somewhere between zero and maximum solar cell output, the transistor will be in its active mode, and the output voltage will be somewhere between zero and full battery voltage. An important quality to note here about the common-emitter configuration is that the output voltage is *inversely proportional* to the input signal strength. That is, the output voltage decreases as the input signal increases. For this reason, the common-emitter amplifier configuration is referred to as an *inverting* amplifier.

So far, we've seen the transistor used as an amplifier for DC signals. In the solar cell light meter example, we were interested in amplifying the DC output of the solar cell to drive a DC meter movement, or to produce a DC output voltage.

However, this is not the only way in which a transistor may be employed as an amplifier. In many cases, what is desired is an *AC* amplifier for amplifying *alternating* current and voltage signals. One common application of this is in audio electronics (radios, televisions, and public-address systems). Earlier, we saw an example where the audio output of a tuning fork could be used to activate a transistor as a switch. Let's see if we can modify that circuit to send power to a speaker rather than to a lamp:

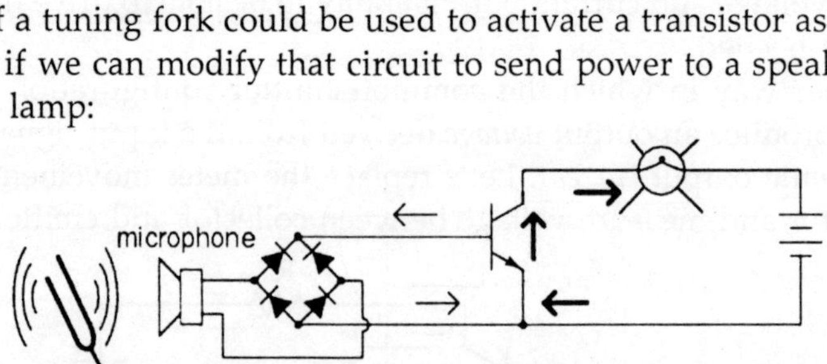

In the original circuit, a full-wave bridge rectifier was used to convert the microphone's AC output signal into a DC voltage to drive the input of the transistor. All we cared about here was turning the lamp on with a sound signal from the microphone, and this arrangement sufficed for that purpose. But now we want to actually reproduce the AC signal and drive a speaker. This means we cannot rectify the microphone's output anymore, because we need undistorted AC signal to drive the transistor! Let's remove the bridge rectifier and replace the lamp with a speaker:

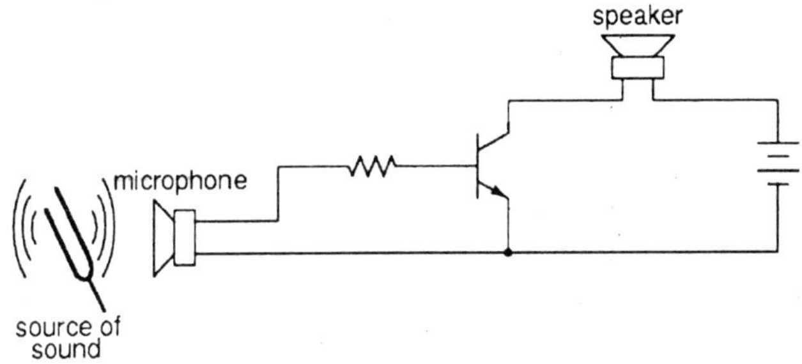

The simulation plots both the input voltage (an AC signal of 1.5 volt peak amplitude and 2000 Hz frequency) and the current through the 15 volt battery, which is the same as the current through the speaker. What we see here is a full AC sine wave alternating in both positive and negative directions, and a half-wave output current waveform that only pulses in one direction. If we were actually driving a speaker with this waveform, the sound produced would be horribly distorted.

What's wrong with the circuit? Why won't it faithfully reproduce the entire AC waveform from the microphone? The answer to this question is found by close inspection of the transistor diode-regulating diode model:

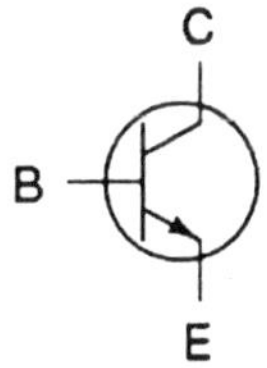

NPN diode-regulating diode model

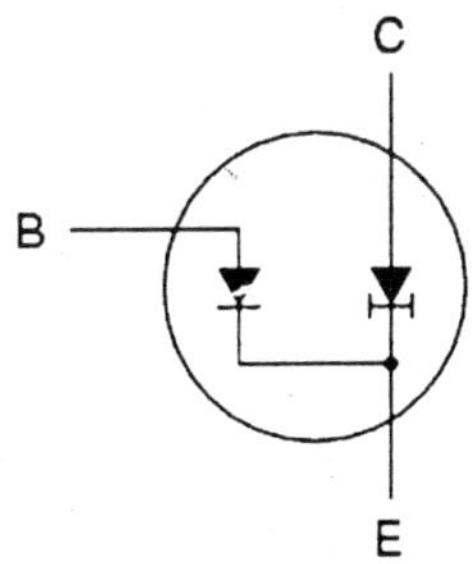

Collector current is controlled, or regulated, through the constant-current mechanism according to the pace set by the current through the base-emitter diode.

Note that both current paths through the transistor are monodirectional: *one way only!* Despite our intent to use the transistor to amplify an *AC* signal, it is essentially a *DC* device, capable of handling currents in a single direction only. We may apply an AC voltage input signal between the base and emitter, but electrons cannot flow in that circuit during the part of the cycle that reverse-biases the base-emitter diode junction.

Therefore, the transistor will remain in cutoff mode throughout that portion of the cycle. It will "turn on" in its active mode only when the input voltage is of the correct polarity to forward-bias the base-emitter diode, and only when that voltage is sufficiently high to overcome the diode's forward voltage drop.

Remember that bipolar transistors are *current-controlled devices*: they regulate collector current based on the existence of base-to-emitter *current*, not base-to-emitter *voltage*.

The only way we can get the transistor to reproduce the entire waveform as current through the speaker is to keep the transistor in its active mode the entire time.

This means we must maintain current through the base during the entire input waveform cycle.

Consequently, the base-emitter diode junction must be kept forward-biased at all times. Fortunately, this can be accomplished with the aid of a DC *bias voltage* added to the input signal. By connecting a sufficient DC voltage in series with the AC signal source, forward-bias can be maintained at all points throughout the wave cycle:

- *Common-emitter* transistor amplifiers are so-called because the input and output voltage points share the emitter lead of the transistor in common with each other, not considering any power supplies.
- Transistors are essentially DC devices: they cannot directly handle voltages or currents that reverse direction. In order to make them work for amplifying AC signals, the input signal must be offset with a DC voltage to keep the transistor in its active mode throughout the entire cycle of the wave. This is called *biasing*.
- If the output voltage is measured between emitter and collector on a common-emitter amplifier, it will be 180° out of phase with the input voltage waveform. For this reason, the common-emitter amplifier is called an *inverting* amplifier circuit.
- The current gain of a common-emitter transistor amplifier with the load connected in series with the collector is equal to b. The

voltage gain of a common-emitter transistor amplifier is approximately given here:

- $A_V = \beta \frac{R_{out}}{R_{in}}$
- Where "R_{out}" is the resistor connected in series with the collector and "R_{in}" is the resistor connected in series with the base.

THE COMMON-COLLECTOR AMPLIFIER

Our next transistor configuration to study is a bit simpler in terms of gain calculations. Called the *common-collector* configuration, its schematic diagram looks like this:

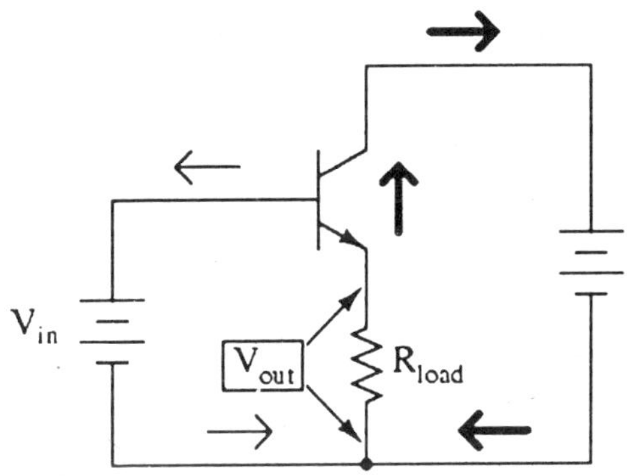

It is called the *common-collector* configuration because (ignoring the power supply battery) both the signal source and the load share the collector lead as a common connection point:

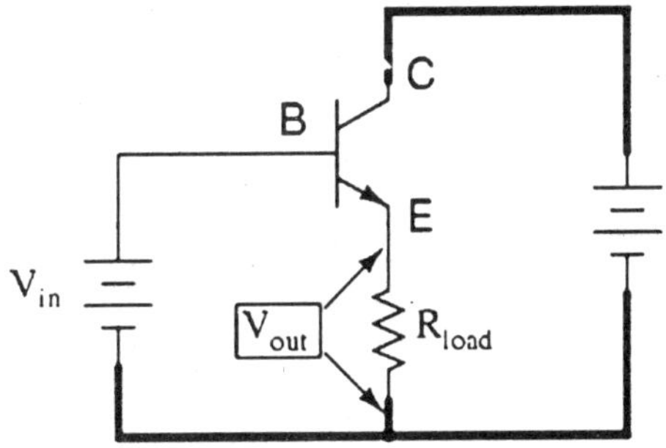

It should be apparent that the load resistor in the common-collector amplifier circuit receives both the base and collector currents, being placed in series with the emitter. Since the emitter lead of a transistor is the one handling the most current (the sum of base and collector currents, since base and collector currents always mesh together to form the emitter current), it would be reasonable to presume that this amplifier will have a very large current gain (maximum output current for minimum input current). This presumption is indeed correct: the current gain for a common-collector amplifier is quite large, larger than any other transistor amplifier configuration. However, this is not necessarily what sets it apart from other amplifier designs.

Unlike the common-emitter amplifier from the previous section, the common-collector produces an output voltage in *direct* rather than *inverse* proportion to the rising input voltage. As the input voltage increases, so does the output voltage. More than that, a close examination reveals that the output voltage is nearly *identical* to the input voltage, lagging behind only about 0.77 volts.

This is the unique quality of the common-collector amplifier: an output voltage that is nearly equal to the input voltage. Examined from the perspective of output voltage *change* for a given amount of input voltage *change*, this amplifier has a voltage gain of almost exactly unity (1), or 0 dB. This holds true for transistors of any β value, and for load resistors of any resistance value.

It is simple to understand why the output voltage of a common-collector amplifier is always nearly equal to the input voltage. Referring back to the diode-regulating diode transistor model, we see that the base current must go through the base-emitter PN junction, which is equivalent to a normal rectifying diode. So long as this junction is forward-biased (the transistor conducting current in either its active or saturated modes), it will have a voltage drop of approximately 0.7 volts, assuming silicon construction. This 0.7 volt drop is largely irrespective of the actual magnitude of base current, so we can regard it as being constant:

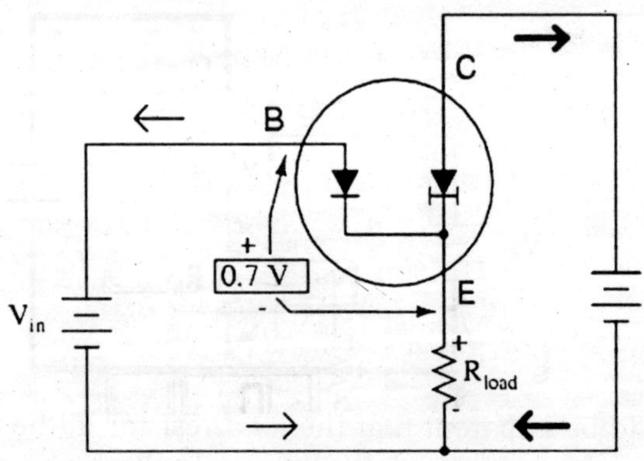

Given the voltage polarities across the base-emitter PN junction and the load resistor, we see that they *must* add together to equal the input voltage, in accordance with Kirchhoff's Voltage Law. In other words, the load voltage will always be about 0.7 volts less than the input voltage for all conditions where the transistor is conducting. Cutoff occurs at input voltages below 0.7 volts, and saturation at input voltages in excess of battery (supply) voltage plus 0.7 volts.

Because of this behaviour, the common-collector amplifier circuit is also known as the *voltage-follower* or *emitter-follower* amplifier, in reference to the fact that the input and load voltages follow each other so closely.

Applying the common-collector circuit to the amplification of AC signals requires the same input "biasing" used in the common-emitter circuit: a DC voltage must be added to the AC input signal to keep the transistor in its active mode during the entire cycle. When this is done, the result is a non-inverting amplifier:

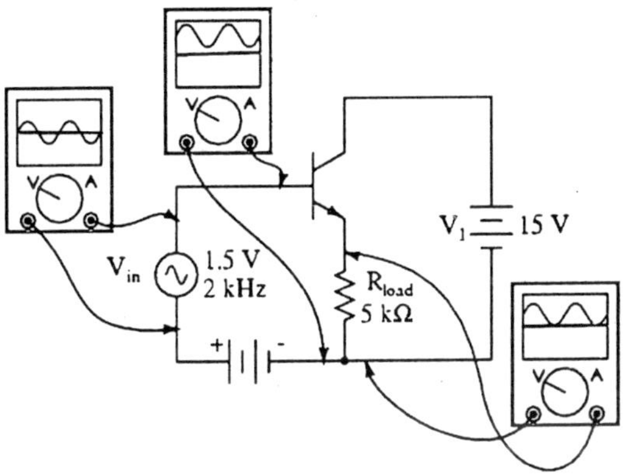

Since this amplifier configuration doesn't provide any voltage gain (in fact, in practice it actually has a voltage gain of slightly *less* than 1), its only amplifying factor is current.

The common-emitter amplifier configuration examined in the previous section had a current gain equal to the β of the transistor, being that the input current went through the base and the output (load) current went through the collector, and β by definition is the ratio between the collector and base currents. In the common-collector configuration, though, the load is situated in series with the emitter, and thus its current is equal to the emitter current. With the emitter carrying collector current *and* base current, the load in this type of amplifier has all the current of the collector running through it *plus* the input current of the base. This yields a current gain of β plus 1:

$$A_I = \frac{I_{emitter}}{I_{base}}$$

$$A_I = \frac{I_{collector} + I_{base}}{I_{base}}$$

$$A_1 = \frac{I_{collector}}{I_{base}} + 1$$

$$A_1 = \beta + 1$$

Once again, PNP transistors are just as valid to use in the common-collector configuration as NPN transistors. The gain calculations are all the same, as is the non-inverting behaviour of the amplifier. The only difference is in voltage polarities and current directions:

A popular application of the common-collector amplifier is for regulated DC power supplies, where an unregulated (varying) source of DC voltage is clipped at a specified level to supply regulated (steady) voltage to a load. Of course, zener diodes already provide this function of voltage regulation:

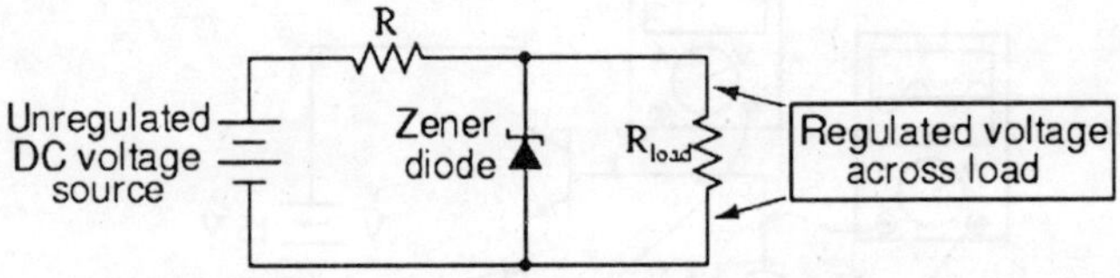

However, when used in this direct fashion, the amount of current that may be supplied to the load is usually quite limited. In essence, this circuit regulates voltage across the load by keeping current through the series resistor at a high enough level to drop all the excess power source voltage across it, the zener diode drawing more or less current as necessary to keep the voltage across itself steady.

For high-current loads, an plain zener diode voltage regulator would have to be capable of shunting a lot of current through the diode in order to be effective at regulating load voltage in the event of large load resistance or voltage source changes.

One popular way to increase the current-handling ability of a regulator circuit like this is to use a common-collector transistor to amplify current to the load, so that the zener diode circuit only has to handle the amount of current necessary to drive the base of the transistor:

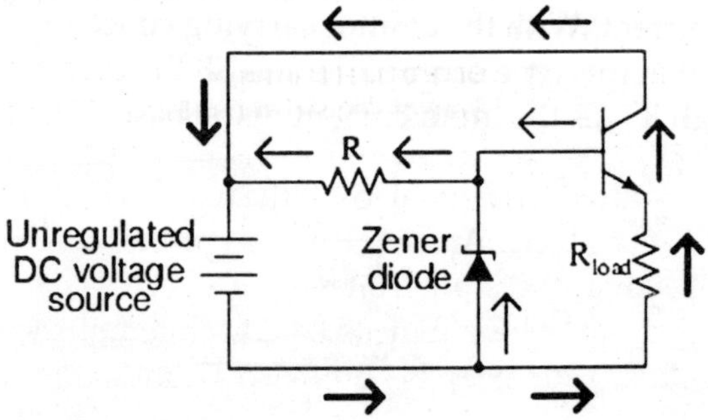

There's really only one caveat to this approach: the load voltage will be approximately 0.7 volts less than the zener diode voltage, due to the transistor's 0.7 volt base-emitter drop. However, since this 0.7 volt difference is fairly constant over a wide range of load currents, a zener diode with a 0.7 volt higher rating can be chosen for the application.

Sometimes the high current gain of a single-transistor, common-collector configuration isn't enough for a particular application. If this is the case, multiple transistors may be staged together in a popular configuration known as a *Darlington pair*, just an extension of the common-collector concept:

An NPN "Darlington pair"

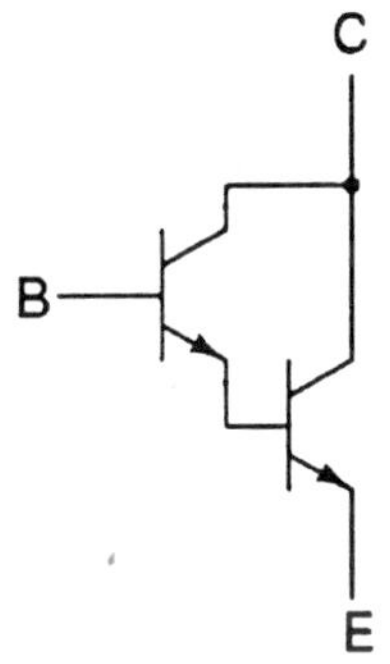

Darlington pairs essentially place one transistor as the common-collector load for another transistor, thus multiplying their individual current gains.

Base current through the upper-left transistor is amplified through that transistor's emitter, which is directly connected to the base of the lower-right transistor, where the current is again amplified. The overall current gain is as follows:

Darlington pair current gain

$$A_1 = (\beta_1 + 1)(\beta_2 + 1)$$

Where,

β_1 = Beta of first transistor

β_2 = Beta of second transistor

Voltage gain is still nearly equal to 1 if the entire assembly is connected to a load in common-collector fashion, although the load voltage will be a full 1.4 volts less than the input voltage:

Darlington pairs may be purchased as discrete units (two transistors in the same package), or may be built up from a pair of individual transistors.

Of course, if even more current gain is desired than what may be obtained with a pair, Darlington triplet or quadruplet assemblies may be constructed.

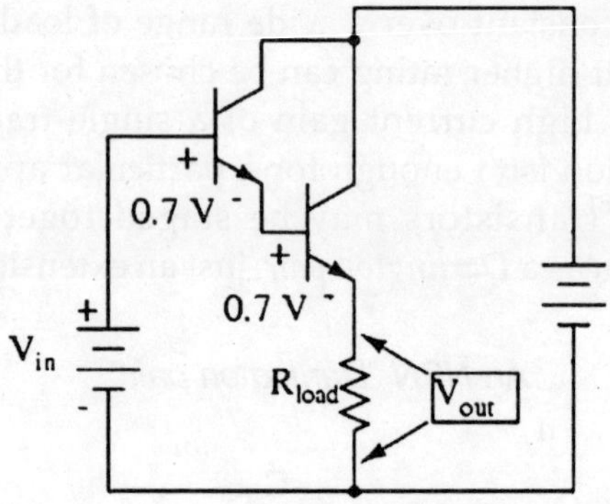

$$V_{out} = V_{in} - 1.4$$

- *Common-collector* transistor amplifiers are so-called because the input and output voltage points share the collector lead of the transistor in common with each other, not considering any power supplies.
- The output voltage on a common-collector amplifier will be in phase with the input voltage, making the common-collector a *non-inverting* amplifier circuit.
- The current gain of a common-collector amplifier is equal to b plus 1. The voltage gain is approximately equal to 1 (in practice, just a little bit less).
- A *Darlington pair* is a pair of transistors "piggybacked" on one another so that the emitter of one feeds current to the base of the other in common-collector form. The result is an overall current gain equal to the product (multiplication) of their individual common-collector current gains (b plus 1).

FEEDBACK

If some percentage of an amplifier's output signal is connected to the input, so that the amplifier amplifies part of its own output signal, we have what is known as *feedback*. Feedback comes in two varieties: *positive* (also called *regenerative*), and *negative* (also called *degenerative*). Positive feedback reinforces the direction of an amplifier's output voltage change, while negative feedback does just the opposite.

A familiar example of feedback happens in public-address ("PA") systems where someone holds the microphone too close to a speaker: a high-pitched "whine" or "howl" ensues, because the audio amplifier system is detecting and amplifying its own noise. Specifically, this is an example of *positive* or *regenerative* feedback, as any sound detected by the microphone is amplified and turned into a louder sound by the speaker, which is then detected by the microphone again, and so on... the result being a noise of steadily increasing volume until the system becomes "saturated" and cannot produce any more volume.

One might wonder what possible benefit feedback is to an amplifier circuit, given such an annoying example as PA system "howl." If we introduce positive, or regenerative, feedback into an amplifier circuit, it has the tendency of creating and sustaining oscillations, the frequency of which determined by the values of components handling the feedback signal from output to input. This is one way to make an *oscillator* circuit to produce AC from a DC power supply. Oscillators are very useful circuits, and so feedback has a definite, practical application for us.

Negative feedback, on the other hand, has a "dampening" effect on an amplifier: if the output signal happens to increase in magnitude, the feedback signal introduces a decreasing influence into the input of the amplifier, thus opposing the change in output signal. While positive feedback drives an amplifier circuit toward a point of instability (oscillations), negative feedback drives it the opposite direction: toward a point of stability.

An amplifier circuit equipped with some amount of negative feedback is not only more stable, but it tends to distort the input waveform to a lesser degree and is generally capable of amplifying a wider range of frequencies. The tradeoff for these advantages (there just *has* to be a disadvantage to negative feedback, right?) is decreased gain.

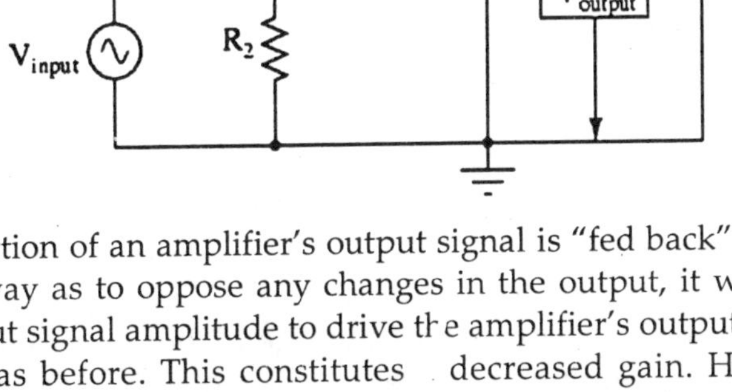

If a portion of an amplifier's output signal is "fed back" to the input in such a way as to oppose any changes in the output, it will require a greater input signal amplitude to drive the amplifier's output to the same amplitude as before. This constitutes decreased gain. However, the advantages of stability, lower distortion, and greater bandwidth are worth the tradeoff in reduced gain for many applications.

Let's examine a simple amplifier circuit and see how we might introduce negative feedback into it: The amplifier configuration shown here is a common-emitter, with a resistor bias network formed by R_1 and R_2. The capacitor couples V_{input} to the amplifier so that the signal source doesn't have a DC voltage imposed on it by the R_1/R_2 divider network. Resistor R_3 serves the purpose of controlling voltage gain. We could omit if for maximum voltage gain, but since base resistors like this are common in common-emitter amplifier circuits, we'll keep it in this schematic.

Like all common-emitter amplifiers, this one *inverts* the input signal as it is amplified. In other words, a positive-going input voltage causes the output voltage to decrease, or go in the direction of negative, and vice versa. If we were to examine the waveforms with oscilloscopes, it would look something like this:

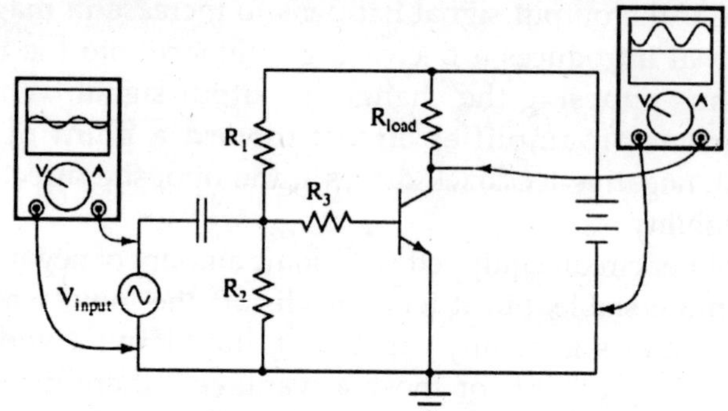

Because the output is an inverted, or mirror-image, reproduction of the input signal, any connection between the output (collector) wire and the input (base) wire of the transistor will result in *negative* feedback:

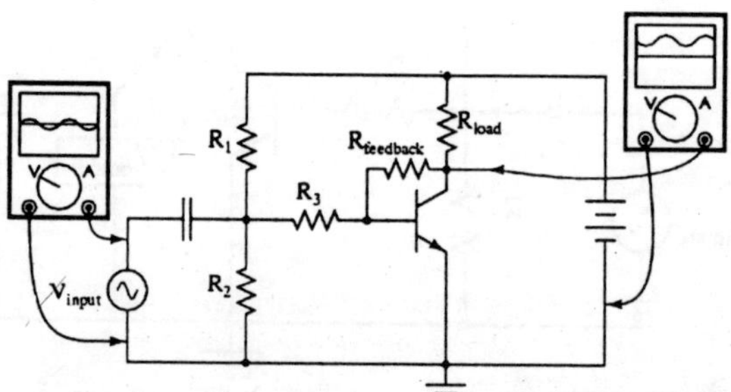

The resistances of R_1, R_2, R_3, and $R_{feedback}$ function together as a signal-mixing network so that the voltage seen at the base of the transistor (in reference to ground) is a weighted average of the input voltage and the feedback voltage, resulting in signal of reduced amplitude going into the transistor. The amplifier circuit will have reduced voltage gain, but improved linearity (reduced distortion) and increased bandwidth.

A resistor connecting collector to base is not the only way to introduce negative feedback into this amplifier circuit, though. Another method, although more difficult to understand at first, involves the placement of a resistor between the transistor's emitter terminal and circuit ground, like this:

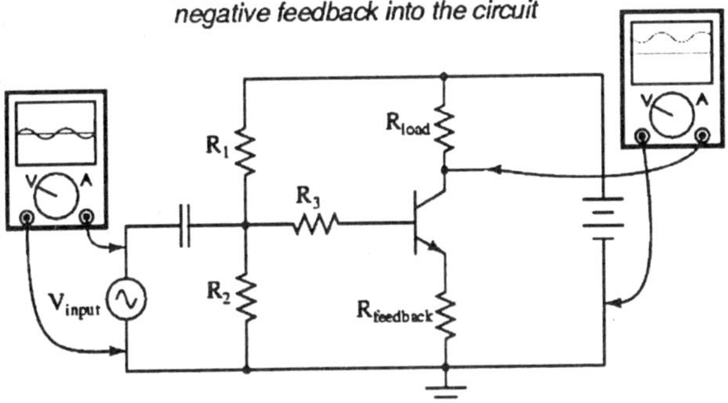

This new feedback resistor drops voltage proportional to the emitter current through the transistor, and it does so in such a way as to oppose the input signal's influence on the base-emitter junction of the transistor. Let's take a closer look at the emitter-base junction and see what difference this new resistor makes:

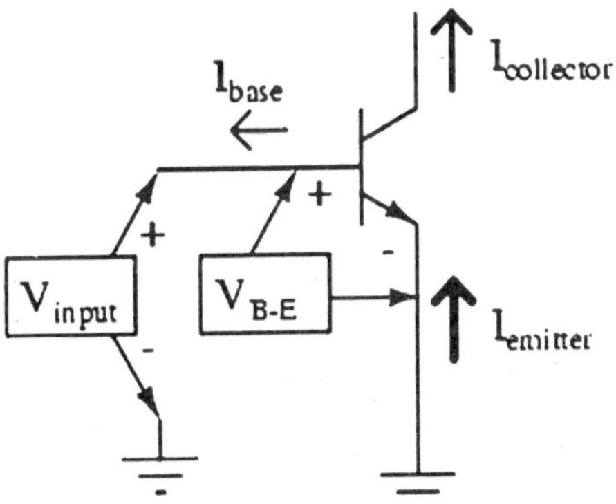

With no feedback resistor connecting the emitter to ground, whatever level of input signal (V_{input}) makes it through the coupling capacitor and $R_1/R_2/R_3$ resistor network will be impressed directly across the base-emitter junction as the transistor's input voltage (V_{B-E}). In other words, with no feedback resistor, V_{B-E} equals V_{input}. Therefore, if V_{input} increases by 100 mV, then V_{B-E} likewise increases by 100 mV: a change in one is the same as a change in the other, since the two voltages are equal to each other. Now let's consider the effects of inserting a resistor ($R_{feedback}$) between the transistor's emitter lead and ground:

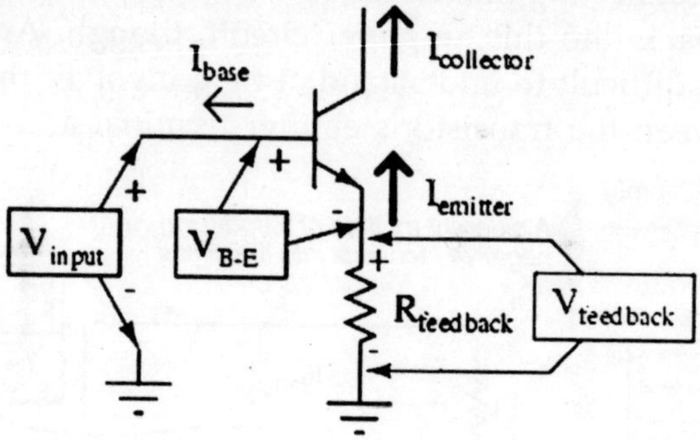

Note how the voltage dropped across $R_{feedback}$ adds with V_{B-E} to equal V_{input}. With $R_{feedback}$ in the $V_{input} - V_{B-E}$ loop, V_{B-E} will no longer be equal to V_{input}. We know that $R_{feedback}$ will drop a voltage proportional to emitter current, which is in turn controlled by the base current, which is in turn controlled by the voltage dropped across the base-emitter junction of the transistor (V_{B-E}).

Thus, if V_{input} were to increase in a positive direction, it would increase V_{B-E}, causing more base current, causing more collector (load) current, causing more emitter current, and causing more feedback voltage to be dropped across $R_{feedback}$.

This increase of voltage drop across the feedback resistor, though, *subtracts* from V_{input} to reduce the V_{B-E}, so that the actual voltage increase for V_{B-E} will be less than the voltage increase of V_{input}.

No longer will a 100 mV increase in V_{input} result in a full 100 mV increase for V_{B-E}, because the two voltages are *not* equal to each other.

Consequently, the input voltage has less control over the transistor than before, and the voltage gain for the amplifier is reduced: just what we expected from negative feedback.

In practical common-emitter circuits, negative feedback isn't just a luxury; it's a necessity for stable operation. In a perfect world, we could build and operate a common-emitter transistor amplifier with no negative feedback, and have the full amplitude of V_{input} impressed across the transistor's base-emitter junction.

This would give us a large voltage gain. Unfortunately, though, the relationship between base-emitter voltage and base-emitter current changes with temperature, as predicted by the "diode equation." As the transistor heats up, there will be less of a forward voltage drop across the base-emitter junction for any given current. This causes a problem for us, as the R_1/R_2 voltage divider network is designed to provide the correct quiescent current through the base of the transistor so that it will operate in whatever class of operation we desire (in this example, I've shown the amplifier working in class-A mode). If the transistor's voltage/current relationship changes with temperature, the amount of DC bias voltage necessary for the desired class of operation will change.

In this case, a hot transistor will draw more bias current for the same amount of bias voltage, making it heat up even more, drawing even more bias current. The result, if unchecked, is called *thermal runaway.*

Common-collector amplifiers, however, do not suffer from thermal runaway. Why is this? The answer has everything to do with negative feedback:

A common-collector amplifier

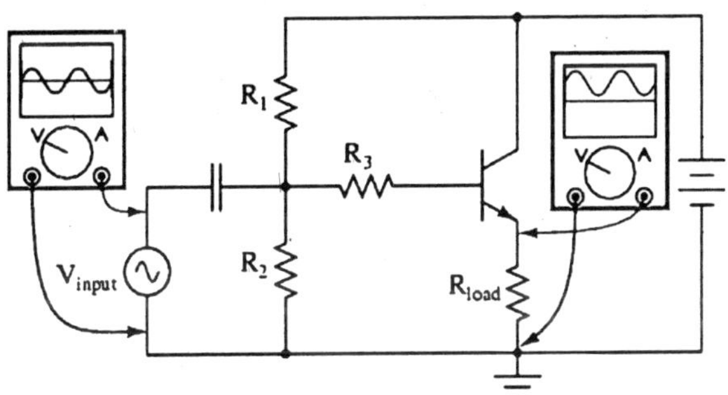

Note that the common-collector amplifier has its load resistor placed in exactly the same spot as we had the $R_{feedback}$ resistor in the last circuit: between emitter and ground. This means that the only voltage impressed across the transistor's base-emitter junction is the *difference* between V_{input} and V_{output}, resulting in a very low voltage gain (usually close to 1 for a common-collector amplifier).

Thermal runaway is impossible for this amplifier: if base current happens to increase due to transistor heating, emitter current will likewise increase, dropping more voltage across the load, which in turn *subtracts* from V_{input} to reduce the amount of voltage dropped between base and emitter.

In other words, the negative feedback afforded by placement of the load resistor makes the problem of thermal runaway *self-correcting*. In exchange for a greatly reduced voltage gain, we get superb stability and immunity from thermal runaway.

By adding a "feedback" resistor between emitter and ground in a common-emitter amplifier, we make the amplifier behave a little less like an "ideal" common-emitter and a little more like a common-collector. The feedback resistor value is typically quite a bit less than the load, minimizing the amount of negative feedback and keeping the voltage gain fairly high.

Another benefit of negative feedback, seen clearly in the common-collector circuit, is that it tends to make the voltage gain of the amplifier less dependent on the characteristics of the transistor. Note that in a common-collector amplifier, voltage gain is nearly equal to unity (1), regardless of the transistor's β.

This means, among other things, that we could replace the transistor in a common-collector amplifier with one having a different β and not see any significant changes in voltage gain. In a common-emitter circuit, the voltage gain is highly dependent on β.

If we were to replace the transistor in a common-emitter circuit with another of differing β, the voltage gain for the amplifier would change significantly. In a common-emitter amplifier equipped with negative feedback, the voltage gain will still be dependent upon transistor β to some degree, but not as much as before, making the circuit more predictable despite variations in transistor β.

The fact that we have to introduce negative feedback into a common-emitter amplifier to avoid thermal runaway is an unsatisfying solution. It would be nice, after all, to avoid thermal runaway without having to suppress the amplifier's inherently high voltage gain. A best-of-both-worlds solution to this dilemma is available to us if we closely examine the nature of the problem: the voltage gain that we have to minimize in order to avoid thermal runaway is the *DC* voltage gain, not the *AC* voltage gain. After all, it isn't the AC input signal that fuels thermal runaway: it's the DC bias voltage required for a certain class of operation: that quiescent DC signal that we use to "trick" the transistor (fundamentally a DC device) into amplifying an AC signal.

We can suppress DC voltage gain in a common-emitter amplifier circuit without suppressing AC voltage gain if we figure out a way to make the negative feedback function with DC only. That is, if we only feed back an inverted DC signal from output to input, but not an inverted AC signal.

The $R_{feedback}$ emitter resistor provides negative feedback by dropping a voltage proportional to load current. In other words, negative feedback is accomplished by inserting an impedance into the emitter current path. If we want to feed back DC but not AC, we need an impedance that is high for DC but low for AC. What kind of circuit presents a high impedance to DC but a low impedance to AC? A high-pass filter, of course!

By connecting a capacitor in parallel with the feedback resistor, we create the very situation we need: a path from emitter to ground that is easier for AC than it is for DC:

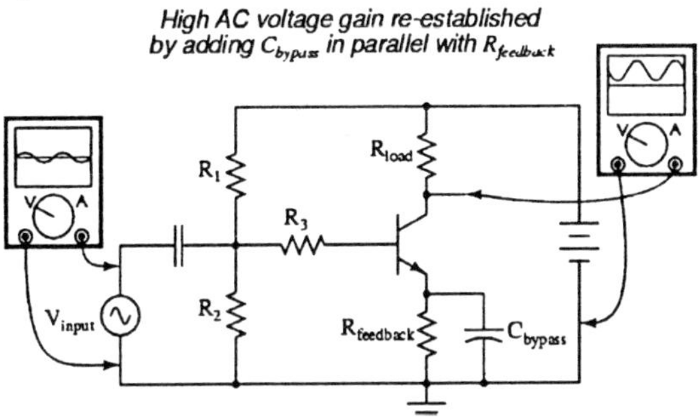

The new capacitor "bypasses" AC from the transistor's emitter to ground, so that no appreciable AC voltage will be dropped from emitter to ground to "feed back" to the input and suppress voltage gain. Direct current, on the other hand, cannot go through the bypass capacitor, and so must travel through the feedback resistor, dropping a DC voltage between emitter and ground which lowers the DC voltage gain and stabilizes the amplifier's DC response, preventing thermal runaway. Because we want the reactance of this capacitor (X_C) to be as low as possible, C_{bypass} should be sized relatively large. Because the polarity across this capacitor will never change, it is safe to use a polarized (electrolytic) capacitor for the task.

Another approach to the problem of negative feedback reducing voltage gain is to use multi-stage amplifiers rather than single-transistor amplifiers. If the attenuated gain of a single transistor is insufficient for the task at hand, we can use more than one transistor to make up for the reduction caused by feedback. Here is an example circuit showing negative feedback in a three-stage common-emitter amplifier:

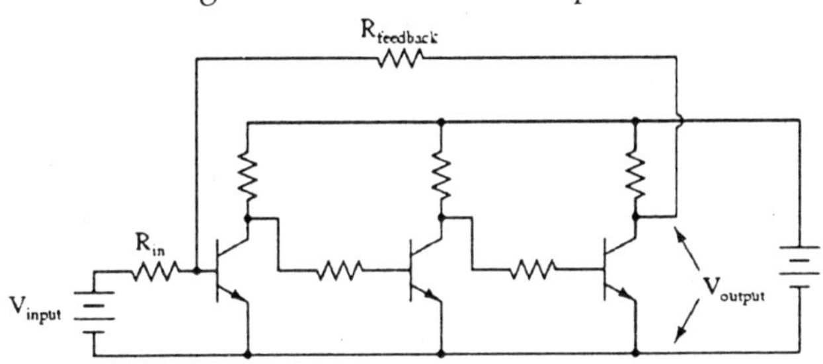

Note how there is but one "path" for feedback, from the final output to the input through a single resistor, $R_{feedback}$. Since each stage is a common-emitter amplifier — and thus inverting in nature — and there are an odd number of stages from input to output, the output signal will be inverted with respect to the input signal, and the feedback will be negative (degenerative).

Relatively large amounts of feedback may be used without sacrificing voltage gain, because the three amplifier stages provide so much gain to begin with.

At first, this design philosophy may seem inelegant and perhaps even counter-productive. Isn't this a rather crude way to overcome the loss in gain incurred through the use of negative feedback, to simply recover gain by adding stage after stage? What is the point of creating a huge voltage gain using three transistor stages if we're just going to attenuate all that gain anyway with negative feedback?

The point, though perhaps not apparent at first, is increased predictability and stability from the circuit as a whole. If the three transistor stages are designed to provide an arbitrarily high voltage gain (in the tens of thousands, or greater) with no feedback, it will be found that the addition of negative feedback causes the overall voltage gain to become less dependent of the individual stage gains, and approximately equal to the simple ratio $R_{feedback}/R_{in}$.

The more voltage gain the circuit has (without feedback), the more closely the voltage gain will approximate $R_{feedback}/R_{in}$ once feedback is established. In other words, voltage gain in this circuit is fixed by the values of two resistors, and nothing more.

This advantage has profound impact on mass-production of electronic circuitry: if amplifiers of predictable gain may be constructed using transistors of widely varied β values, it makes the selection and replacement of components very easy and inexpensive. It also means the amplifier's gain varies little with changes in temperature. This principle of stable gain control through a high-gain amplifier "tamed" by negative feedback is elevated almost to an art form in electronic circuits called *operational amplifiers*, or *op-amps*.

- *Feedback* is the coupling of an amplifier's output to its input.
- *Positive*, or *regenerative* feedback has the tendency of making an amplifier circuit unstable, so that it produces oscillations (AC). The frequency of these oscillations is largely determined by the components in the feedback network.
- *Negative*, or *degenerative* feedback has the tendency of making an amplifier circuit more stable, so that its output changes *less* for a given input signal than without feedback. This reduces the gain of the amplifier, but has the advantage of decreasing distortion and increasing bandwidth (the range of frequencies the amplifier can handle).
- Negative feedback may be introduced into a common-emitter

circuit by coupling collector to base, or by inserting a resistor between emitter and ground.

- An emitter-to-ground "feedback" resistor is usually found in common-emitter circuits as a preventative measure against *thermal runaway.*
- Negative feedback also has the advantage of making amplifier voltage gain more dependent on resistor values and less dependent on the transistor's characteristics.
- Common-collector amplifiers have a lot of negative feedback, due to the placement of the load resistor between emitter and ground. This feedback accounts for the extremely stable voltage gain of the amplifier, as well as its immunity against thermal runaway.
- Voltage gain for a common-emitter circuit may be re-established without sacrificing immunity to thermal runaway, by connecting a *bypass capacitor* in parallel with the emitter "feedback resistor."
- If the voltage gain of an amplifier is arbitrarily high, and negative feedback is used to reduce the gain to reasonable levels, it will be found that the gain will approximately equal $R_{feedback}/R_{in}$. Changes in transistor b or other internal component values will have comparatively little effect on voltage gain with feedback in operation, resulting in an amplifier that is stable and easy to design.

Chapter 6

Resonance

The inductance reactance vector and the capacitive reactance vector point in the opposite directions. Thus, when an inductor and capacitor are placed in series, their reactances subtract from one and other. At high frequencies the series circuit will appear as an inductive reactance. At low frequencies the inductor and capacitor series circuit will appear as a capacitive reactance. At some in between frequency the inductance and the capacitance will be equal.

The sum of these vector quantities will be zero. This frequency is called the resonant frequency. $w = 1/(LC)^{1/2}$ where w = 2*Pif or simply w = angular velocity. Pi = 3.141..., and f = frequency.Note that at frequency of about 700 kHz, the voltage at TP1 dips to 2.23 V. At 700 kHz the series circuit formed by C1 and L1 exhibits resonance. The impedance of a resonant circuit is low compared to Rs, and therefore a dip in signal amplitude occurs at TP1. At this resonant frequency the current in the series circuit Rs/C1/L1 is maximum.

Thus, the voltage across L1 peaks to 16.29 volts. This is characteristic of resonance.If you calculated the resonant frequency for L1 and C1 you would get 877 kHz, not the 700 kHz we observed in the bottom animation. That is because the second stage of this filter effects the resonant frequency of the first stage. In the top animation the first stage of the filter has been isolated. Note that the voltage at TP1 dips to 43 milli-volts at 877 kHz. The voltage at TP1 would dip even lower if L1 were an ideal inductor. L1 has a resistance of.1 ohms. This prevents the impedance of L1/C1 series circuit from going to zero at the resonant frequency.

AN ELECTRIC PENDULUM

Capacitors store energy in the form of an electric field, and electrically manifest that stored energy as a potential: *static voltage*. Inductors store energy in the form of a magnetic field, and electrically manifest that stored energy as a kinetic motion of electrons: *current*.

Capacitors and inductors are flip-sides of the same reactive coin, storing and releasing energy in complementary modes. When these two types of reactive components are directly connected together, their complementary tendencies to store energy will produce an unusual result.

If either the capacitor or inductor starts out in a charged state, the two components will exchange energy between them, back and forth, creating their own AC voltage and current cycles.

If we assume that both components are subjected to a sudden application of voltage (say, from a momentarily connected battery), the capacitor will very quickly charge and the inductor will oppose change in current, leaving the capacitor in the charged state and the inductor in the discharged state:

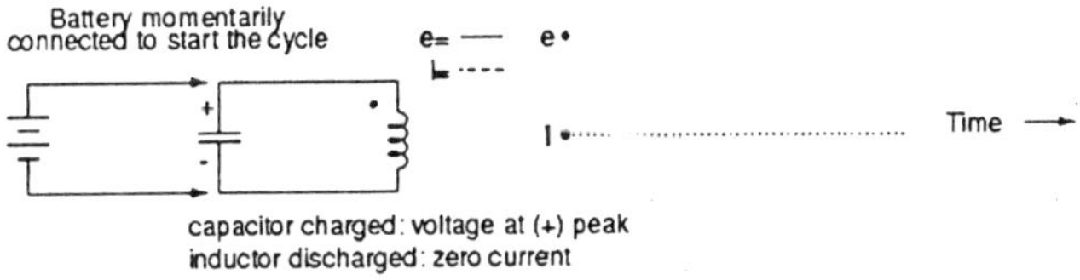

Fig. Capacitor Charged: Voltage at (+) Peak, Inductor Discharged: Zero Current.

The capacitor will begin to discharge, its voltage decreasing. Meanwhile, the inductor will begin to build up a "charge" in the form of a magnetic field as current increases in the circuit:

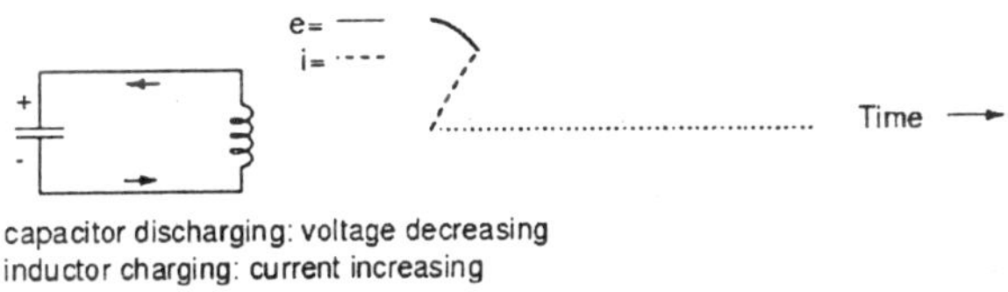

Fig. Capacitor Discharging: Voltage Decreasing, Inductor Charging: Current Increasing.

The inductor, still charging, will keep electrons flowing in the circuit until the capacitor has been completely discharged, leaving zero voltage across it:

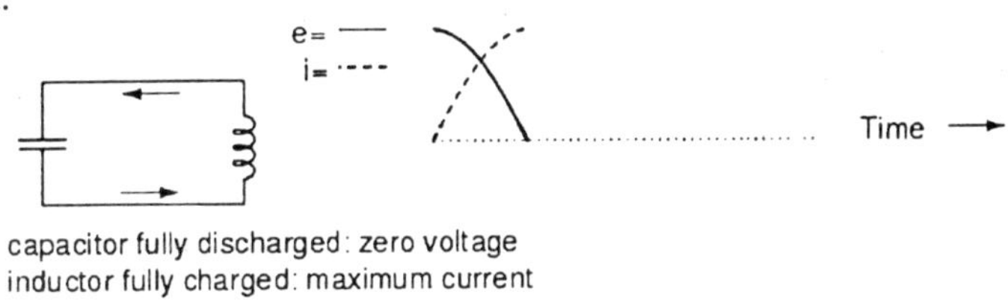

Fig. Capacitor Fully Discharged: Zero Voltage, Inductor Fully charged: Maximum Current.

The inductor will maintain current flow even with no voltage applied. In fact, it will generate a voltage (like a battery) in order to keep current in the same direction.

The capacitor, being the recipient of this current, will begin to accumulate a charge in the opposite polarity as before: When the inductor is finally depleted of its energy reserve and the electrons come to a halt, the capacitor will have reached full (voltage) charge in the opposite polarity as when it started:

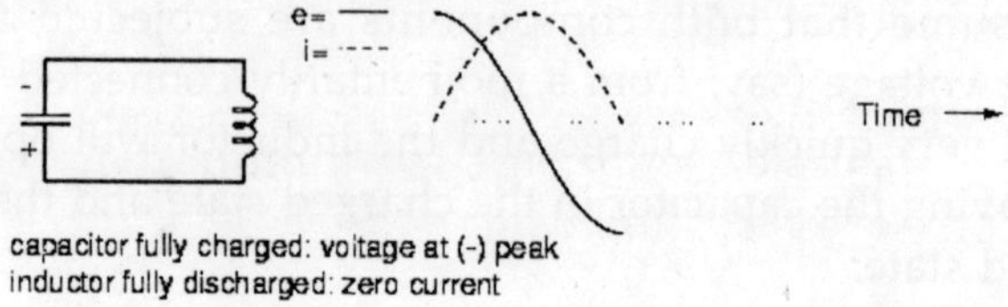

Capacitor fully charged: voltage at (-) peak, inductor fully discharged: zero current.

Now we're at a condition very similar to where we started: the capacitor at full charge and zero current in the circuit. The capacitor, as before, will begin to discharge through the inductor, causing an increase in current (in the opposite direction as before) and a decrease in voltage as it depletes its own energy reserve:

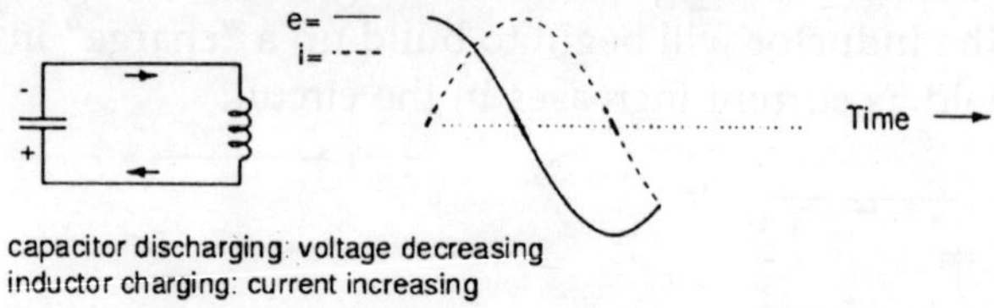

Capacitor discharging: voltage decreasing, inductor charging: current increasing. Eventually the capacitor will discharge to zero volts, leaving the inductor fully charged with full current through it:

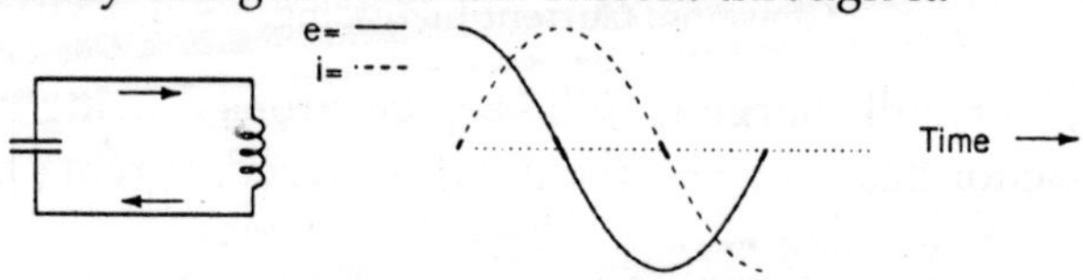

Fig. Capacitor Fully Discharged: Zero Voltage, Inductor Fully Charged: Current at (–) Peak.

The inductor, desiring to maintain current in the same direction, will act like a source again, generating a voltage like a battery to continue the flow. In doing so, the capacitor will begin to charge up and the current will decrease in magnitude: Eventually the capacitor will become fully charged again as the inductor expends all of its energy reserves trying to maintain current. The voltage will once again be at its positive peak and the current at zero. This completes one full cycle of the energy exchange between the capacitor and inductor:

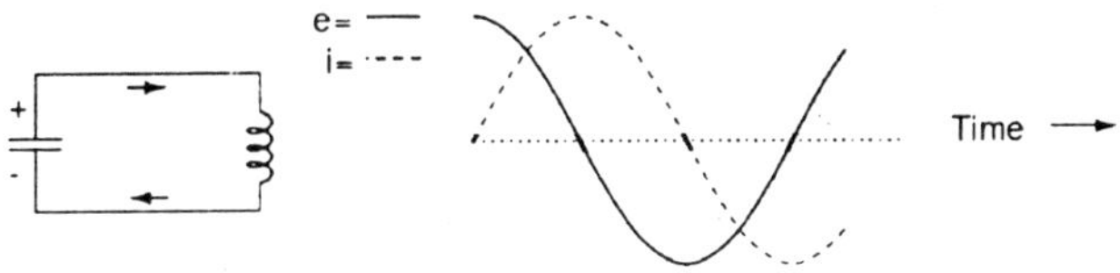

Fig. Capacitor Charging: Voltage Increasing, Inductor Discharging: Current Decreasing.

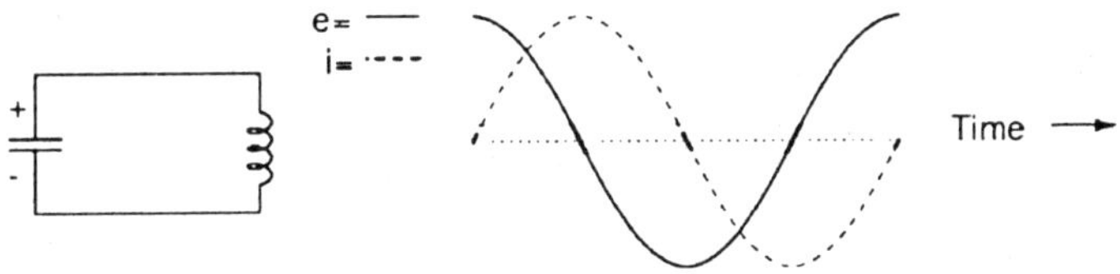

Fig. Capacitor Fully Charged: Voltage at (+) Peak, Inductor Fully Discharged: Zero Current.

This oscillation will continue with steadily decreasing amplitude due to power losses from stray resistances in the circuit, until the process stops altogether. Overall, this behaviour is akin to that of a pendulum: as the pendulum mass swings back and forth, there is a transformation of energy taking place from kinetic (motion) to potential (height), in a similar fashion to the way energy is transferred in the capacitor/inductor circuit back and forth in the alternating forms of current (kinetic motion of electrons) and voltage (potential electric energy). At the peak height of each swing of a pendulum, the mass briefly stops and switches directions. It is at this point that potential energy (height) is at a maximum and kinetic energy (motion) is at zero. As the mass swings back the other way, it passes quickly through a point where the string is pointed straight down. At this point, potential energy (height) is at zero and kinetic energy (motion) is at maximum.

Like the circuit, a pendulum's back-and-forth oscillation will continue with a steadily dampened amplitude, the result of air friction (resistance) dissipating energy.

Also like the circuit, the pendulum's position and velocity measurements trace two sine waves (90 degrees out of phase) over time: In physics, this kind of natural sine-wave oscillation for a mechanical system is called *Simple Harmonic Motion* (often abbreviated as "SHM"). The same underlying principles govern both the oscillation of a capacitor/inductor circuit and the action of a pendulum, hence the similarity in effect. It is an interesting property of any pendulum that its periodic time is governed by the length of the string holding the mass, and not the weight of the mass itself. That is why a pendulum will keep swinging at the same frequency as the oscillations decrease in amplitude. The oscillation rate is independent of the *amount* of energy stored in it.

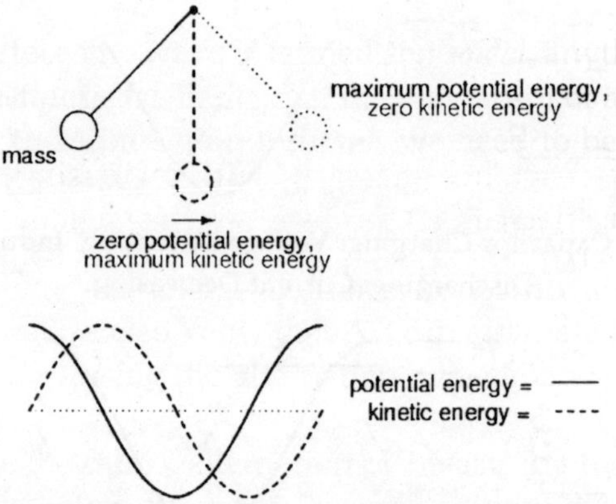

Fig. Pendelum Transfers Energy between Kinetic and Potential Energy as it Swings Low to High.

The same is true for the capacitor/inductor circuit. The rate of oscillation is strictly dependent on the sizes of the capacitor and inductor, not on the amount of voltage (or current) at each respective peak in the waves. The ability for such a circuit to store energy in the form of oscillating voltage and current has earned it the name *tank circuit*. Its property of maintaining a single, natural frequency regardless of how much or little energy is actually being stored in it gives it special significance in electric circuit design.

However, this tendency to oscillate, or *resonate*, at a particular frequency is not limited to circuits exclusively designed for that purpose. In fact, nearly any AC circuit with a combination of capacitance and inductance (commonly called an "LC circuit") will tend to manifest unusual effects when the AC power source frequency approaches that natural frequency. This is true regardless of the circuit's intended purpose.

If the power supply frequency for a circuit exactly matches the natural frequency of the circuit's LC combination, the circuit is said to be in a state of *resonance*. The unusual effects will reach maximum in this condition of resonance. For this reason, we need to be able to predict what the resonant frequency will be for various combinations of L and C, and be aware of what the effects of resonance are.

- A capacitor and inductor directly connected together form something called a *tank circuit*, which oscillates (or *resonates*) at one particular frequency. At that frequency, energy is alternately shuffled between the capacitor and the inductor in the form of alternating voltage and current 90 degrees out of phase with each other.
- When the power supply frequency for an AC circuit exactly matches that circuit's natural oscillation frequency as set by the L and C components, a condition of *resonance* will have been reached.

SIMPLE PARALLEL (TANK CIRCUIT) RESONANCE

A condition of resonance will be experienced in a tank circuit when the reactances of the capacitor and inductor are equal to each other. Because inductive reactance increases with increasing frequency and capacitive reactance decreases with increasing frequency, there will only be one frequency where these two reactances will be equal.

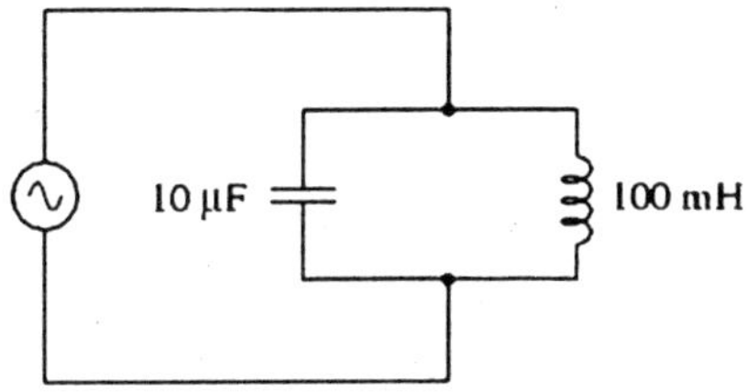

Fig. Simple Parallel Resonant Circuit (Tank Circuit).

In the above circuit, we have a 10 μF capacitor and a 100 mH inductor. Since we know the equations for determining the reactance of each at a given frequency, and we're looking for that point where the two reactances are equal to each other, we can set the two reactance formulae equal to each other and solve for frequency algebraically:

$$X_L = 2\pi fL \qquad X_C = \frac{1}{2\pi fC}$$

...setting the two equal to each other, representing a condition of eaqual reactance (resonance)...

$$2\pi fL = \frac{1}{2\pi fC}$$

Multiplying both sides by f eliminates the f term in the denominator of the fraction...

$$2\pi f^2 L = \frac{1}{2\pi C}$$

Dividing both sides by $2\pi L$ leaves f^2 by itself on the left – hand side of the equation...

$$f^2 = \frac{1}{2\pi 2\pi LC}$$

Taking the square root of both sides of the equation leaves f by itself on the left side...

$$f = \frac{\sqrt{1}}{\sqrt{2\pi 2\pi LC}}$$

...simplifying...

$$\boxed{f = \frac{1}{2\pi\sqrt{LC}}}$$

So there we have it: a formula to tell us the resonant frequency of a tank circuit, given the values of inductance (L) in Henrys and capacitance (C) in Farads. Plugging in the values of L and C in our example circuit, we arrive at a resonant frequency of 159.155 Hz.

What happens at resonance is quite interesting. With capacitive and inductive reactances equal to each other, the total impedance increases to infinity, meaning that the tank circuit draws no current from the AC power source! We can calculate the individual impedances of the 10 μF capacitor and the 100 mH inductor and work through the parallel impedance formula to demonstrate this mathematically:

$$X_L = 2\pi fL$$

$$X_L = (2)(\pi)(159.155\ \text{Hz})(100\ \text{mH})$$

$$X_L = 100\,\Omega$$

$$X_C = \frac{1}{2\pi fC}$$

$$X_C = \frac{1}{(2)(\pi)(159.155\ \text{Hz})(10\ \mu\text{F})}$$

$$X_C = 100\,\Omega$$

These component values to give resonance impedances that were easy to work with (100 Ω even). Now, we use the parallel impedance formula to see what happens to total Z:

$$Z_{parallel} = \frac{1}{\dfrac{1}{Z_L + \dfrac{1}{Z_C}}}$$

$$Z_{parallel} = \frac{1}{\dfrac{1}{100\,\Omega\angle 90^\circ} + \dfrac{1}{100\,\Omega\angle -90^\circ}}$$

$$Z_{parallel} = \frac{1}{0.01\angle -90^\circ + 0.01\angle 90^\circ}$$

$$Z_{parallel} = \frac{1}{0}\ \text{Undefined!}$$

- Resonance occurs when capacitive and inductive reactances are equal to each other.
- For a tank circuit with no resistance (R), resonant frequency can be calculated with the following formula:

$$f_{Resonant} = \frac{1}{2\pi\sqrt{LC}}$$

- The total impedance of a parallel LC circuit approaches infinity as the power supply frequency approaches resonance.
- A *Bode plot* is a graph plotting waveform amplitude or phase on one axis and frequency on the other.

SIMPLE SERIES RESONANCE

A similar effect happens in series inductive/capacitive circuits. When a state of resonance is reached (capacitive and inductive reactances equal), the two impedances cancel each other out and the total impedance drops to zero!

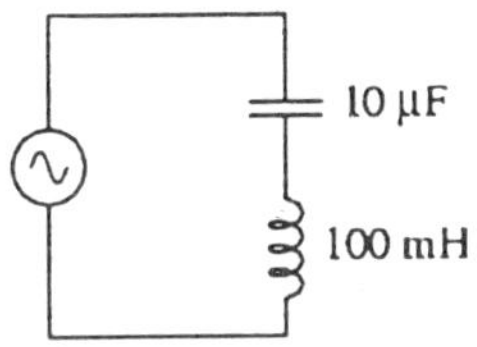

Fig. Simple Series Resonant Circuit.

APPLICATIONS OF RESONANCE

So far, the phenomenon of resonance appears to be a useless curiosity, or at most a nuisance to be avoided (especially if series resonance makes for a short-circuit across our AC voltage source!). However, this is not the case. Resonance is a very valuable property of reactive AC circuits, employed in a variety of applications.

One use for resonance is to establish a condition of stable frequency in circuits designed to produce AC signals. Usually, a parallel (tank) circuit is used for this purpose, with the capacitor and inductor directly connected together, exchanging energy between each other. Just as a pendulum can be used to stabilize the frequency of a clock mechanism's oscillations, so can a tank circuit be used to stabilize the electrical frequency of an AC *oscillator* circuit.

The frequency set by the tank circuit is solely dependent upon the values of L and C, and not on the magnitudes of voltage or current present in the oscillations:

At 159.155 Hz:

$$Z_L = 0 + j\,100\ \Omega \qquad Z_C = 0 - j\,100\ \Omega$$

$$Z_{Series} = Z_L + Z_C$$

$$Z_{Series} = (0 + j\,100\ \Omega) + (0 - j\,100\ \Omega)$$

$$Z_{Series} = 0\ \Omega$$

Fig. Resonant Circuit Serves as Stable Frequency Source.

Another use for resonance is in applications where the effects of greatly increased or decreased impedance at a particular frequency is desired. A resonant circuit can be used to "block" (present high impedance toward) a frequency or range of frequencies, thus acting as a sort of frequency "filter" to strain certain frequencies out of a mix of others. In fact, these particular circuits are called *filters,* and their design constitutes a discipline of study all by itself:

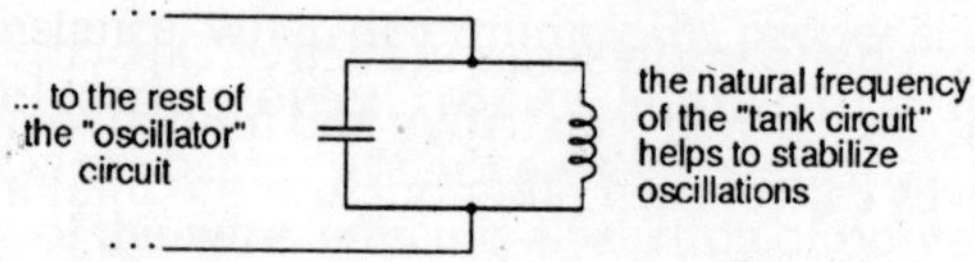

Fig. Resonant Circuit Serves as Filter

In essence, this is how analog radio receiver tuner circuits work to filter, or select, one station frequency out of the mix of different radio station frequency signals intercepted by the antenna.

- Resonance can be employed to maintain AC circuit oscillations at a constant frequency, just as a pendulum can be used to maintain constant oscillation speed in a timekeeping mechanism.
- Resonance can be exploited for its impedance properties: either dramatically increasing or decreasing impedance for certain frequencies.

  Circuits designed to screen certain frequencies out of a mix of different frequencies are called *filters.*

Resonance in Series-parallel Circuits

In simple reactive circuits with little or no resistance, the effects of radically altered impedance will manifest at the resonance frequency predicted by the equation given earlier. In a parallel (tank) LC circuit, this means infinite impedance at resonance. In a series LC circuit, it means zero impedance at resonance:

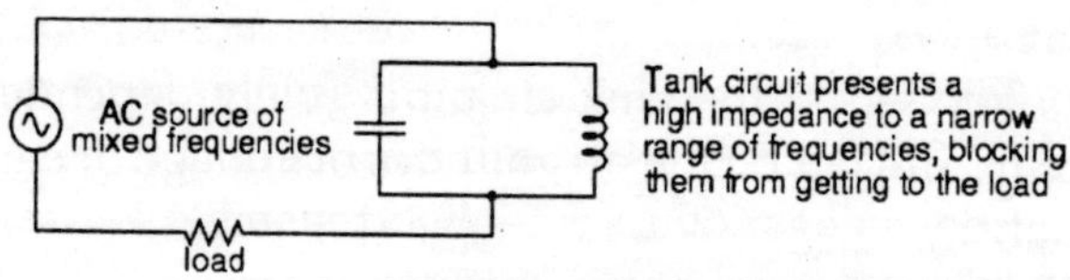

Fig. Q and Bandwidth of a Resonant Circuit

The Q, *quality factor,* of a resonant circuit is a measure of the "goodness" or quality of a resonant circuit. A higher value for this figure of merit correspondes to a more narrow bandwith, which is desirable in many applications.

More formally, Q is the ration of power stored to power dissipated in the circuit reactance and resistance, respectively:

$$Q = P_{stored}/P_{dissipated} = I^2X/I^2R$$

$$Q = X/R$$

where: X = Capacitive or Inductive reactance at resonance

R = Series resistance.

This formula is applicable to series resonant circuits, and also parallel resonant ciruits if the resistance is in series with the inductor. This is the case in practical applications, as we are mostly concerned with the resistance of the inductor limiting the Q.

Note: Some text may show X and R interchanged in the "Q" formula for a parallel resonant circuit.

This is correct for a large value of R in parallel with C and L. Our formula is correct for a small R in series with L.

A practical application of "Q" is that voltage across L or C in a series resonant circuit is Q times total applied voltage. In a parallel resonant circuit, current through L or C is Q times the total applied current.

Chapter 7

Alternating Current

WHAT IS ALTERNATING CURRENT (AC)?

Most students of electricity begin their study with what is known as *direct current* (DC), which is electricity flowing in a constant direction, and/ or possessing a voltage with constant polarity. DC is the kind of electricity made by a battery (with definite positive and negative terminals), or the kind of charge generated by rubbing certain types of materials against each other.

As useful and as easy to understand as DC is, it is not the only "kind" of electricity in use. Certain sources of electricity (most notably, rotary electro-mechanical generators) naturally produce voltages alternating in polarity, reversing positive and negative over time. Either as a voltage switching polarity or as a current switching direction back and forth, this "kind" of electricity is known as Alternating Current (AC):

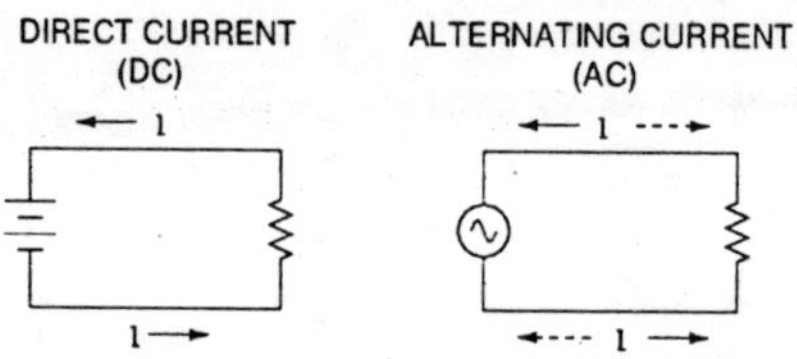

Fig. Direct vs Alternating Current

Whereas the familiar battery symbol is used as a generic symbol for any DC voltage source, the circle with the wavy line inside is the generic symbol for any AC voltage source.

One might wonder why anyone would bother with such a thing as AC. It is true that in some cases AC holds no practical advantage over DC. In applications where electricity is used to dissipate energy in the form of heat, the polarity or direction of current is irrelevant, so long as there is enough voltage and current to the load to produce the desired heat (power dissipation).

However, with AC it is possible to build electric generators, motors and power distribution systems that are far more efficient than DC, and so we find AC used predominately across the world in high power applications. To explain the details of why this is so, a bit of background knowledge about AC is necessary.

If a machine is constructed to rotate a magnetic field around a set of stationary wire coils with the turning of a shaft, AC voltage will be produced across the wire coils as that shaft is rotated, in accordance with Faraday's Law of electromagnetic induction. This is the basic operating principle of an AC generator, also known as an *alternator*:

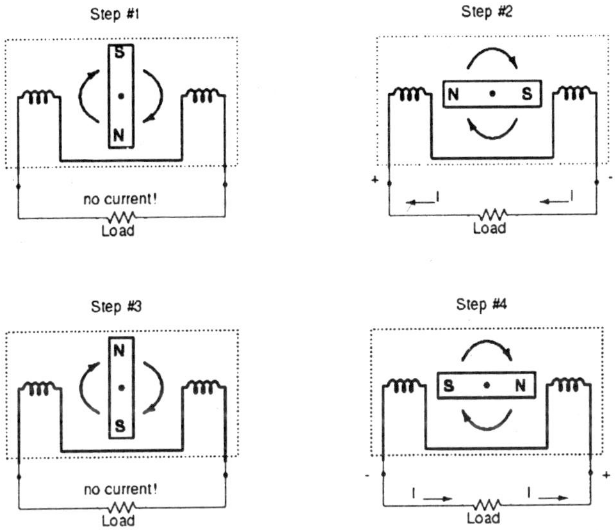

Fig. Alterantor Operation

Notice how the polarity of the voltage across the wire coils reverses as the opposite poles of the rotating magnet pass by. Connected to a load, this reversing voltage polarity will create a reversing current direction in the circuit. The faster the alternator's shaft is turned, the faster the magnet will spin, resulting in an alternating voltage and current that switches directions more often in a given amount of time.

While DC generators work on the same general principle of electromagnetic induction, their construction is not as simple as their AC counterparts.

With a DC generator, the coil of wire is mounted in the shaft where the magnet is on the AC alternator, and electrical connections are made to this spinning coil via stationary carbon "brushes" contacting copper strips on the rotating shaft. All this is necessary to switch the coil's changing output polarity to the external circuit so the external circuit sees a constant polarity:

The generator shown above will produce two pulses of voltage per revolution of the shaft, both pulses in the same direction (polarity). In order for a DC generator to produce *constant* voltage, rather than brief pulses of voltage once every 1/2 revolution, there are multiple sets of coils making intermittent contact with the brushes. The diagram shown above is a bit more simplified than what you would see in real life.

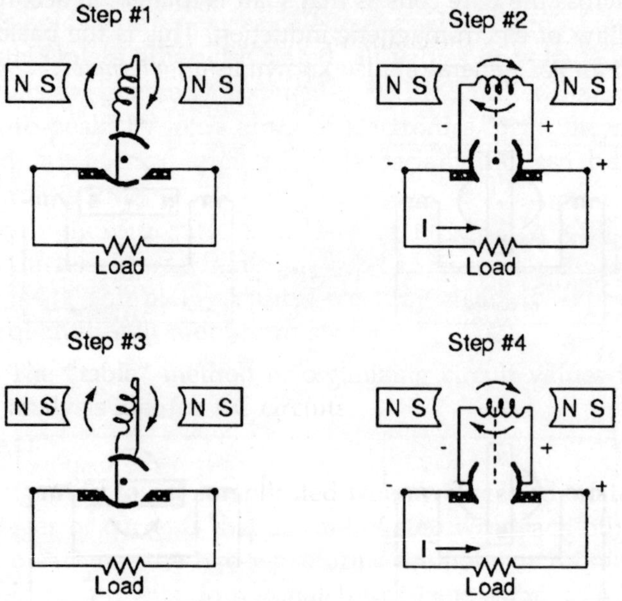

Fig. DC Generator Operation

The problems involved with making and breaking electrical contact with a moving coil should be obvious (sparking and heat), especially if the shaft of the generator is revolving at high speed. If the atmosphere surrounding the machine contains flammable or explosive vapors, the practical problems of spark-producing brush contacts are even greater. An AC generator (alternator) does not require brushes and commutators to work, and so is immune to these problems experienced by DC generators.

The benefits of AC over DC with regard to generator design is also reflected in electric motors. While DC motors require the use of brushes to make electrical contact with moving coils of wire, AC motors do not.

In fact, AC and DC motor designs are very similar to their generator counterparts (identical for the sake of this tutorial), the AC motor being dependent upon the reversing magnetic field produced by alternating current through its stationary coils of wire to rotate the rotating magnet around on its shaft, and the DC motor being dependent on the brush contacts making and breaking connections to reverse current through the rotating coil every 1/2 rotation (180 degrees). So we know that AC generators and AC motors tend to be simpler than DC generators and DC motors. This relative simplicity translates into greater reliability and lower cost of manufacture.

But what else is AC good for? Surely there must be more to it than design details of generators and motors! Indeed there is. There is an effect of electromagnetism known as *mutual induction,* whereby two or more coils of wire placed so that the changing magnetic field created by one induces a voltage in the other. If we have two mutually inductive coils and we energize one coil with AC, we will create an AC voltage in the other coil. When used as such, this device is known as a *transformer*:

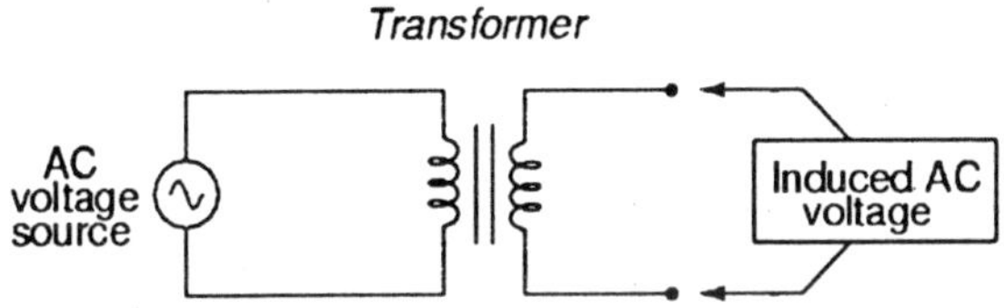

Fig. Transformer "Transforms" AC Voltage and Current.

The fundamental significance of a transformer is its ability to step voltage up or down from the powered coil to the unpowered coil. The AC voltage induced in the unpowered ("secondary") coil is equal to the AC voltage across the powered ("primary") coil multiplied by the ratio of secondary coil turns to primary coil turns.

If the secondary coil is powering a load, the current through the secondary coil is just the opposite: primary coil current multiplied by the ratio of primary to secondary turns. This relationship has a very close mechanical analogy, using torque and speed to represent voltage and current, respectively:

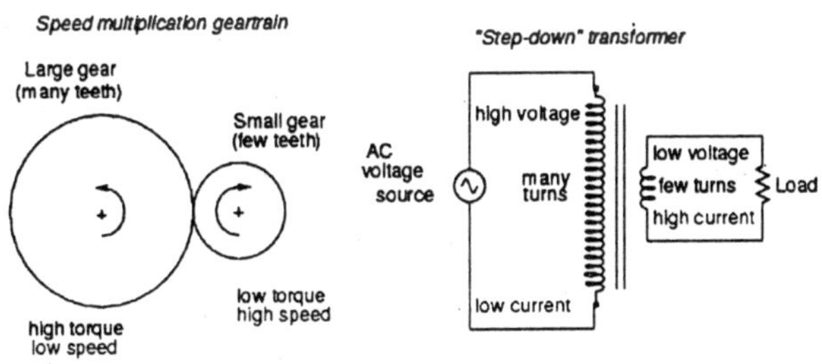

Speed multiplication gear train steps torque down and speed up. Step-down transformer steps voltage down and current up. If the winding ratio is reversed so that the primary coil has less turns than the secondary coil, the transformer "steps up" the voltage from the source level to a higher level at the load:

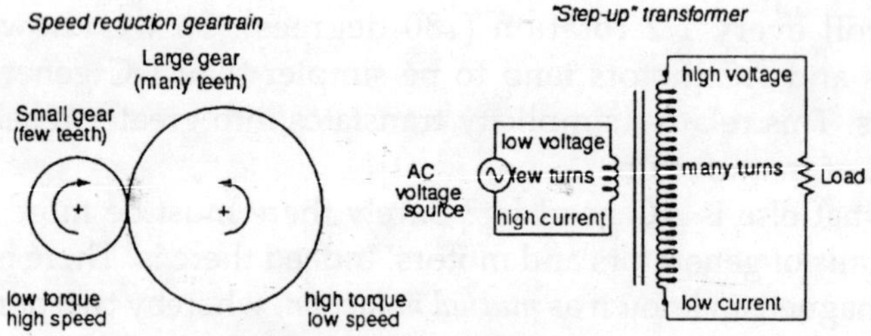

Fig. Speed Reduction Gear Train Steps Torque up and Speed Down. Step-up Transformer Steps Voltage up and Current Down.

The transformer's ability to step AC voltage up or down with ease gives AC an advantage unmatched by DC in the realm of power distribution in figure. When transmitting electrical power over long distances, it is far more efficient to do so with stepped-up voltages and stepped-down currents (smaller-diameter wire with less resistive power losses), then step the voltage back down and the current back up for industry, business, or consumer use use.

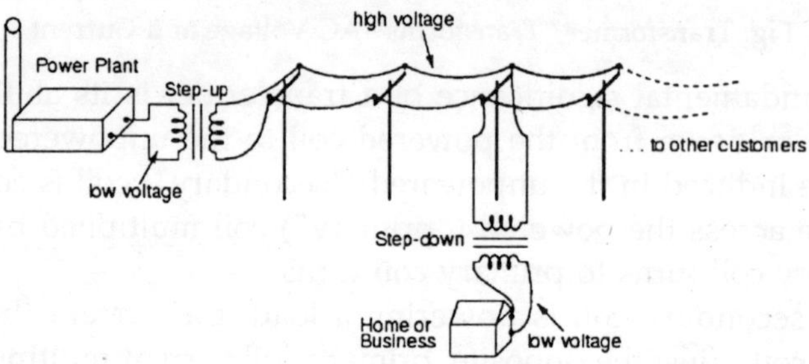

Fig. Transformers Enable Efficient Long Distance high Voltage Transmission of Electric Energy.

Transformer technology has made long-range electric power distribution practical. Without the ability to efficiently step voltage up and down, it would be cost-prohibitive to construct power systems for anything but close-range (within a few miles at most) use.

As useful as transformers are, they only work with AC, not DC. Because the phenomenon of mutual inductance relies on *changing* magnetic fields, and direct current (DC) can only produce steady magnetic fields, transformers simply will not work with direct current.

Of course, direct current may be interrupted (pulsed) through the primary winding of a transformer to create a changing magnetic field (as is done in automotive ignition systems to produce high-voltage spark plug power from a low-voltage DC battery), but pulsed DC is not that different from AC. Perhaps more than any other reason, this is why AC finds such widespread application in power systems.

- DC stands for "Direct Current," meaning voltage or current that maintains constant polarity or direction, respectively, over time.
- AC stands for "Alternating Current," meaning voltage or current that changes polarity or direction, respectively, over time.
- AC electromechanical generators, known as *alternators*, are of simpler construction than DC electromechanical generators.
- AC and DC motor design follows respective generator design principles very closely.
- A *transformer* is a pair of mutually-inductive coils used to convey AC power from one coil to the other. Often, the number of turns in each coil is set to create a voltage increase or decrease from the powered (primary) coil to the unpowered (secondary) coil.
- Secondary voltage = Primary voltage (secondary turns/ primary turns)
- Secondary current = Primary current (primary turns/ secondary turns)

AC WAVEFORMS

When an alternator produces AC voltage, the voltage switches polarity over time, but does so in a very particular manner. When graphed over time, the "wave" traced by this voltage of alternating polarity from an alternator takes on a distinct shape, known as a *sine wave*:

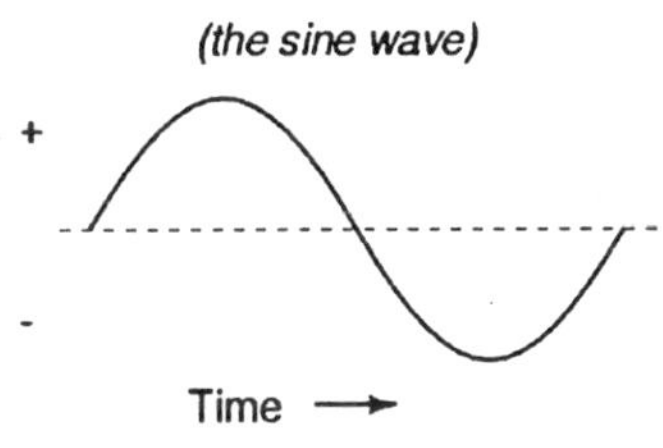

Fig. Graph of AC Voltage over time (The Sine Wave).

In the voltage plot from an electromechanical alternator, the change from one polarity to the other is a smooth one, the voltage level changing most rapidly at the zero ("crossover") point and most slowly at its peak.

The reason why an electromechanical alternator outputs sine-wave AC is due to the physics of its operation.

The voltage produced by the stationary coils by the motion of the rotating magnet is proportional to the rate at which the magnetic flux is changing perpendicular to the coils (Faraday's Law of Electromagnetic Induction). That rate is greatest when the magnet poles are closest to the coils, and least when the magnet poles are furthest away from the coils.

Mathematically, the rate of magnetic flux change due to a rotating magnet follows that of a sine function, so the voltage produced by the coils follows that same function.

If we were to follow the changing voltage produced by a coil in an alternator from any point on the sine wave graph to that point when the wave shape begins to repeat itself, we would have marked exactly one *cycle* of that wave.

This is most easily shown by spanning the distance between identical peaks, but may be measured between any corresponding points on the graph. The degree marks on the horizontal axis of the graph represent the domain of the trigonometric sine function, and also the angular position of our simple two-pole alternator shaft as it rotates:

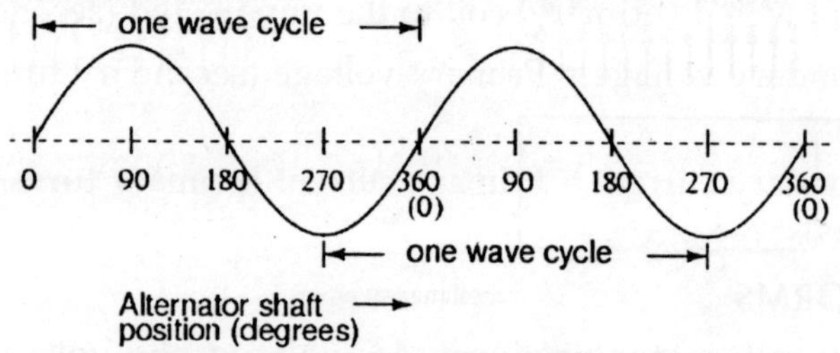

Fig. Alternator Voltage as Function of Shaft Position (time).

Since the horizontal axis of this graph can mark the passage of time as well as shaft position in degrees, the dimension marked for one cycle is often measured in a unit of time, most often seconds or fractions of a second. When expressed as a measurement, this is often called the *period* of a wave. The period of a wave in degrees is *always* 360, but the amount of time one period occupies depends on the rate voltage oscillates back and forth.

A more popular measure for describing the alternating rate of an AC voltage or current wave than *period* is the rate of that back-and-forth oscillation. This is called *frequency*. The modern unit for frequency is the Hertz (abbreviated Hz), which represents the number of wave cycles completed during one second of time.

In the United States of America, the standard power-line frequency is 60 Hz, meaning that the AC voltage oscillates at a rate of 60 complete back-and-forth cycles every second. In Europe, where the power system frequency is 50 Hz, the AC voltage only completes 50 cycles every second. A radio station transmitter broadcasting at a frequency of 100 MHz generates an AC voltage oscillating at a rate of 100 *million* cycles every second.

Prior to the canonization of the Hertz unit, frequency was simply expressed as "cycles per second." Older meters and electronic equipment often bore frequency units of "CPS" (Cycles Per Second) instead of Hz. Many people believe the change from self-explanatory units like CPS to Hertz constitutes a step backward in clarity. A similar change occurred when the unit of "Celsius" replaced that of "Centigrade" for metric temperature measurement. The name Centigrade was based on a 100-count ("Centi-") scale ("-grade") representing the melting and boiling points of H_2O, respectively. The name Celsius, on the other hand, gives no hint as to the unit's origin or meaning.

Period and frequency are mathematical reciprocals of one another. That is to say, if a wave has a period of 10 seconds, its frequency will be 0.1 Hz, or 1/10 of a cycle per second:

Frequency in HertZ =

An instrument called an *oscilloscope*, Figure, is used to display a changing voltage over time on a graphical screen. You may be familiar with the appearance of an *ECG* or *EKG* (electrocardiograph) machine, used by physicians to graph the oscillations of a patient's heart over time. The ECG is a special-purpose oscilloscope expressly designed for medical use. General-purpose oscilloscopes have the ability to display voltage from virtually any voltage source, plotted as a graph with time as the independent variable. The relationship between period and frequency is very useful to know when displaying an AC voltage or current waveform on an oscilloscope screen. By measuring the period of the wave on the horizontal axis of the oscilloscope screen and reciprocating that time value (in seconds), you can determine the frequency in Hertz.

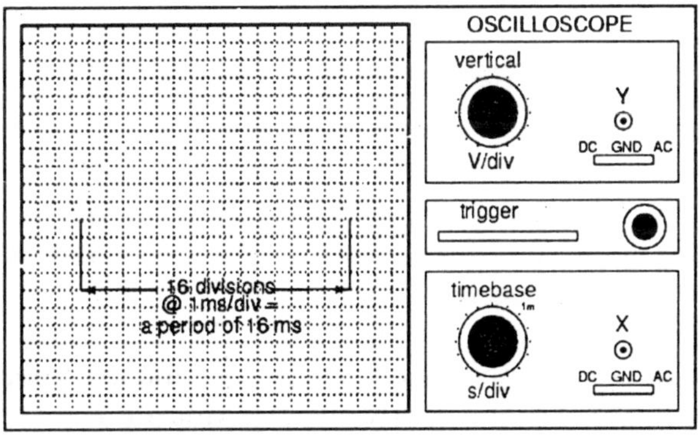

$$\text{Frequency} = \frac{1}{\text{period}} = \frac{1}{16\text{ ms}} = 62.5\text{ Hz}$$

Fig. Time Period of Sinewave is Shown on Oscilloscope.

Voltage and current are by no means the only physical variables subject to variation over time. Much more common to our everyday experience is *sound*, which is nothing more than the alternating compression and decompression (pressure waves) of air molecules, interpreted by our ears as a physical sensation. Because alternating current is a wave phenomenon, it shares many of the properties of other wave phenomena, like sound. For this reason, sound (especially structured music) provides an excellent analogy for relating AC concepts.

In musical terms, frequency is equivalent to *pitch*. Low-pitch notes such as those produced by a tuba or bassoon consist of air molecule vibrations that are relatively slow (low frequency).

High-pitch notes such as those produced by a flute or whistle consist of the same type of vibrations in the air, only vibrating at a much faster rate (higher frequency).

Note	Musical designation	Frequency (in hertz)
A	A_1	220.00
A sharp (or B flat)	$A^\#$ or B^b	233.08
B	B_1	246.94
C (middle)	C	261.63
C sharp (or D flat)	$C^\#$ or D^b	277.18
D	D	293.66
D sharp (or E flat)	$D^\#$ or E^b	311.13
E	E	329.63
F	F	349.23
F sharp (or G flat)	$F^\#$ or G^b	369.99
G	G	392.00
G sharp (or A flat)	$G^\#$ or A^b	415.30
A	A	440.00
A sharp (or B flat)	$A^\#$ or B^b	466.16
B	B	493.88
C	C^1	523.25

Fig. The frequency in Hertz (Hz) is Shown for Various Musical Notes.

Astute observers will notice that all notes on the table bearing the same letter designation are related by a frequency ratio of 2:1. For example, the first frequency shown (designated with the letter "A") is 220 Hz. The next highest "A" note has a frequency of 440 Hz — exactly twice as many sound wave cycles per second.

The same 2:1 ratio holds true for the first A sharp (233.08 Hz) and the next A sharp (466.16 Hz), and for all note pairs found in the table. Audibly, two notes whose frequencies are exactly double each other sound remarkably similar.

This similarity in sound is musically recognized, the shortest span on a musical scale separating such note pairs being called an *octave*.

Following this rule, the next highest "A" note (one octave above 440 Hz) will be 880 Hz, the next lowest "A" (one octave below 220 Hz) will be 110 Hz. A view of a piano keyboard helps to put this scale into perspective:

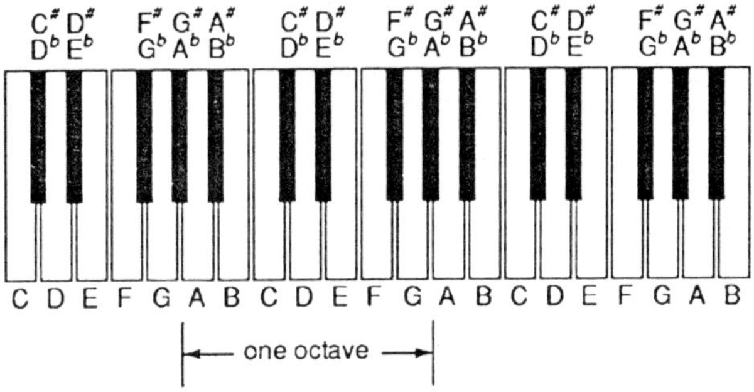

Fig. An Octave is Shown on a Musical Keyboard.

One octave is equal to *eight* white keys' worth of distance on a piano keyboard. The familiar musical mnemonic (doe-ray-mee-fah-so-lah-tee-doe) — yes, the same pattern immortalized in the whimsical Rodgers and Hammerstein song sung in The Sound of Music — covers one octave from C to C.

While electromechanical alternators and many other physical phenomena naturally produce sine waves, this is not the only kind of alternating wave in existence. Other "waveforms" of AC are commonly produced within electronic circuitry. Here are but a few sample waveforms and their common designations in figure.

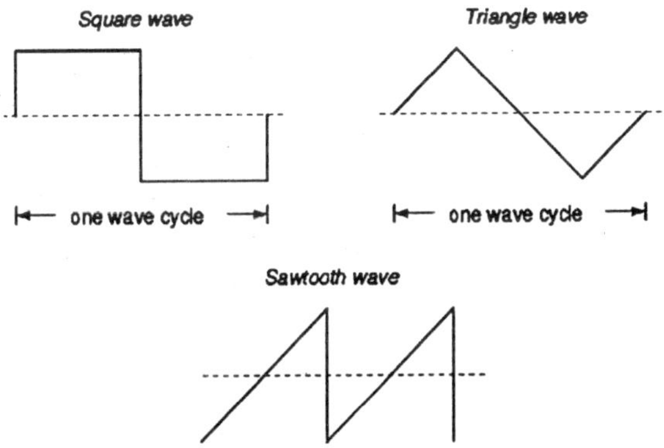

Fig. Some Common Waveshapes (waveforms).

These waveforms are by no means the only kinds of waveforms in existence. They're simply a few that are common enough to have been given distinct names. Even in circuits that are supposed to manifest "pure" sine, square, triangle, or sawtooth voltage/current waveforms, the real-life result is often a distorted version of the intended waveshape.

Some waveforms are so complex that they defy classification as a particular "type" (including waveforms associated with many kinds of musical instruments). Generally speaking, any waveshape bearing close resemblance to a perfect sine wave is termed *sinusoidal*, anything different being labeled as *non-sinusoidal*. Being that the waveform of an AC voltage or current is crucial to its impact in a circuit, we need to be aware of the fact that AC waves come in a variety of shapes.

MEASUREMENTS OF AC MAGNITUDE

So far we know that AC voltage alternates in polarity and AC current alternates in direction. We also know that AC can alternate in a variety of different ways, and by tracing the alternation over time we can plot it as a "waveform."

We can measure the rate of alternation by measuring the time it takes for a wave to evolve before it repeats itself (the "period"), and express this as cycles per unit time, or "frequency." In music, frequency is the same as *pitch*, which is the essential property distinguishing one note from another.

However, we encounter a measurement problem if we try to express how large or small an AC quantity is. With DC, where quantities of voltage and current are generally stable, we have little trouble expressing how much voltage or current we have in any part of a circuit. But how do you grant a single measurement of magnitude to something that is constantly changing?

One way to express the intensity, or magnitude (also called the *amplitude*), of an AC quantity is to measure its peak height on a waveform graph. This is known as the *peak* or *crest* value of an AC waveform:

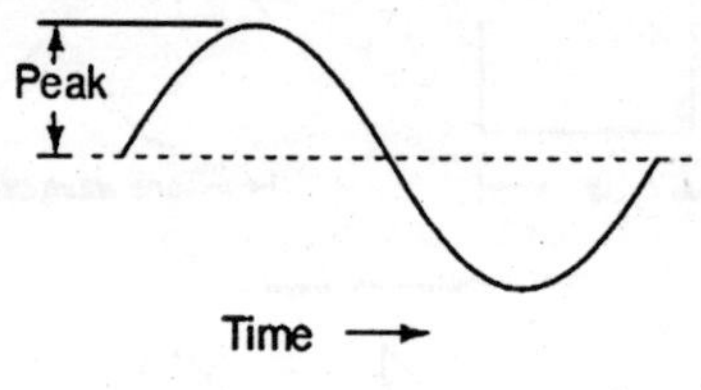

Fig. Peak voltage of a waveform.

Another way is to measure the total height between opposite peaks. This is known as the *peak-to-peak* (P-P) value of an AC waveform:

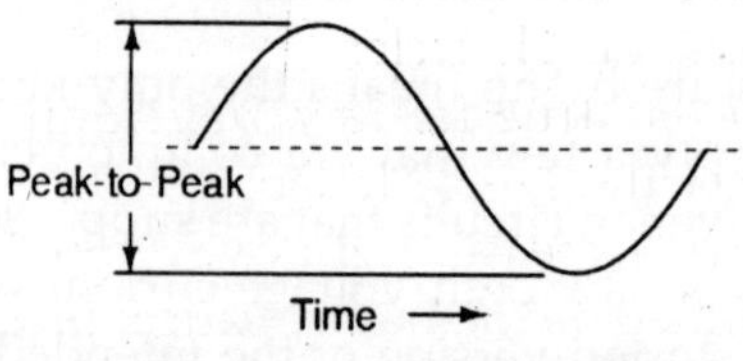

Fig. Peak-to-peak Voltage of a Waveform.

Unfortunately, either one of these expressions of waveform amplitude can be misleading when comparing two different types of waves. For example, a square wave peaking at 10 volts is obviously a greater amount of voltage for a greater amount of time than a triangle wave peaking at 10 volts. The effects of these two AC voltages powering a load would be quite different:

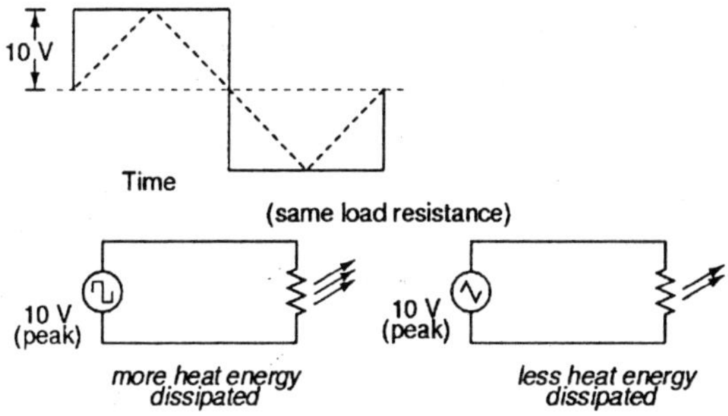

Fig. A Square Wave Produces a Greater Heating Effect than the Same Peak Voltage Triangle Wave.

One way of expressing the amplitude of different waveshapes in a more equivalent fashion is to mathematically average the values of all the points on a waveform's graph to a single, aggregate number.

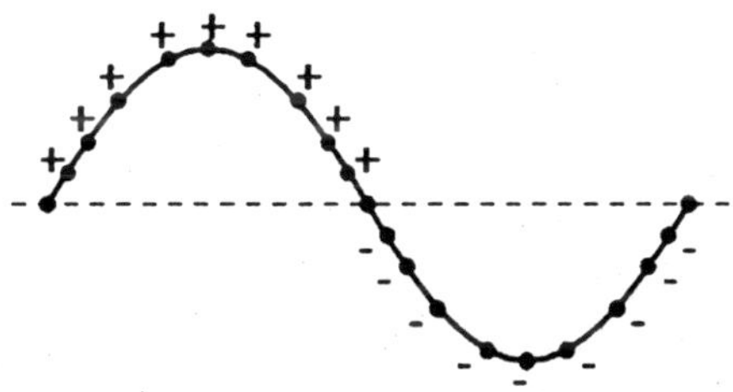

Fig. The Average Value of a Sinewave is Zero.

This amplitude measure is known simply as the *average* value of the waveform.

If we average all the points on the waveform algebraically (that is, to consider their *sign,* either positive or negative), the average value for most waveforms is technically zero, because all the positive points cancel out all the negative points over a full cycle:

This, of course, will be true for any waveform having equal-area portions above and below the "zero" line of a plot.

However, as a *practical* measure of a waveform's aggregate value, "average" is usually defined as the mathematical mean of all the points' *absolute values* over a cycle.

In other words, we calculate the practical average value of the waveform by considering all points on the wave as positive quantities, as if the waveform looked like this:

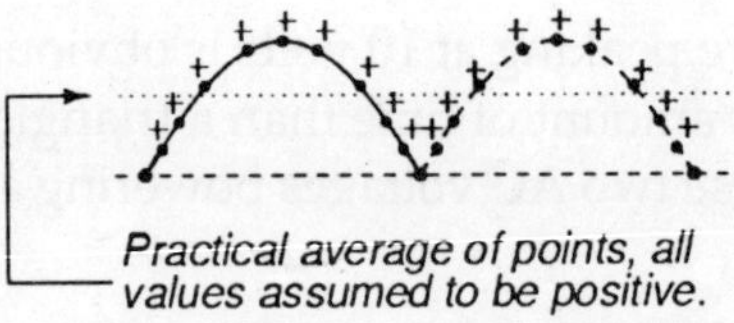

Fig. Waveform Seen by AC "Average Responding" Meter.

Polarity-insensitive mechanical meter movements (meters designed to respond equally to the positive and negative half-cycles of an alternating voltage or current) register in proportion to the waveform's (practical) average value, because the inertia of the pointer against the tension of the spring naturally averages the force produced by the varying voltage/ current values over time.

Conversely, polarity-sensitive meter movements vibrate uselessly if exposed to AC voltage or current, their needles oscillating rapidly about the zero mark, indicating the true (algebraic) average value of zero for a symmetrical waveform.

When the "average" value of a waveform is referenced in this text, it will be assumed that the "practical" definition of average is intended unless otherwise specified.

Another method of deriving an aggregate value for waveform amplitude is based on the waveform's ability to do useful work when applied to a load resistance.

Unfortunately, an AC measurement based on work performed by a waveform is not the same as that waveform's "average" value, because the *power* dissipated by a given load (work performed per unit time) is not directly proportional to the magnitude of either the voltage or current impressed upon it. Rather, power is proportional to the *square* of the voltage or current applied to a resistance ($P = E^2/R$, and $P = I^2R$). Although the mathematics of such an amplitude measurement might not be straightforward, the utility of it is.

Consider a bandsaw and a jigsaw, two pieces of modern woodworking equipment. Both types of saws cut with a thin, toothed, motor-powered metal blade to cut wood.

But while the bandsaw uses a continuous motion of the blade to cut, the jigsaw uses a back-and-forth motion. The comparison of alternating current (AC) to direct current (DC) may be likened to the comparison of these two saw types:

The problem of trying to describe the changing quantities of AC voltage or current in a single, aggregate measurement is also present in this saw analogy: how might we express the speed of a jigsaw blade? A bandsaw blade moves with a constant speed, similar to the way DC voltage pushes or DC current moves with a constant magnitude.

A jigsaw blade, on the other hand, moves back and forth, its blade speed constantly changing. What is more, the back-and-forth motion of any two jigsaws may not be of the same type, depending on the mechanical design of the saws. One jigsaw might move its blade with a sine-wave motion, while another with a triangle-wave motion.

To rate a jigsaw based on its *peak* blade speed would be quite misleading when comparing one jigsaw to another (or a jigsaw with a bandsaw!). Despite the fact that these different saws move their blades in different manners, they are equal in one respect: they all cut wood, and a quantitative comparison of this common function can serve as a common basis for which to rate blade speed.

Picture a jigsaw and bandsaw side-by-side, equipped with identical blades (same tooth pitch, angle, etc.), equally capable of cutting the same thickness of the same type of wood at the same rate. We might say that the two saws were equivalent or equal in their cutting capacity. Might this comparison be used to assign a "bandsaw equivalent" blade speed to the jigsaw's back-and-forth blade motion; to relate the wood-cutting effectiveness of one to the other?

This is the general idea used to assign a "DC equivalent" measurement to any AC voltage or current: whatever magnitude of DC voltage or current would produce the same amount of heat energy dissipation through an equal resistance:

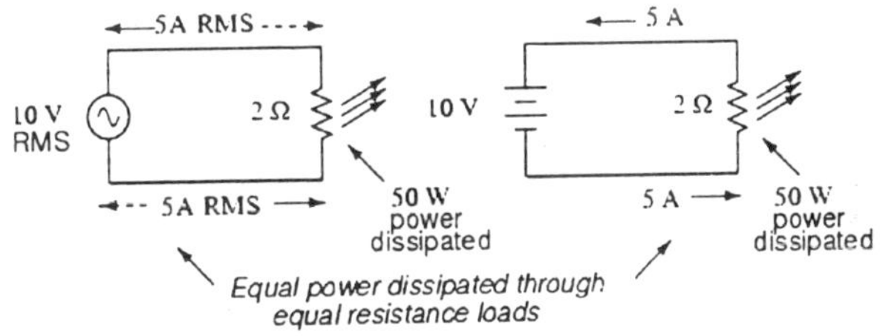

Fig. An RMS Voltage Produces the same Heating Effect as a the same DC Voltage

In the two circuits above, we have the same amount of load resistance (2 Ω) dissipating the same amount of power in the form of heat (50 watts), one powered by AC and the other by DC. Because the AC voltage source pictured above is equivalent (in terms of power delivered to a load) to a 10 volt DC battery, we would call this a "10 volt" AC source.

More specifically, we would denote its voltage value as being 10 volts *RMS*. The qualifier "RMS" stands for *Root Mean Square*, the algorithm used to obtain the DC equivalent value from points on a graph (essentially, the procedure consists of squaring all the positive and negative points on a waveform graph, averaging those squared values, then taking the square root of that average to obtain the final answer). Sometimes the alternative terms *equivalent* or *DC equivalent* are used instead of "RMS," but the quantity and principle are both the same.

RMS amplitude measurement is the best way to relate AC quantities to DC quantities, or other AC quantities of differing waveform shapes, when dealing with measurements of electric power. For other considerations, peak or peak-to-peak measurements may be the best to employ. For instance, when determining the proper size of wire (ampacity) to conduct electric power from a source to a load, RMS current measurement is the best to use, because the principal concern with current is overheating of the wire, which is a function of power dissipation caused by current through the resistance of the wire.

However, when rating insulators for service in high-voltage AC applications, peak voltage measurements are the most appropriate, because the principal concern here is insulator "flashover" caused by brief spikes of voltage, irrespective of time.

Peak and peak-to-peak measurements are best performed with an oscilloscope, which can capture the crests of the waveform with a high degree of accuracy due to the fast action of the cathode-ray-tube in response to changes in voltage.

For RMS measurements, analog meter movements (D'Arsonval, Weston, iron vane, electrodynamometer) will work so long as they have been calibrated in RMS figures.

Because the mechanical inertia and dampening effects of an electromechanical meter movement makes the deflection of the needle naturally proportional to the *average* value of the AC, not the true RMS value, analog meters must be specifically calibrated (or mis-calibrated, depending on how you look at it) to indicate voltage or current in RMS units. The accuracy of this calibration depends on an assumed waveshape, usually a sine wave.

Electronic meters specifically designed for RMS measurement are best for the task. Some instrument manufacturers have designed ingenious methods for determining the RMS value of any waveform.

One such manufacturer produces "True-RMS" meters with a tiny resistive heating element powered by a voltage proportional to that being measured. The heating effect of that resistance element is measured thermally to give a true RMS value with no mathematical calculations whatsoever, just the laws of physics in action in fulfillment of the definition of RMS. The accuracy of this type of RMS measurement is independent of waveshape.

For "pure" waveforms, simple conversion coefficients exist for equating Peak, Peak-to-Peak, Average (practical, not algebraic), and RMS measurements to one another:

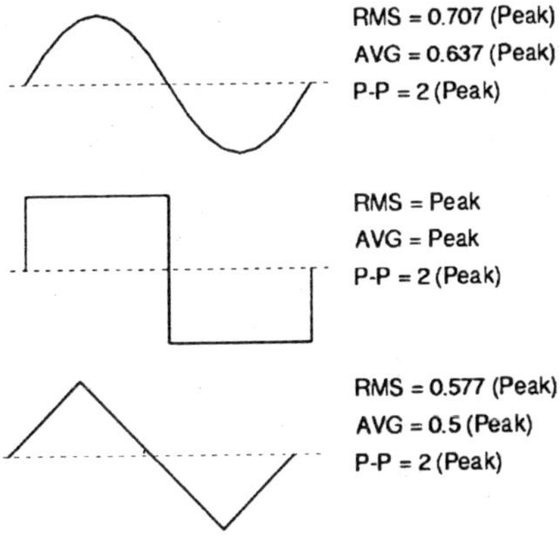

Fig. Conversion Factors for Common Waveforms.

In addition to RMS, average, peak (crest), and peak-to-peak measures of an AC waveform, there are ratios expressing the proportionality between some of these fundamental measurements. The *crest factor* of an AC waveform, for instance, is the ratio of its peak (crest) value divided by its RMS value. The *form factor* of an AC waveform is the ratio of its RMS value divided by its average value.

Square-shaped waveforms always have crest and form factors equal to 1, since the peak is the same as the RMS and average values. Sinusoidal waveforms have an RMS value of 0.707 (the reciprocal of the square root of 2) and a form factor of 1.11 (0.707/0.636). Triangle- and sawtooth-shaped waveforms have RMS values of 0.577 (the reciprocal of square root of 3) and form factors of 1.15 (0.577/0.5).

Bear in mind that the conversion constants shown here for peak, RMS, and average amplitudes of sine waves, square waves, and triangle waves hold true only for *pure* forms of these waveshapes. The RMS and average values of distorted waveshapes are not related by the same ratios:

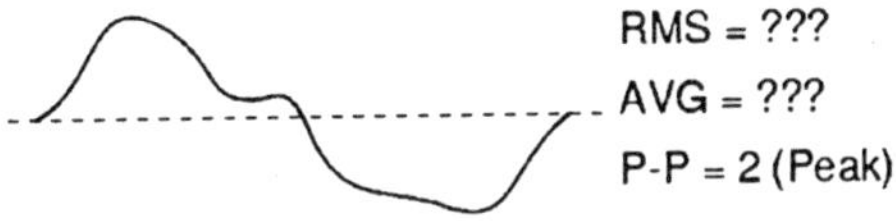

Fig. Arbitrary Waveforms have no Simple Conversions.

This is a very important concept to understand when using an analog meter movement to measure AC voltage or current. An analog movement, calibrated to indicate sine-wave RMS amplitude, will only be accurate when measuring pure sine waves.

If the waveform of the voltage or current being measured is anything but a pure sine wave, the indication given by the meter will not be the true RMS value of the waveform, because the degree of needle deflection in an analog meter movement is proportional to the *average* value of the waveform, not the RMS.

RMS meter calibration is obtained by "skewing" the span of the meter so that it displays a small multiple of the average value, which will be equal to be the RMS value for a particular waveshape and *a particular waveshape only*.

Since the sine-wave shape is most common in electrical measurements, it is the waveshape assumed for analog meter calibration, and the small multiple used in the calibration of the meter is 1.1107 (the form factor: 0.707/0.636: the ratio of RMS divided by average for a sinusoidal waveform). Any waveshape other than a pure sine wave will have a different ratio of RMS and average values, and thus a meter calibrated for sine-wave voltage or current will not indicate true RMS when reading a non-sinusoidal wave. Bear in mind that this limitation applies only to simple, analog AC meters not employing "True-RMS" technology.

- The *amplitude* of an AC waveform is its height as depicted on a graph over time. An amplitude measurement can take the form of peak, peak-to-peak, average, or RMS quantity.
- *Peak* amplitude is the height of an AC waveform as measured from the zero mark to the highest positive or lowest negative point on a graph. Also known as the *crest* amplitude of a wave.
- *Peak-to-peak* amplitude is the total height of an AC waveform as measured from maximum positive to maximum negative peaks on a graph. Often abbreviated as "P-P".
- *Average* amplitude is the mathematical "mean" of all a waveform's points over the period of one cycle. Technically, the average amplitude of any waveform with equal-area portions above and below the "zero" line on a graph is zero. However, as a practical measure of amplitude, a waveform's average value is often calculated as the mathematical mean of all the points' *absolute values* (taking all the negative values and considering them as positive). For a sine wave, the average value so calculated is approximately 0.637 of its peak value.
- "RMS" stands for *Root Mean Square,* and is a way of expressing an AC quantity of voltage or current in terms functionally equivalent to DC. For example, 10 volts AC RMS is the amount of voltage that would produce the same amount of heat dissipation across a resistor of given value as a 10 volt DC power supply. Also known as the "equivalent" or "DC equivalent" value of an AC voltage or current. For a sine wave, the RMS value is approximately 0.707 of its peak value.

- The *crest factor* of an AC waveform is the ratio of its peak (crest) to its RMS value.
- The *form factor* of an AC waveform is the ratio of its RMS value to its average value.
- Analog, electromechanical meter movements respond proportionally to the *average* value of an AC voltage or current. When RMS indication is desired, the meter's calibration must be "skewed" accordingly. This means that the accuracy of an electromechanical meter's RMS indication is dependent on the purity of the waveform: whether it is the exact same waveshape as the waveform used in calibrating.

SIMPLE AC CIRCUIT CALCULATIONS

You will learn that AC circuit measurements and calculations can get very complicated due to the complex nature of alternating current in circuits with inductance and capacitance.

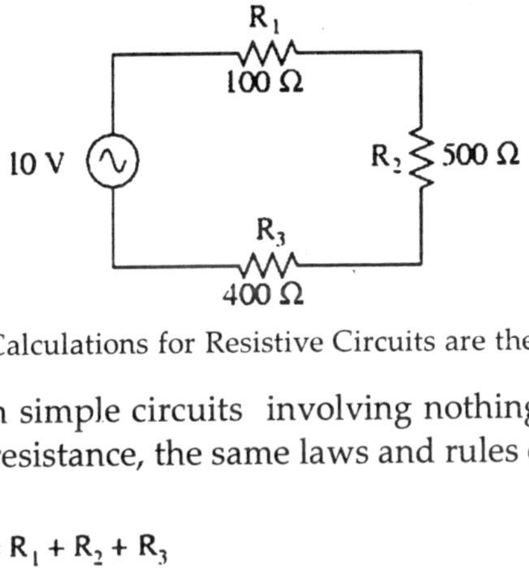

Fig. AC Circuit Calculations for Resistive Circuits are the Same as for DC.

However, with simple circuits involving nothing more than an AC power source and resistance, the same laws and rules of DC apply simply and directly.

$$R_{total} = R_1 + R_2 + R_3$$

$$R_{total} = 1\ k\Omega$$

$$I_{total} = \frac{E_{total}}{R_{total}} \qquad I_{total} = \frac{10\ V}{1\ k\Omega} \qquad I_{total} = 10\ mA$$

$$E_{R1} = I_{total}R_1 \qquad E_{R2} = I_{total}R_2 \qquad E_{R3} = I_{total}R_3$$

$$E_{R1} = 1\ V \qquad E_{R2} = 5\ V \qquad E_{R3} = 4\ V$$

Series resistances still add, parallel resistances still diminish, and the Laws of Kirchhoff and Ohm still hold true.Because all these mathematical relationships still hold true, we can make use of our familiar "table" method of organizing circuit values just as with DC:

One major caveat needs to be given here: all measurements of AC voltage and current must be expressed in the same terms (peak, peak-to-peak, average, or RMS).

If the source voltage is given in peak AC volts, then all currents and voltages subsequently calculated are cast in terms of peak units. If the source voltage is given in AC RMS volts, then all calculated currents and voltages are cast in AC RMS units as well.

This holds true for *any* calculation based on Ohm's Laws, Kirchhoff's Laws, etc. Unless otherwise stated, all values of voltage and current in AC circuits are generally assumed to be RMS rather than peak, average, or peak-to-peak. In some areas of electronics, peak measurements are assumed, but in most applications (especially industrial electronics) the assumption is RMS.

- All the old rules and laws of DC (Kirchhoff's Voltage and Current Laws, Ohm's Law) still hold true for AC. However, with more complex circuits, we may need to represent the AC quantities in more complex form.
- The "table" method of organizing circuit values is still a valid analysis tool for AC circuits.

AC PHASE

Things start to get complicated when we need to relate two or more AC voltages or currents that are out of step with each other. By "out of step," I mean that the two waveforms are not synchronized: that their peaks and zero points do not match up at the same points in time. The graph in illustrates an example of this.

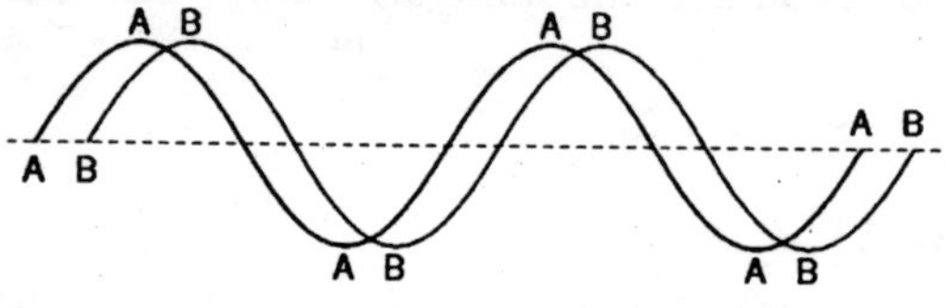

Fig. Out of Phase Waveforms

The two waves shown above (A versus B) are of the same amplitude and frequency, but they are out of step with each other. In technical terms, this is called a *phase shift*. Earlier we saw how we could plot a "sine wave" by calculating the trigonometric sine function for angles ranging from 0 to 360 degrees, a full circle.

The starting point of a sine wave was zero amplitude at zero degrees, progressing to full positive amplitude at 90 degrees, zero at 180 degrees, full negative at 270 degrees, and back to the starting point of zero at 360 degrees. We can use this angle scale along the horizontal axis of our waveform plot to express just how far out of step one wave is with another:

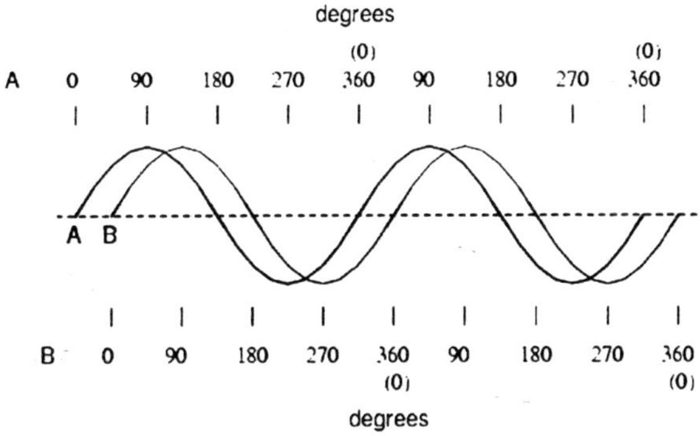

Fig. Wave A Leads Wave B by 45°

The shift between these two waveforms is about 45 degrees, the "A" wave being ahead of the "B" wave. A sampling of different phase shifts is given in the following graphs to better illustrate this concept:

Because the waveforms in the above examples are at the same frequency, they will be out of step by the same angular amount at every point in time.

For this reason, we can express phase shift for two or more waveforms of the same frequency as a constant quantity for the entire wave, and not just an expression of shift between any two particular points along the waves.

That is, it is safe to say something like, "voltage 'A' is 45 degrees out of phase with voltage 'B'." Whichever waveform is ahead in its evolution is said to be *leading* and the one behind is said to be *lagging*.

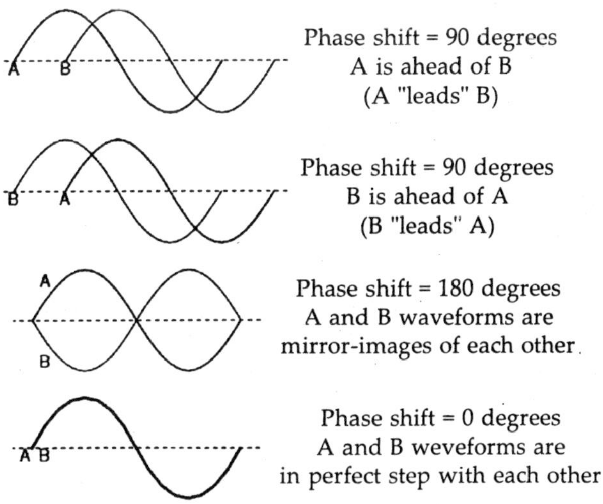

Fig. Examples of Phase Shifts.

Phase shift, like voltage, is always a measurement relative between two things. There's really no such thing as a waveform with an *absolute* phase measurement because there's no known universal reference for phase. Typically in the analysis of AC circuits, the voltage waveform of the power supply is used as a reference for phase, that voltage stated as "xxx volts at 0 degrees." Any other AC voltage or current in that circuit will have its phase shift expressed in terms relative to that source voltage.

This is what makes AC circuit calculations more complicated than DC. When applying Ohm's Law and Kirchhoff's Laws, quantities of AC voltage and current must reflect phase shift as well as amplitude. Mathematical operations of addition, subtraction, multiplication, and division must operate on these quantities of phase shift as well as amplitude. Fortunately, there is a mathematical system of quantities called *complex numbers* ideally suited for this task of representing amplitude and phase.

Because the subject of complex numbers is so essential to the understanding of AC circuits.

- *Phase shift* is where two or more waveforms are out of step with each other.
- The amount of phase shift between two waves can be expressed in terms of degrees, as defined by the degree units on the horizontal axis of the waveform graph used in plotting the trigonometric sine function.
- A *leading* waveform is defined as one waveform that is ahead of another in its evolution. A *lagging* waveform is one that is behind another. Example:
- Calculations for AC circuit analysis must take into consideration both amplitude and phase shift of voltage and current waveforms to be completely accurate. This requires the use of a mathematical system called *complex numbers*.

PRINCIPLES OF RADIO

One of the more fascinating applications of electricity is in the generation of invisible ripples of energy called *radio waves*. The limited scope of this lesson on alternating current does not permit full exploration of the concept, some of the basic principles will be covered.

With Oersted's accidental discovery of electromagnetism, it was realized that electricity and magnetism were related to each other. When an electric current was passed through a conductor, a magnetic field was generated perpendicular to the axis of flow.

Likewise, if a conductor was exposed to a change in magnetic flux perpendicular to the conductor, a voltage was produced along the length of that conductor. So far, scientists knew that electricity and magnetism always seemed to affect each other at right angles. However, a major discovery lay hidden just beneath this seemingly simple concept of related perpendicularity, and its unveiling was one of the pivotal moments in modern science.

This breakthrough in physics is hard to overstate. The man responsible for this conceptual revolution was the Scottish physicist James Clerk Maxwell (1831-1879), who "unified" the study of electricity and magnetism in four relatively tidy equations. In essence, what he discovered was that electric and magnetic *fields* were intrinsically related to one another, with or without the presence of a conductive path for electrons to flow. Stated more formally, Maxwell's discovery was this:

A changing electric field produces a perpendicular magnetic field, and A changing magnetic field produces a perpendicular electric field.

All of this can take place in open space, the alternating electric and magnetic fields supporting each other as they travel through space at the speed of light. This dynamic structure of electric and magnetic fields propagating through space is better known as an *electromagnetic wave.*

There are many kinds of natural radiative energy composed of electromagnetic waves. Even light is electromagnetic in nature. So are X-rays and "gamma" ray radiation.

The only difference between these kinds of electromagnetic radiation is the frequency of their oscillation (alternation of the electric and magnetic fields back and forth in polarity). By using a source of AC voltage and a special device called an *antenna,* we can create electromagnetic waves (of a much lower frequency than that of light) with ease.

An antenna is nothing more than a device built to produce a dispersing electric or magnetic field. Two fundamental types of antennae are the *dipole* and the *loop*:

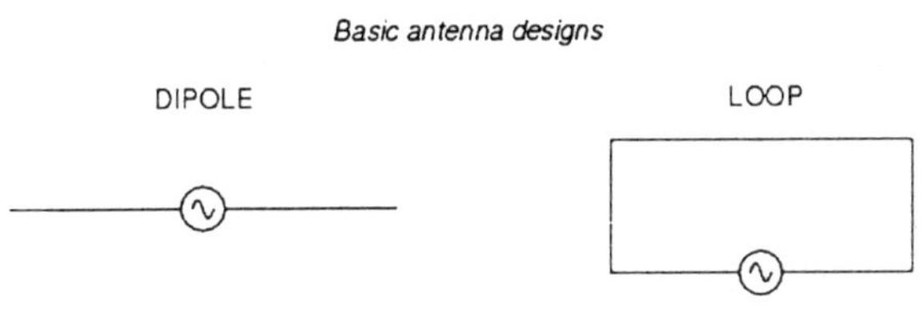

Fig. Dipole and Loop Antennae

While the dipole looks like nothing more than an open circuit, and the loop a short circuit, these pieces of wire are effective radiators of electromagnetic fields when connected to AC sources of the proper frequency. The two open wires of the dipole act as a sort of capacitor (two conductors separated by a dielectric), with the electric field open to dispersal instead of being concentrated between two closely-spaced plates.

The closed wire path of the loop antenna acts like an inductor with a large air core, again providing ample opportunity for the field to disperse away from the antenna instead of being concentrated and contained as in a normal inductor.

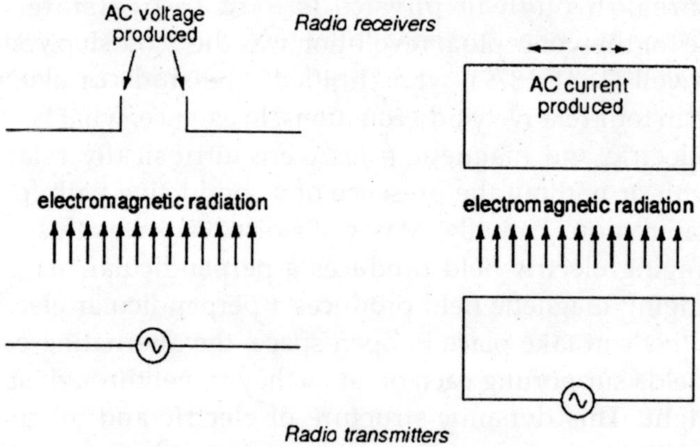

Fig. Basic Radio Transmitter and Receiverm

As the powered dipole radiates its changing electric field into space, a changing magnetic field is produced at right angles, thus sustaining the electric field further into space, and so on as the wave propagates at the speed of light. As the powered loop antenna radiates its changing magnetic field into space, a changing electric field is produced at right angles, with the same end-result of a continuous electromagnetic wave sent away from the antenna. Either antenna achieves the same basic task: the controlled production of an electromagnetic field.

When attached to a source of high-frequency AC power, an antenna acts as a *transmitting* device, converting AC voltage and current into electromagnetic wave energy. Antennas also have the ability to intercept electromagnetic waves and convert their energy into AC voltage and current. In this mode, an antenna acts as a *receiving* device:

While there is *much* more that may be said about antenna technology, this brief introduction is enough to give you the general idea of what's going on (and perhaps enough information to provoke a few experiments).

- James Maxwell discovered that changing electric fields produce perpendicular magnetic fields, and vice versa, even in empty space.
- A twin set of electric and magnetic fields, oscillating at right angles to each other and traveling at the speed of light, constitutes an *electromagnetic wave.*

- An *antenna* is a device made of wire, designed to radiate a changing electric field or changing magnetic field when powered by a high-frequency AC source, or intercept an electromagnetic field and convert it to an AC voltage or current.
- The *dipole* antenna consists of two pieces of wire (not touching), primarily generating an electric field when energized, and secondarily producing a magnetic field in space.
- The *loop* antenna consists of a loop of wire, primarily generating a magnetic field when energized, and secondarily producing an electric field in space.

VECTORS AND AC WAVEFORMS

The length of the vector represents the magnitude (or amplitude) of the waveform.

The greater the amplitude of the waveform, the greater the length of its corresponding vector.

The angle of the vector, however, represents the phase shift in degrees between the waveform in question and another waveform acting as a "reference" in time.

Usually, when the phase of a waveform in a circuit is expressed, it is referenced to the power supply voltage waveform (arbitrarily stated to be "at" 0°). Remember that phase is always a *relative* measurement between two waveforms rather than an absolute property.

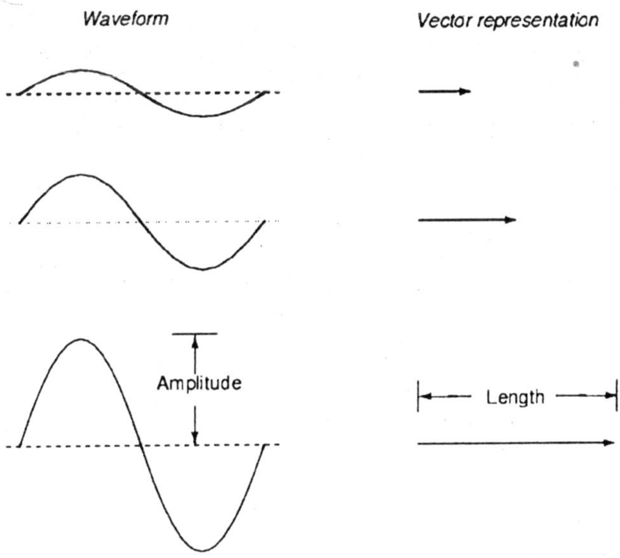

Fig. Vector Length Represents AC Voltage Magnitude.

The greater the phase shift in degrees between two waveforms, the greater the angle difference between the corresponding vectors. Being a relative measurement, like voltage, phase shift (vector angle) only has meaning in reference to some standard waveform. Generally this "reference" waveform is the main AC power supply voltage in the circuit. If there is more than one AC voltage source, then one of those sources is arbitrarily chosen to be the phase reference for all other measurements in the circuit.

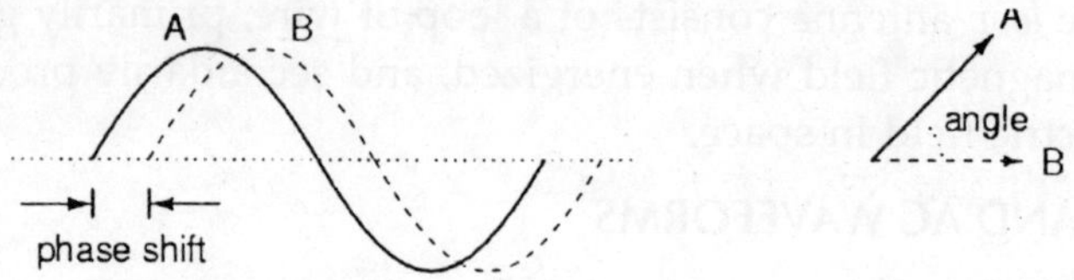

Fig. Phase Shift between Waves and Vector Phase Angle

This concept of a reference point is not unlike that of the "ground" point in a circuit for the benefit of voltage reference. With a clearly defined point in the circuit declared to be "ground," it becomes possible to talk about voltage "on" or "at" single points in a circuit, being understood that those voltages (always relative between *two* points) are referenced to "ground." Correspondingly, with a clearly defined point of reference for phase it becomes possible to speak of voltages and currents in an AC circuit having definite phase angles. For example, if the current in an AC circuit is described as "24.3 milliamps at -64 degrees," it means that the current waveform has an amplitude of 24.3 mA, and it lags 64° behind the reference waveform, usually assumed to be the main source voltage waveform.

When used to describe an AC quantity, the length of a vector represents the amplitude of the wave while the angle of a vector represents the phase angle of the wave relative to some other waveform.

SIMPLE VECTOR ADDITION

Remember that vectors are mathematical objects just like numbers on a number line: they can be added, subtracted, multiplied, and divided. Addition is perhaps the easiest vector operation to visualize, so we'll begin with that.

If vectors with common angles are added, their magnitudes (lengths) add up just like regular scalar quantities:

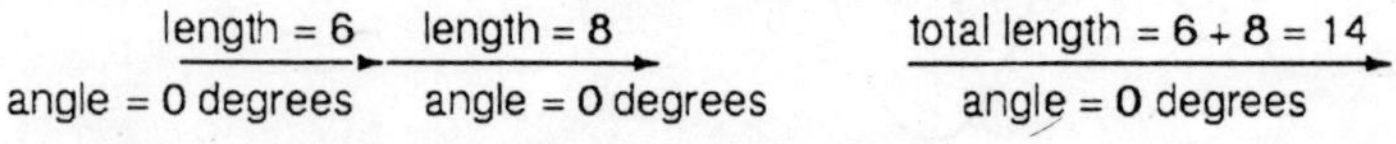

Fig. Vector Magnitudes Add Like Scalars for a Common Angle.

Similarly, if AC voltage sources with the same phase angle are connected together in series, their voltages add just as you might expect with DC batteries:

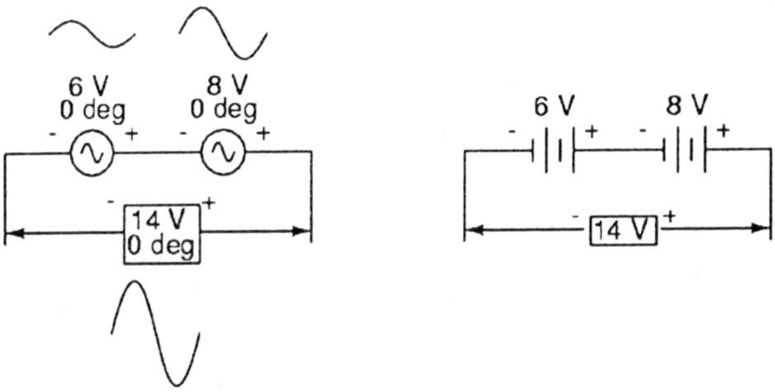

Fig. "In Phase" AC Voltages Add Like DC Battery Voltages.

Please note the (+) and (-) polarity marks next to the leads of the two AC sources.

Even though we know AC doesn't have "polarity" in the same sense that DC does, these marks are essential to knowing how to reference the given phase angles of the voltages. This will become more apparent in the next example.

If vectors directly opposing each other (180° out of phase) are added together, their magnitudes (lengths) subtract just like positive and negative scalar quantities subtract when added:

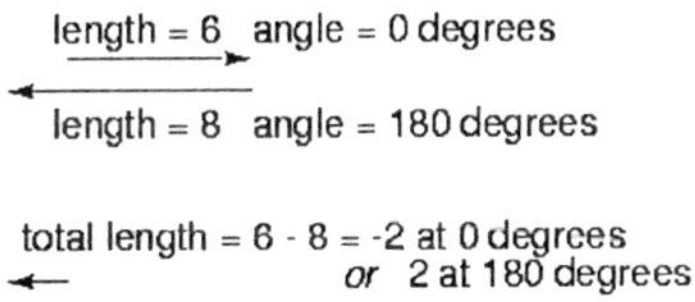

Fig. Directly Opposing Vector Magnitudes Subtract.

Similarly, if opposing AC voltage sources are connected in series, their voltages subtract as you might expect with DC batteries connected in an opposing fashion:

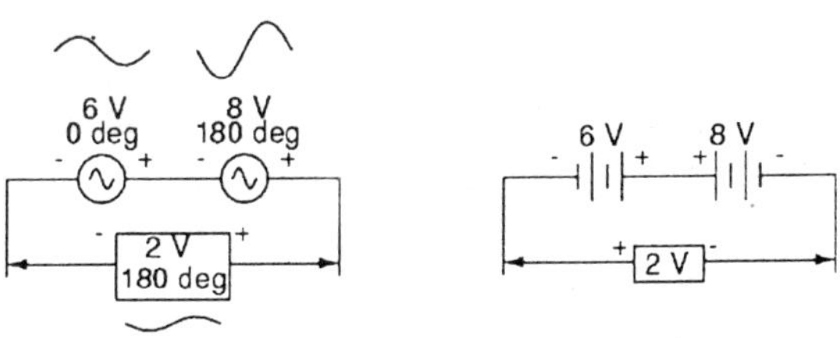

Fig. Opposing AC Voltages Subtract Like Opposing Battery Voltages.

Determining whether or not these voltage sources are opposing each other requires an examination of their polarity markings *and* their phase angles. Notice how the polarity markings in the above diagram seem to indicate additive voltages (from left to right, we see - and + on the 6 volt source, - and + on the 8 volt source).

Even though these polarity markings would normally indicate an *additive* effect in a DC circuit (the two voltages working together to produce a greater total voltage), in this AC circuit they're actually pushing in opposite directions because one of those voltages has a phase angle of 0° and the other a phase angle of 180°. The result, of course, is a total voltage of 2 volts. We could have just as well shown the opposing voltages subtracting in series like this:

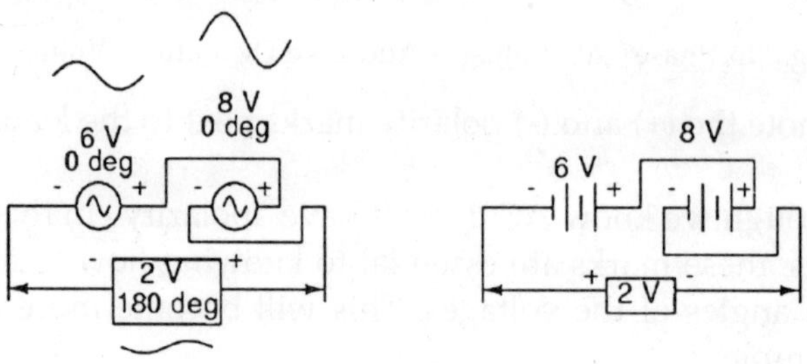

Fig. Opposing Voltages in Spite of Equal phase Angles.

Note how the polarities appear to be opposed to each other now, due to the reversal of wire connections on the 8 volt source. Since both sources are described as having equal phase angles (0°), they truly are opposed to one another, and the overall effect is the same as the former scenario with "additive" polarities and differing phase angles: a total voltage of only 2 volts.

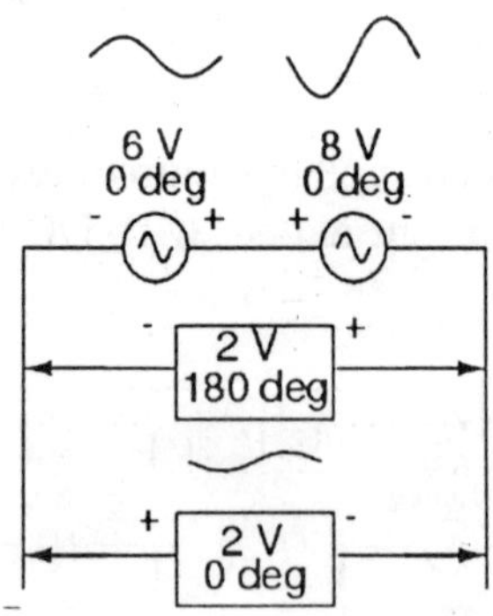

Fig. Just as There are two ways to Express the Phase of the Sources, There are two ways to Express the Resultant Their Sum.

The resultant voltage can be expressed in two different ways: 2 volts at 180° with the (-) symbol on the left and the (+) symbol on the right, or

2 volts at 0° with the (+) symbol on the left and the (-) symbol on the right. A reversal of wires from an AC voltage source is the same as phase-shifting that source by 180°.

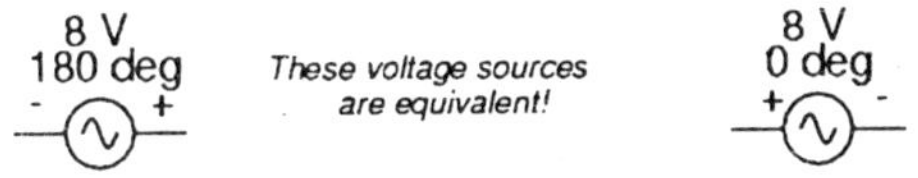

Fig. Example of Equivalent Voltage Sources.

COMPLEX VECTOR ADDITION

If vectors with uncommon angles are added, their magnitudes (lengths) add up quite differently than that of scalar magnitudes:

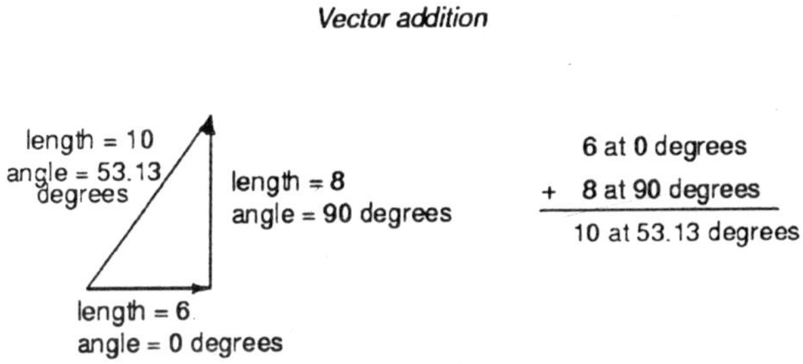

Fig. Vector Magnitudes do not Directly Add for Unequal Angles.

If two AC voltages — 90° out of phase — are added together by being connected in series, their voltage magnitudes do not directly add or subtract as with scalar voltages in DC. Instead, these voltage quantities are complex quantities, and just like the above vectors, which add up in a trigonometric fashion, a 6 volt source at 0° added to an 8 volt source at 90° results in 10 volts at a phase angle of 53.13°:

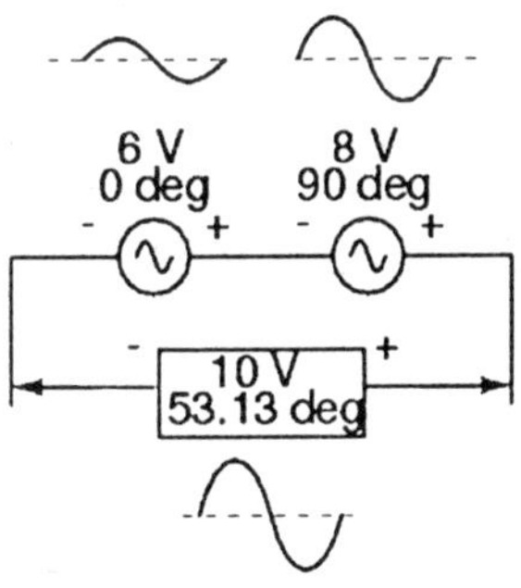

Fig. The 6V and 8V Sources Add to 10V with the Help of Trigonometry.

Compared to DC circuit analysis, this is very strange indeed. Note that it's possible to obtain voltmeter indications of 6 and 8 volts, respectively, across the two AC voltage sources, yet only read 10 volts for a total voltage! There is no suitable DC analogy for what we're seeing here with two AC voltages slightly out of phase. DC voltages can only directly aid or directly oppose, with nothing in between. With AC, two voltages can be aiding or opposing one another *to any degree* between fully-aiding and fully-opposing, inclusive.

Without the use of vector (complex number) notation to describe AC quantities, it would be *very* difficult to perform mathematical calculations for AC circuit analysis.

In the next section, we'll learn how to represent vector quantities in symbolic rather than graphical form. Vector and triangle diagrams suffice to illustrate the general concept, but more precise methods of symbolism must be used if any serious calculations are to be performed on these quantities.

DC voltages can only either directly aid or directly oppose each other when connected in series. AC voltages may aid or oppose *to any degree* depending on the phase shift between them.

POLAR AND RECTANGULAR NOTATION

In order to work with these complex numbers without drawing vectors, we first need some kind of standard mathematical notation. There are two basic forms of complex number notation: *polar* and *rectangular*.

Polar form is where a complex number is denoted by the *length* (otherwise known as the *magnitude, absolute value,* or *modulus*) and the *angle* of its vector (usually denoted by an angle symbol that looks like this: "). To use the map analogy, polar notation for the vector from New York City to San Diego would be something like "2400 miles, southwest." Here are two examples of vectors and their polar notations:

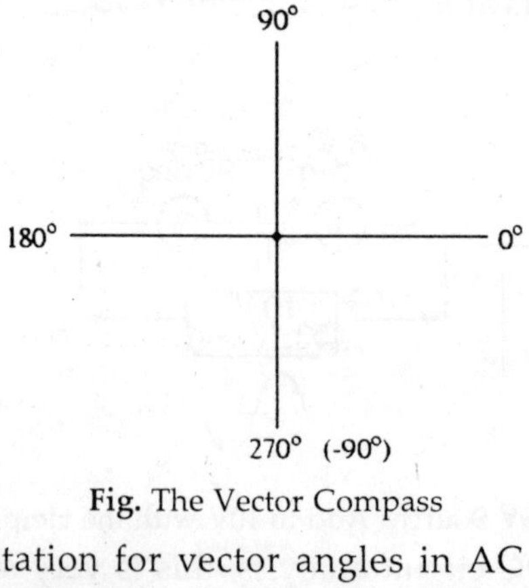

Fig. The Vector Compass

Standard orientation for vector angles in AC circuit calculations defines 0° as being to the right (horizontal), making 90° straight up, 180° to the left, and 270° straight down. Please note that vectors angled "down" can have angles represented in polar form as positive numbers in excess of 180, or negative numbers less than 180.

For example, a vector angled "270° (straight down) can also be said to have an angle of -90°. The above vector on the right (7.81 "230.19°) can also be denoted as 7.81 " -129.81°.

Rectangular form, on the other hand, is where a complex number is denoted by its respective horizontal and vertical components. In essence, the angled vector is taken to be the hypotenuse of a right triangle, described by the lengths of the adjacent and opposite sides.

Rather than describing a vector's length and direction by denoting magnitude and angle, it is described in terms of "how far left/right" and "how far up/down."

These two dimensional figures (horizontal and vertical) are symbolized by two numerical figures. In order to distinguish the horizontal and vertical dimensions from each other, the vertical is prefixed with a lower-case "i" (in pure mathematics) or "j" (in electronics). These lower-case letters do not represent a physical variable (such as instantaneous current, also symbolized by a lower-case letter "i"), but rather are mathematical *operators* used to distinguish the vector's vertical component from its horizontal component. As a complete complex number, the horizontal and vertical quantities are written as a sum:

In "rectangular" form the vector's length and direction are denoted in terms of its horizontal and vertical span, the first number representing the the horizontal ("real") and the second number (with the "j" prefix) representing the vertical ("imaginary") dimensions. The horizontal component is referred to as the *real* component, since that dimension is compatible with normal, scalar ("real") numbers.

The vertical component is referred to as the *imaginary* component, since that dimension lies in a different direction, totally alien to the scale of the real numbers.

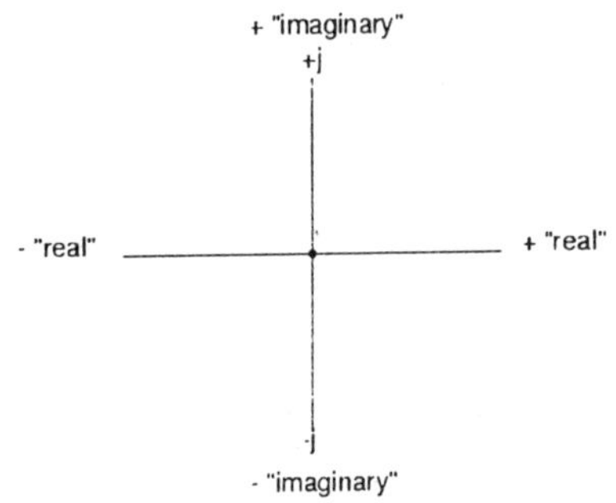

Fig. Vector Compass Showing Real and Imaginary Axes

The "real" axis of the graph corresponds to the familiar number line we saw earlier: the one with both positive and negative values on it. The "imaginary" axis of the graph corresponds to another number line situated at 90° to the "real" one.

Vectors being two-dimensional things, we must have a two-dimensional "map" upon which to express them, thus the two number lines perpendicular to each other:

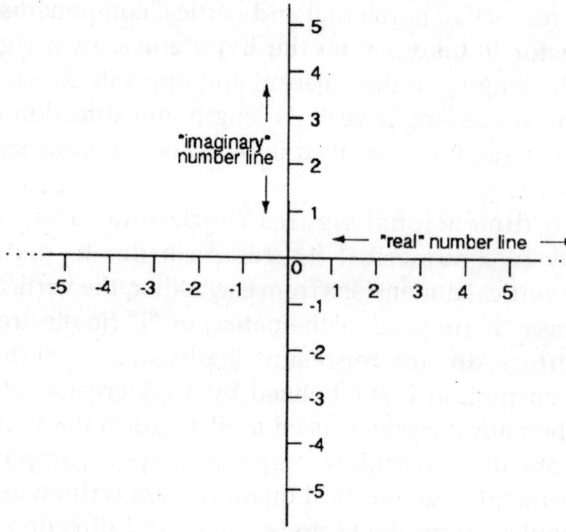

Fig. Vector Compass with Real and Imaginary ("j") Number Lines.

Either method of notation is valid for complex numbers. The primary reason for having two methods of notation is for ease of longhand calculation, rectangular form lending itself to addition and subtraction, and polar form lending itself to multiplication and division.

Conversion between the two notational forms involves simple trigonometry. To convert from polar to rectangular, find the real component by multiplying the polar magnitude by the cosine of the angle, and the imaginary component by multiplying the polar magnitude by the sine of the angle.

This may be understood more readily by drawing the quantities as sides of a right triangle, the hypotenuse of the triangle representing the vector itself (its length and angle with respect to the horizontal constituting the polar form), the horizontal and vertical sides representing the "real" and "imaginary" rectangular components.

To convert from rectangular to polar, find the polar magnitude through the use of the Pythagorean Theorem (the polar magnitude is the hypotenuse of a right triangle, and the real and imaginary components are the adjacent and opposite sides, respectively), and the angle by taking the arctangent of the imaginary component divided by the real component:

AC "POLARITY"

Complex numbers are useful for AC circuit analysis because they provide a convenient method of symbolically denoting phase shift between AC quantities like voltage and current.

However, for most people the equivalence between abstract vectors and real circuit quantities is not an easy one to grasp. AC voltage sources are given voltage figures in complex form (magnitude *and* phase angle), as well as polarity markings.

Being that alternating current has no set "polarity" as direct current does, these polarity markings and their relationship to phase angle tends to be confusing. This section is written in the attempt to clarify some of these issues. Voltage is an inherently *relative* quantity. When we measure a voltage, we have a choice in how we connect a voltmeter or other voltage-measuring instrument to the source of voltage, as there are two points between which the voltage exists, and two test leads on the instrument with which to make connection.

In DC circuits, we denote the polarity of voltage sources and voltage drops explicitly, using "+" and "-" symbols, and use colour-coded meter test leads (red and black).

If a digital voltmeter indicates a negative DC voltage, we know that its test leads are connected "backward" to the voltage (red lead connected to the "-" and black lead to the "+").

Batteries have their polarity designated by way of intrinsic symbology: the short-line side of a battery is always the negative (-) side and the long-line side always the positive (+):

Although it would be mathematically correct to represent a battery's voltage as a negative figure with reversed polarity markings, it would be decidedly unconventional:

Interpreting such notation might be easier if the "+" and "-" polarity markings were viewed as reference points for voltmeter test leads, the "+" meaning "red" and the "-" meaning "black."

A voltmeter connected to the above battery with red lead to the bottom terminal and black lead to the top terminal would indeed indicate a negative voltage (-6 volts). Actually, this form of notation and interpretation is not as unusual as you might think: it's commonly encountered in problems of DC network analysis where "+" and "-" polarity marks are initially drawn according to educated guess, and later interpreted as correct or "backward" according to the mathematical sign of the figure calculated.

In AC circuits, though, we don't deal with "negative" quantities of voltage. Instead, we describe to what degree one voltage aids or opposes another by *phase*: the time-shift between two waveforms. We never describe an AC voltage as being negative in sign, because the facility of polar notation allows for vectors pointing in an opposite direction. If one AC voltage directly opposes another AC voltage, we simply say that one is 180° out of phase with the other.

Still, voltage is relative between two points, and we have a choice in how we might connect a voltage-measuring instrument between those two points. The mathematical sign of a DC voltmeter's reading has meaning only in the context of its test lead connections: which terminal the red lead is touching, and which terminal the black lead is touching. Likewise, the phase angle of an AC voltage has meaning only in the context of knowing which of the two points is considered the "reference" point.

Because of this fact, "+" and "-" polarity marks are often placed by the terminals of an AC voltage in schematic diagrams to give the stated phase angle a frame of reference.

Let's review these principles with some graphical aids. First, the principle of relating test lead connections to the mathematical sign of a DC voltmeter indication:

The mathematical sign of a digital DC voltmeter's display has meaning only in the context of its test lead connections. Consider the use of a DC voltmeter in determining whether or not two DC voltage sources are aiding or opposing each other, assuming that both sources are unlabeled as to their polarities. Using the voltmeter to measure across the first source:

This first measurement of +24 across the left-hand voltage source tells us that the black lead of the meter really is touching the negative side of voltage source #1, and the red lead of the meter really is touching the positive. Thus, we know source #1 is a battery facing in this orientation:

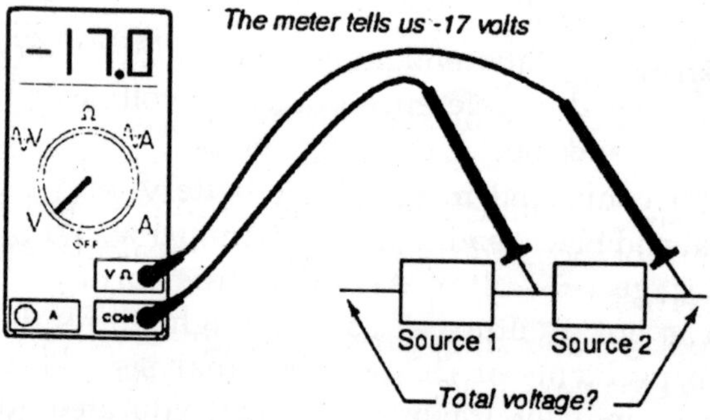

24V source is polarized (-) to (+).

Measuring the other unknown voltage source:

This second voltmeter reading, however, is a *negative* (-) 17 volts, which tells us that the black test lead is actually touching the positive side of voltage source #2, while the red test lead is actually touching the negative.

Thus, we know that source #2 is a battery facing in the *opposite* direction:

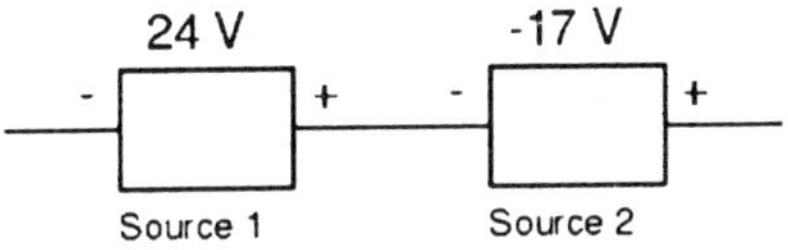

Fig. 17V Source is Polarized (+) to (-)

It should be obvious to any experienced student of DC electricity that these two batteries are opposing one another. By definition, opposing voltages *subtract* from one another, so we subtract 17 volts from 24 volts to obtain the total voltage across the two: 7 volts.

We could, however, draw the two sources as nondescript boxes, labeled with the exact voltage figures obtained by the voltmeter, the polarity marks indicating voltmeter test lead placement:

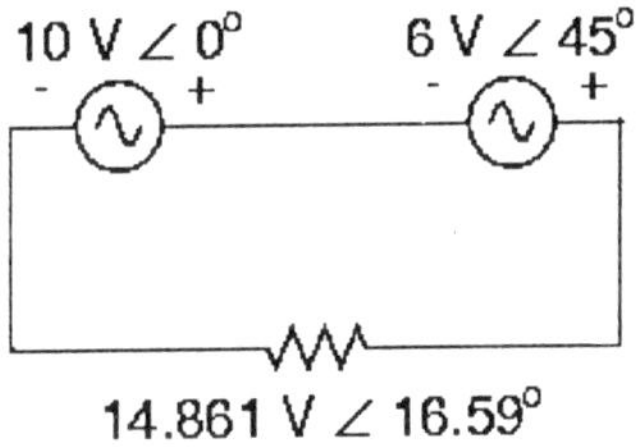

Fig. Voltmeter Readings as Read from Meters.

According to this diagram, the polarity marks (which indicate meter test lead placement) indicate the sources *aiding* each other. By definition, aiding voltage sources *add* with one another to form the total voltage, so we add 24 volts to -17 volts to obtain 7 volts: still the correct answer. If we let the polarity markings guide our decision to either add or subtract voltage figures — whether those polarity markings represent the *true* polarity or just the meter test lead orientation — and include the mathematical signs of those voltage figures in our calculations, the result will always be correct.

Again, the polarity markings serve as *frames of reference* to place the voltage figures' mathematical signs in proper context.

The same is true for AC voltages, except that *phase angle* substitutes for mathematical *sign*. In order to relate multiple AC voltages at different phase angles to each other, we need polarity markings to provide frames of reference for those voltages' phase angles.

Take for example the following circuit: The polarity markings show these two voltage sources aiding each other, so to determine the total voltage across the resistor we must *add* the voltage figures of 10 V " 0° and 6 V " 45° together to obtain 14.861 V " 16.59°.

However, it would be perfectly acceptable to represent the 6 volt source as 6 V " 225°, with a reversed set of polarity markings, and still arrive at the same total voltage:

6 V " 45° with negative on the left and positive on the right is exactly the same as 6 V " 225° with positive on the left and negative on the right: the reversal of polarity markings perfectly complements the addition of 180° to the phase angle designation:

Unlike DC voltage sources, whose symbols intrinsically define polarity by means of short and long lines, AC voltage symbols have no intrinsic polarity marking. Therefore, any polarity marks must by included as additional symbols on the diagram, and there is no one "correct" way in which to place them. They must, however, correlate with the given phase angle to represent the true phase relationship of that voltage with other voltages in the circuit.

Polarity markings are sometimes given to AC voltages in circuit schematics in order to provide a frame of reference for their phase angles.

LC CIRCUITS

The easiest-to-understand system that demonstrates electromagnetic oscillations is the combination of a capacitor and an inductor. Typically, we begin with a fully charged capacitor which is then linked to an inductor by a switch. The basic circuit is shown in figure.

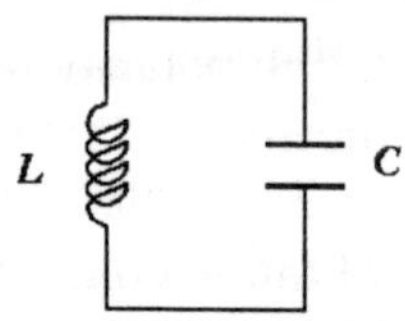

A combination of an inductor and a capacitor forms an LC circuit.

For this circuit, the behaviour is characterized by *oscillating* current, charge, and voltages. This is in sharp contrast to the exponential behaviour of the current, voltages, and charges for the case of RL or RC circuits. If we assume we have initially charged the capacitor to a voltage V_{max} before connecting it to the inductor, we can ask what the behaviour will be.

While it is possible to give a qualitative picture of what happens when the capacitor and inductor are conjoined, it might be insightful to go through the quantitative determination first, then see if that adequately meets what we would naively expect. Before starting even the quantitative approach though, we need to state that this is a case in which the energy approach for looking at the circuit is more immediately obvious than the voltage approach. The main reason is that we can clearly see that energy is conserved for this case since capacitors and inductors involve only electric and magnetic fields which are conservative. We can use Maple to handle the mathematical details. In dealing with specific cases of circuits, we now need to deal with one of the ever-present "difficult" issues of physics, namely: how do you decide what approach to use for solving any particular problem?

This is especially true for circuits involving inductors, capacitors, and resistors because the formulas which describe the behaviour of the individual circuit elements are difficult to memorize. In such cases, which do arise quite often in physics, you must remember the mantra: memorize behaviors, not formulas!

+Typically, you will remember enough about a specific formula to be able to reconstruct it either through a re-derivation of the formula from first principles, or through consideration of what mathematically is needed to get a correct description of the behaviour combined with dimensional analysis. The physics is in the behaviour

In applying this idea to the case at hand, remember that the behaviour is determined by the same considerations that were important for physics problem in Mechanics: if there are no dissipative elements, use energy conservation! Conservation rules are always simpler to use for most applications than complex formulas describing intermediate behaviour (i.e. what happens between the initial and final states being considered in the problem). In cases where dissipative elements, resistors in this case, are involved, think about the behaviour in terms of the initial and long-time aspects of inductor or capacitor behaviour.

As you have seen in previous lectures, these kinds of considerations are essential to deriving the exact formulas anyway. But, for a large class of problems, these same considerations allow us to dispense with the exact formulas. Let's consider two contrasting cases.

RLC Circuits

Including a non-negligible resistance in our calculation is straightforward. We simply need to account for the effect of resistance in terms of the power lost. In this case, we know that the time rate of change of potential energy in the system is $dU/dt = Li\, di/dt + (q/C)dq/dt$. Since energy is no longer conserved, we can't have $dU/dt = 0$, but rather $dU/dt = P_R = -i^2R$, i.e. the Joule heating in the resistor is the rate at which energy is lost from the system. The differential equation describing the charge as a function of time is then. We once again turn to Maple to get the solution of this equation and it's ramifications.

AC Circuits

We can also consider the case of resistors, inductors, and capacitors connected to an alternating current source. Such a source produces a current that varies with time, usually sinusoidally, that looks very much like the current in an LC circuit. We can easily see the result of introducing such a current into a resistor, capacitor, or inductor in terms of the voltage across each. Suppose that the current has the form of $i(t) = I \cos wt$.

We describe the resistor as having a voltage which is *in phase* with the current source. The voltage across the inductor is *out of phase* because it *leads* the current source by a quarter-cycle.

This is in keeping with our knowledge of the behaviour of inductors in that any attempt to *change* the current leads to an EMF that opposes the change, hence the current itself must "lag" the change to the current. For a capacitor, the voltage is also out of phase with the current source but in this case it *lags* the current source by a quarter-cycle. It is the current that delivers the charge to the capacitor plates in order to create the potential difference that causes the current to change, hence the current in this circuit must *lead* the change in current.

POWER GENERATION

Rotating a coil in a magnetic field creates alternating current, and is the method that is used to create most of the world's electricity. Even DC generators create a sinusoidal voltage that is rectified by a commutator. AC electricity is better suited for long distance power transmission than DC electricity.

Alternating current voltage can be stepped up to over 100,000 volts, making it possible to transmit power for hundreds of miles over High Voltage transmission lines.

The voltage can than be reduced to 120 volts (US single phase) or 220 volts (European single phase) for safe home use. The transformers that you see on poles or behind chain link fences are used to step down high voltage to a safer level. Power Generation is the province of electrical engineering not electronic engineering.

One aspect of electronics deals with AC signal generation, filtering, and amplification. It encompasses audio signals, radio signals, radar signals and the like. Of course most electronic system do have power supplies that convert AC to the DC voltages required by vacuum tube, transistor, and integrated circuits.

Electronic AC Signal Generation: This is one of the simplest ways to generate a sinusoidal signal. The capacitor is charged by DC supply voltage. Potential energy is stored in the electrostatic E field between the plates of the capacitor.

The capacitor discharges through the coil. When the capacitor discharges to zero volts the current in coil reaches a maximum. At this point the potential energy of the E field is reduced to zero and the current is at a maximum. All the potential energy of capacitor is now kinetic energy stored in the coil magnetic H field. As the H field collapses, the change in magnetic flux sustains the current flow through it. This current causes the capacitor to charge to a level opposite in polarity with respect to original voltage. This transfers the energy back to the capacitor E field. This cycle repeats indefinitely unless damped. This oscillating circuit is frequently referred to as a tank circuit.

An ideal tank circuit has no resistance and will oscillate indefinitely. Of course the wire of the inductors has some resistance, which will course some damping of the oscillations. Heat can also be generated in the electrolyte of the capacitor. This will also contribute to oscillation damping. An electromagnetic oscillator can also radiate energy in the form of electromagnetic radiation.

Electromagnetic radiation consists of an oscillating E and H fields in space. This radiation propagates at the speed of light.AC Signals are usually sinusoidal: The equation below defines a sinusoidal waveform.Electromagnetic radiation in space propagates as a sinusoidal wave moving at the speed of light.V = A sin wt where w = 2 *Pi*f or angular velocity. Ohm's and Kirchhoff's Laws: In purely capacitive and purely inductive circuits, Ohm's and Kirchhoff's Laws still apply.

Perform AC calculation in the same manner as you performed DC calculations. Simply substitute Vac for Vdc.

In purely inductive or purely capacitive circuits, you must first calculate the inductive or capacitive reactance before applying Ohm's law. Inductive and capacitive reactance are a function of frequency. Reactance is expressed in ohms.

Chapter 8

Generators and Motors

DIRECT CURRENT GENERATORS

A generator is a machine that converts mechanical energy into electrical energy by using the principle of magnetic induction. This principle is explained as follows: Whenever a conductor is moved within a magnetic field in such a way that the conductor cuts across magnetic lines of flux, voltage is generated in the conductor.

The Amount of voltage generated depends on:

- The strength of the magnetic field,
- The angle at which the conductor cuts the magnetic field,
- The speed at which the conductor is moved, and
- The length of the conductor within the magnetic field.

The polarity of the voltage depends on the direction of the magnetic lines of flux and the direction of movement of the conductor. To determine the direction of current in a given situation, the left-hand rule for generators is used. This rule is explained in the following manner. Extend the thumb, forefinger, and middle finger of your left hand at right angles to one another. Point your thumb in the direction the conductor is being moved. Point your forefinger in the direction of magnetic flux (from north to south). Your middle finger will then point in the direction of current flow in an external circuit to which the voltage is applied.

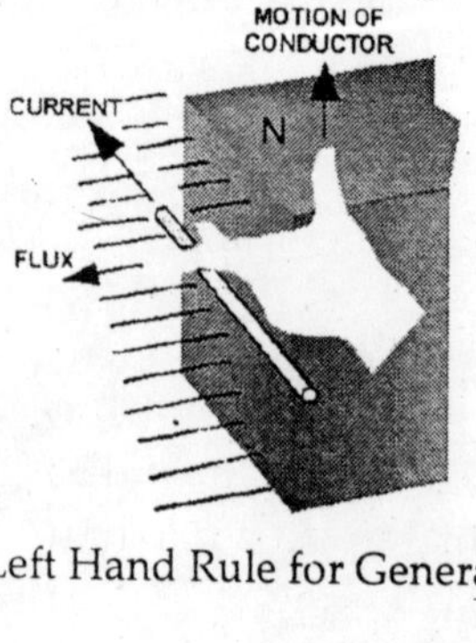

Fig. Left Hand Rule for Generators

The elementary generator the simplest elementary generator that can be built is an ac generator. Basic generating principles are most easily explained through the use of the elementary ac generator.

An elementary generator consists of a wire loop placed so that it can be rotated in a stationary magnetic field. This will produce an induced emf in the loop. Sliding contacts (brushes) connect the loop to an external circuit load in order to pick up or use the induced emf.

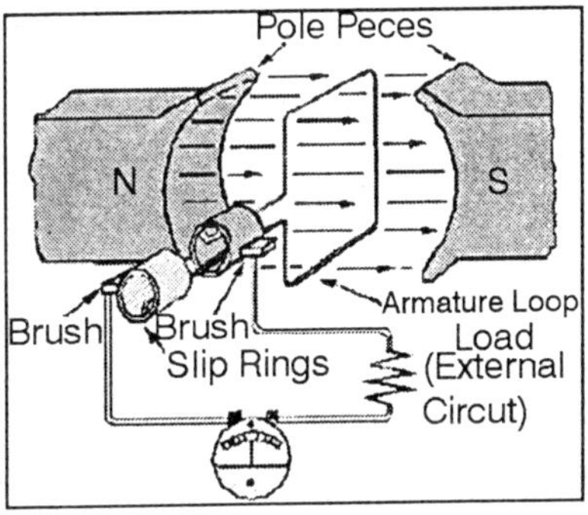

Fig. The Denentary Generator

The pole pieces (marked N and S) provide the magnetic field. The pole pieces are shaped and positioned as shown to concentrate the magnetic field as close as possible to the wire loop.

The loop of wire that rotates through the field is called the Armature. The ends of the armature loop are connected to rings called *slip rings*. They rotate with the armature. The brushes, usually made of carbon, with wires attached to them, ride against the rings. The generated voltage appears across these brushes.

The elementary generator produces a voltage in the following manner. The armature loop is rotated in a clockwise direction. The initial or starting point is shown at position A. (This will be considered the zero-degree position.) At 0° the armature loop is perpendicular to the magnetic field. The black and white conductors of the loop are moving parallel to the field. The instant the conductors are moving parallel to the magnetic field, they do not cut any lines of flux.

Therefore, no emf is induced in the conductors, and the meter at position A indicates zero.

This position is called the neutral plane. As the armature loop rotates from position A (0°) to position B (90°), the conductors cut through more and more lines of flux, at a continually increasing angle. At 90° they are cutting through a maximum number of lines of flux and at maximum angle. The result is that between 0° and 90°, the induced emf in the conductors builds up from zero to a maximum value.

Observe that from 0° to 90°, the black conductor cuts DOWN through the field. At the same time the white conductor cuts UP through the field. The induced emfs in the conductors are series-adding. This means the resultant voltage across the brushes (the terminal voltage) is the sum of the two induced voltages.

The meter at position B reads maximum value. As the armature loop continues rotating from 90° (position B) to 180° (position C), the conductors which were cutting through a maximum number of lines of flux at position B now cut through fewer lines.

They are again moving parallel to the magnetic field at position C. They no longer cut through any lines of flux. As the armature rotates from 90° to 180°, the induced voltage will decrease to zero in the same manner that it increased during the rotation from 0° to 90°.

The meter again reads zero. From 0° to 180° the conductors of the armature loop have been moving in the same direction through the magnetic field. Therefore, the polarity of the induced voltage has remained the same. This is shown by points A through C on the graph. As the loop rotates beyond 180° (position C), through 270° (position D), and back to the initial or starting point (position A), the direction of the cutting action of the conductors through the magnetic field reverses. Now the black conductor cuts UP through the field while the white conductor cuts DOWN through the field.

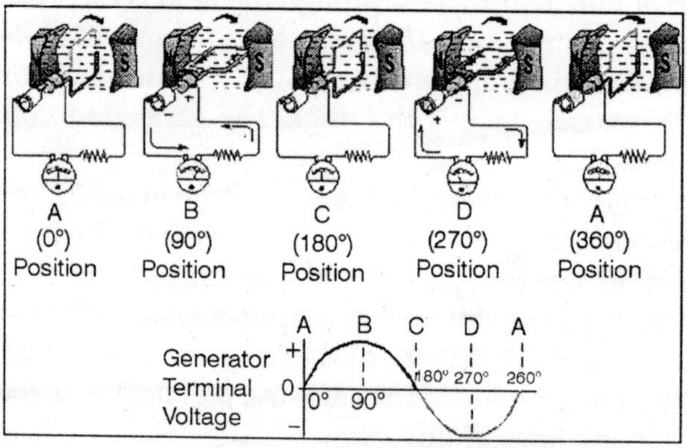

Fig. Output voltage of an Elementary Generator During one Revolution.

As a result, the polarity of the induced voltage reverses. Following the sequence shown by graph points C, D, and back to A, the voltage will be in the direction opposite to that shown from points A, B, and C. The terminal voltage will be the same as it was from A to C except that the polarity is reversed (as shown by the meter deflection at position D). The voltage output waveform for the complete revolution of the loop.

THE ELEMENTARY DC GENERATOR

A single-loop generator with each terminal connected to a segment of a two-segment metal ring is shown in figure. The two segments of the split metal ring are insulated from each other. This forms a simple commutator. The commutator in a dc generator replaces the slip rings of the ac generator.

This is the main difference in their construction. The commutator mechanically reverses the armature loop connections to the external circuit. This occurs at the same instant that the polarity of the voltage in the armature loop reverses. Through this process the commutator changes the generated ac voltage to a pulsating dc voltage. This action is known as commutation.

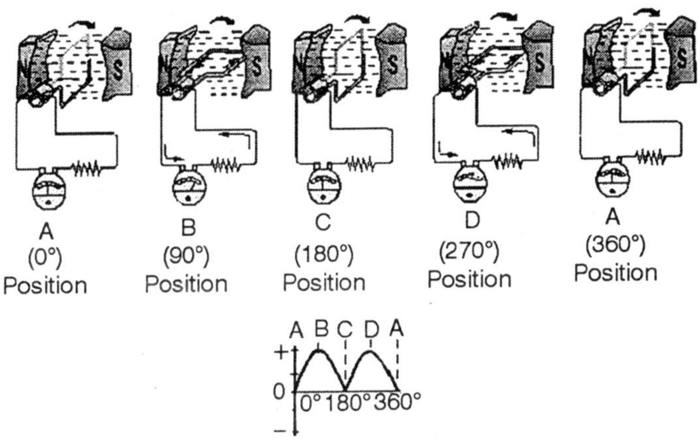

Fig. Effects of Commutation.

The step-by-step description of the operation of a dc generator. When the armature loop rotates clockwise from position A to position B, a voltage is induced in the armature loop which causes a current in a direction that deflects the meter to the right. Current flows through loop, out of the negative brush, through the meter and the load, and back through the positive brush to the loop.

Voltage reaches its maximum value at point B on the graph for reasons explained earlier. The generated voltage and the current fall to zero at position C. At this instant each brush makes contact with both segments of the commutator. As the armature loop rotates to position D, a voltage is again induced in the loop. In this case, however, the voltage is of opposite polarity.

The voltages induced in the two sides of the coil at position D are in the reverse direction to that of the voltages shown at position B. Note that the current is flowing from the black side to the white side in position B and from the white side to the black side in position D. However, because the segments of the commutator have rotated with the loop and are contacted by opposite brushes, the direction of current flow through the brushes and the meter remains the same as at position B.

The voltage developed across the brushes is pulsating and unidirectional (in one direction only). It varies twice during each revolution between zero and maximum. This variation is called RIPPLE. A pulsating voltage, such as that produced in the preceding description, is unsuitable for most applications. Therefore, in practical generators more armature loops (coils) and more commutator segments are used to produce an output voltage waveform with less ripple.

EFFECTS OF ADDING ADDITIONAL COILS AND POLES

The effects of additional coils may be illustrated by the addition of a second coil to the armature. The commutator must now be divided into four parts since there are four coil ends. The coil is rotated in a clockwise direction from the position shown.

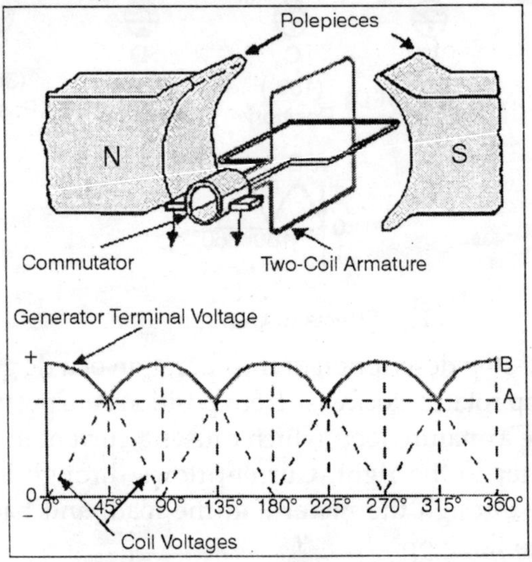

Fig. Effects of Additional Coils.

The voltage induced in the white coil, Decreases for the next 90° of rotation (from maximum to zero). The voltage induced in the black coil increases from zero to maximum at the same time. Since there are four segments in the commutator, a new segment passes each brush every 90° instead of every 180°.

This allows the brush to switch from the white coil to the black coil at the instant the voltages in the two coils are equal. The brush remains in contact with the black coil as its induced voltage increases to maximum, level B in the graph. It then decreases to level A, 90° later. At this point, the brush will contact the white coil again.

The ripple effect of the voltage when two armature coils are used. Since there are now four commutator segments in the commutator and only two brushes, the voltage cannot fall any lower than at point A. Therefore, the ripple is limited to the rise and fall between points A and B on the graph. By adding more armature coils, the ripple effect can be further reduced. Decreasing ripple in this way increases the effective voltage of the output.

Effective voltage is the equivalent level of dc voltage, which will cause the same average current through a given resistance. By using additional armature coils, the voltage across the brushes is not allowed to fall to as low a level between peaks. The ripple has been reduced. Practical generators use many armature coils. They also use more than one pair of magnetic poles. The additional magnetic poles have the same effect on ripple as did the additional armature coils. In addition, the increased number of poles provides a stronger magnetic field (greater number of flux lines). This, in turn, allows an increase in output voltage because the coils cut more lines of flux per revolution.

ELECTROMAGNETIC POLES

Nearly all practical generators use electromagnetic poles instead of the permanent magnets used in our elementary generator. The electromagnetic field poles consist of coils of insulated copper wire wound on soft iron cores.

The main advantages of using electromagnetic poles are:

- Increased field strength and
- A means of controlling the strength of the fields.

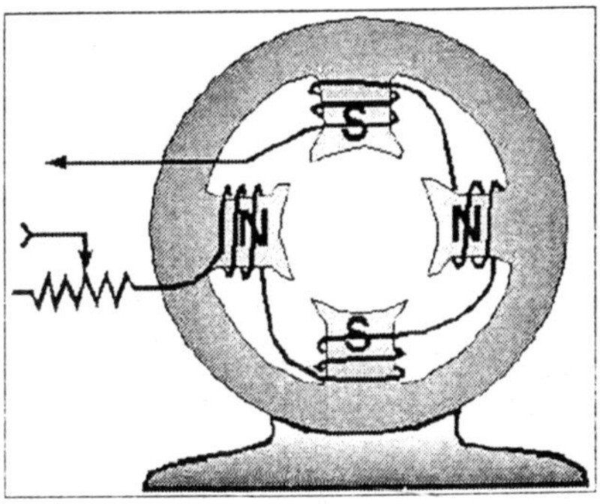

Fig. Four-pole Generator (Without Armature).

By varying the input voltage, the field strength is varied. By varying the field strength, the output voltage of the generator can be controlled.

COMMUTATION

Commutation is the process by which a dc voltage output is taken from an armature that has an ac voltage induced in it. The elementary dc generator that the commutator mechanically reverses the armature loop connections to the external circuit. This occurs at the same instant that the voltage polarity in the armature loop reverses. A dc voltage is applied to the load because the output connections are reversed as each commutator segment passes under a brush. The segments are insulated from each other.

Commutation occurs simultaneously in the two coils that are briefly short-circuited by the brushes. Coil B is short-circuited by the negative brush. Coil Y, the opposite coil, is short-circuited by the positive brush. The brushes are positioned on the commutator so that each coil is short-circuited as it moves through its own electrical neutral plane.

There is no voltage generated in the coil at that time. Therefore, no sparking can occur between the commutator and the brush. Sparking between the brushes and the commutator is an indication of improper commutation. Improper brush placement is the main cause of improper commutation.

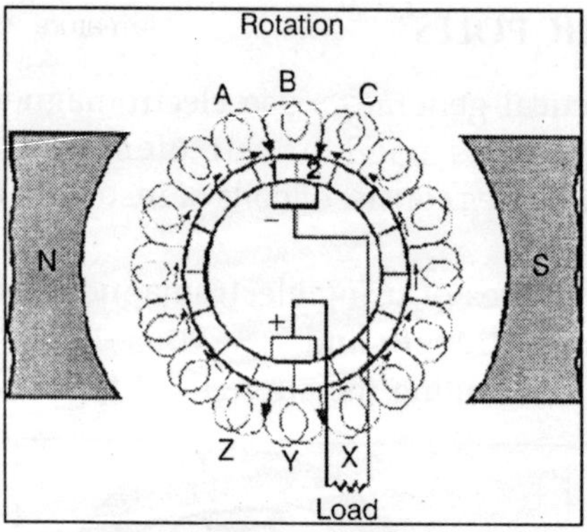

Fig. Commutation of a dc Generator.

ARMATURE REACTION

From previous study, you know that all current-carrying conductors produce magnetic fields. The magnetic field produced by current in the armature of a dc generator affects the flux pattern and distorts the main field. This distortion causes a shift in the neutral plane, which affects commutation. This change in the neutral plane and the reaction of the magnetic field is called *armature reaction.*

You know that for proper commutation, the coil short-circuited by the brushes must be in the neutral plane. Consider the operation of a simple two-pole dc generator. View A of the figure shows the field poles and the main magnetic field. The armature is shown in a simplified view in views B and C with the cross section of its coil represented as little circles. The symbols within the circles represent arrows. The dot represents the point of the arrow coming toward you, and the cross represents the tail, or feathered end, going away from you. When the armature rotates clockwise, the sides of the coil to the left will have current flowing toward you, as indicated by the dot. The side of the coil to the right will have current flowing away from you, as indicated by the cross. The field generated around each side of the coil is shown in view B of figure. This field increases in strength for each wire in the armature coil, and sets up a magnetic field almost perpendicular to the main field.

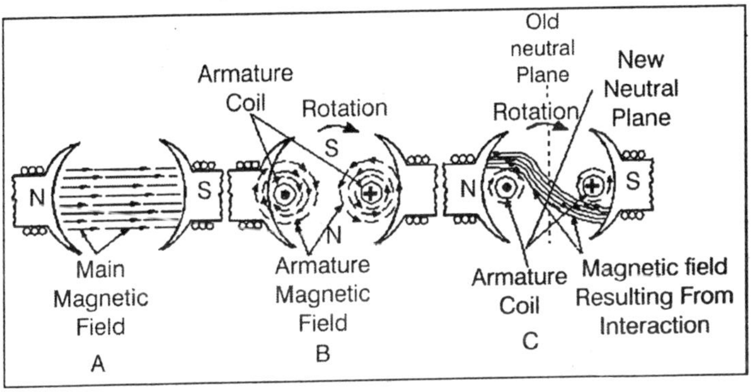

Fig. Armature Reaction.

Now you have two fields — the main field, view A, and the field around the armature coil, view B. View C of figure shows how the armature field distorts the main field and how the neutral plane is shifted in the direction of rotation.

If the brushes remain in the old neutral plane, they will be short-circuiting coils that have voltage induced in them. Consequently, there will be arcing between the brushes and commutator.

COMPENSATING WINDINGS AND INTERPOLES

Shifting the brushes to the advanced position (the new neutral plane) does not completely solve the problems of armature reaction. The effect of armature reaction varies with the load current. Therefore, each time the load current varies, the neutral plane shifts. This means the brush position must be changed each time the load current varies.

In small generators, the effects of armature reaction are reduced by actually mechanically shifting the position of the brushes. The practice of shifting the brush position for each current variation is not practiced except in small generators. In larger generators, other means are taken to eliminate armature reaction. Compensating Windings or Interpoles are used for this purpose.

The compensating windings consist of a series of coils embedded in slots in the pole faces. These coils are connected in series with the armature. The series-connected compensating windings produce a magnetic field, which varies directly with armature current.

Because the compensating windings are wound to produce a field that opposes the magnetic field of the armature, they tend to cancel the effects of the armature magnetic field. The neutral plane will remain stationary and in its original position for all values of armature current. Because of this, once the brushes have been set correctly, they do not have to be moved again.

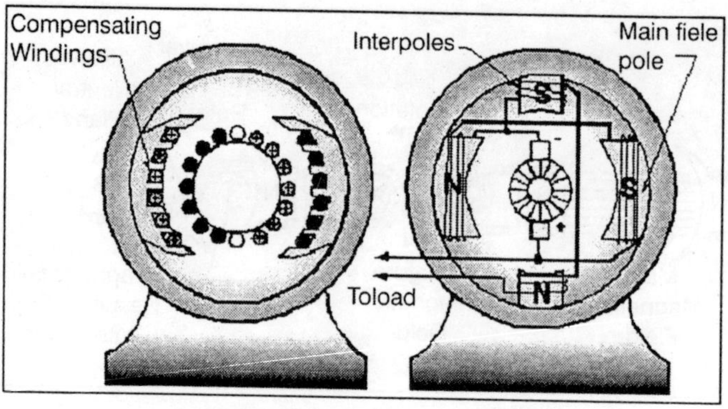

Fig. Compensating Windings and Interpoles.

Another way to reduce the effects of armature reaction is to place small auxiliary poles called "interpoles" between the main field poles. The interpoles have a few turns of large wire and are connected in series with the armature.

Interpoles are wound and placed so that each interpole has the same magnetic polarity as the main pole ahead of it, in the direction of rotation. The field generated by the interpoles produces the same effect as the compensating winding. This field, in effect, cancels the armature reaction for all values of load current by causing a shift in the neutral plane opposite to the shift caused by armature reaction. The amount of shift caused by the interpoles will equal the shift caused by armature reaction since both shifts are a result of armature current.

MOTOR REACTION IN A GENERATOR

When a generator delivers current to a load, the armature current creates a magnetic force that opposes the rotation of the armature. This is called Motor Reaction. A single armature conductor is represented in figure, view A. When the conductor is stationary, no voltage is generated and no current flows.

Therefore, no force acts on the conductor. When the conductor is moved downward and the circuit is completed through an external load, current flows through the conductor in the direction indicated. This sets up lines of flux around the conductor in a clockwise direction.

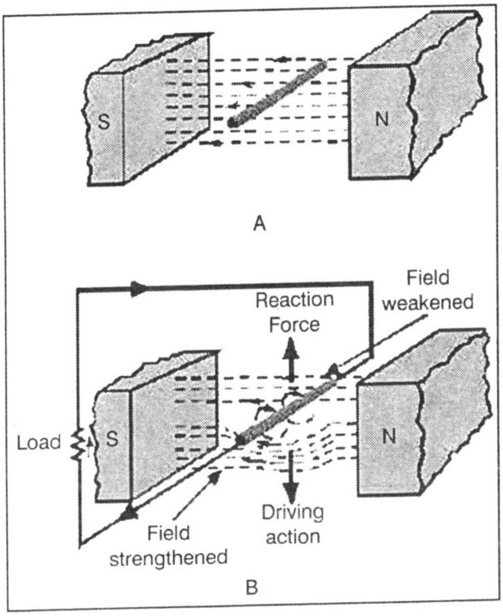

Fig. Motor Reaction in a Generator.

The interaction between the conductor field and the main field of the generator weakens the field above the conductor and strengthens the field below the conductor. The main field consists of lines that now act like stretched rubber bands.

Thus, an upward reaction force is produced that acts in opposition to the downward driving force applied to the armature conductor. If the current in the conductor increases, the reaction force increases. Therefore, more force must be applied to the conductor to keep it moving. With no armature current, there is no magnetic (motor) reaction.

Therefore, the force required to turn the armature is low. As the armature current increases, the reaction of each armature conductor against rotation increases.

The actual force in a generator is multiplied by the number of conductors in the armature. The driving force required to maintain the generator armature speed must be increased to overcome the motor reaction. The force applied to turn the armature must overcome the motor reaction force in all dc generators. The device that provides the turning force applied to the armature is called the PRIME MOVER. The prime mover may be an electric motor, a gasoline engine, a steam turbine, or any other mechanical device that provides turning force.

ARMATURE LOSSES

In dc generators, as in most electrical devices, certain forces act to decrease the efficiency. These forces, as they affect the armature, are considered as losses and may be defined as follows:

- I^2R, or copper loss in the winding
- Eddy current loss in the core
- Hysteresis loss (a sort of magnetic friction)

Copper Losses

The power lost in the form of heat in the armature winding of a generator is known as COPPER LOSS. Heat is generated any time current flows in a conductor. Copper loss is an I^2R loss, which increases as current increases.

The amount of heat generated is also proportional to the resistance of the conductor. The resistance of the conductor varies directly with its length and inversely with its cross- sectional area. Copper loss is minimized in armature windings by using large diameter wire.

Eddy Current Losses

The core of a generator armature is made from soft iron, which is a conducting material with desirable magnetic characteristics. Any conductor will have currents induced in it when it is rotated in a magnetic field. These currents that are induced in the generator armature core are called EDDY CURRENTS. The power dissipated in the form of heat, as a result of the eddy currents, is considered a loss.

Eddy currents, just like any other electrical currents, are affected by the resistance of the material in which the currents flow. The resistance of any material is inversely proportional to its cross-sectional area. The eddy currents induced in an armature core that is a solid piece of soft iron. A soft iron core of the same size, but made up of several small pieces insulated from each other.

This process is called lamination. The currents in each piece of the laminated core are considerably less than in the solid core because the resistance of the pieces is much higher. (Resistance is inversely proportional to cross-sectional area.) The currents in the individual pieces of the laminated core are so small that the sum of the individual currents is much less than the total of eddy currents in the solid iron core.

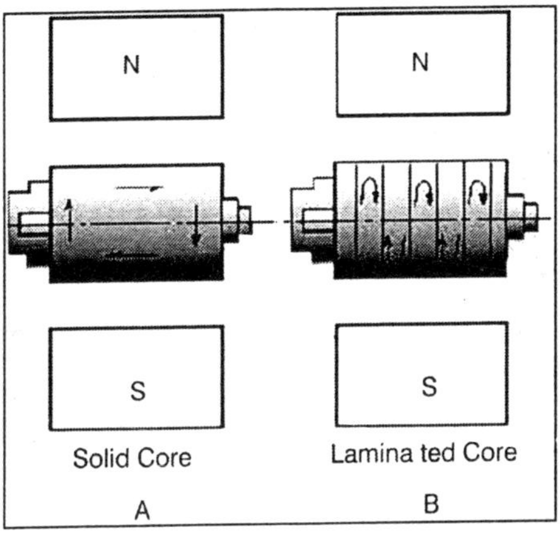

Fig. Eddy Currents in dc Generator Armature Cores.

Eddy current losses are kept low when the core material is made up of many thin sheets of metal. Laminations in a small generator armature may be as thin as 1/64 inch.

The laminations are insulated from each other by a thin coat of lacquer or, in some instances, simply by the oxidation of the surfaces. Oxidation is caused by contact with the air while the laminations are being annealed. The insulation value need not be high because the voltages induced are very small. Most generators use armatures with laminated cores to reduce eddy current losses.

Hysteresis Losses

Hysteresis loss is a heat loss caused by the magnetic properties of the armature. When an armature core is in a magnetic field, the magnetic particles of the core tend to line up with the magnetic field. When the armature core is rotating, its magnetic field keeps changing direction. The continuous movement of the magnetic particles, as they try to align themselves with the magnetic field, produces molecular friction. This, in turn, produces heat.

This heat is transmitted to the armature windings. The heat causes armature resistances to increase. To compensate for hysteresis losses, heat-treated silicon steel laminations are used in most dc generator armatures. After the steel has been formed to the proper shape, the laminations are heated and allowed to cool. This annealing process reduces the hysteresis loss to a low value.

THE PRACTICAL DC GENERATOR

The actual construction and operation of a practical dc generator differs somewhat from our elementary generators. The differences are in the construction of the armature, the manner in which the armature is wound, and the method of developing the main field. A generator that has only one or two armature loops has high ripple voltage. This results in too little current to be of any practical use. To increase the amount of current output, a number of loops of wire are used.

These additional loops do away with most of the ripple. The loops of wire, called windings, are evenly spaced around the armature so that the distance between each winding is the same. The commutator in a practical generator is also different. It has several segments instead of two or four, as in our elementary generators. The number of segments must equal the number of armature coils.

GRAMME-RING ARMATURE

The diagram of a gramme-ring armature. Each coil is connected to two commutator segments. One end of coil 1goes to segment A, and the other end of coil 1 goes to segment B. One end of coil 2 goes to segment C, and the other end of coil 2 goes to segment B. The rest of the coils are connected in a like manner, in series, around the armature. To complete the series arrangement, coil 8 connects to segment A. Therefore, each coil is in series with every other coil.

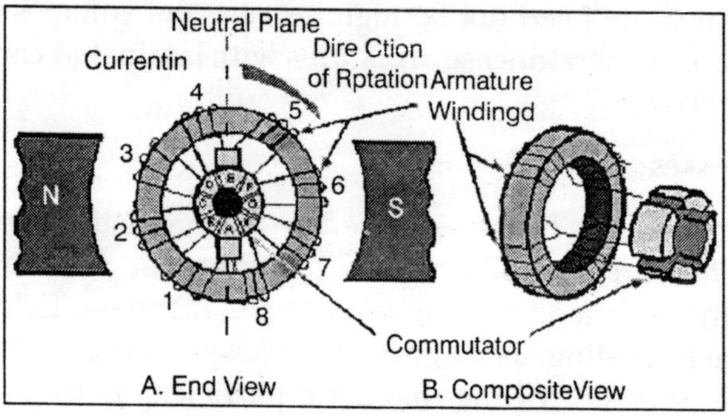

Fig. Gramme-ring Armature.

A composite view of a Gramme-ring armature. It illustrates more graphically the physical relationship of the coils and commutator locations. The windings of a Gramme-ring armature are placed on an iron ring. A disadvantage of this arrangement is that the windings located on the inner side of the iron ring cut few lines of flux. Therefore, they have little, if any, voltage induced in them. For this reason, the Gramme-ring armature is not widely used.

DRUM-TYPE ARMATURE

The armature windings are placed in slots cut in a drum-shaped iron core. Each winding completely surrounds the core so that the entire length of the conductor cuts the main magnetic field. Therefore, the total voltage induced in the armature is greater than in the Gramme-ring. You can see that the drum-type armature is much more efficient than the Gramme-ring. This accounts for the almost universal use of the drum-type armature in modem dc generators.

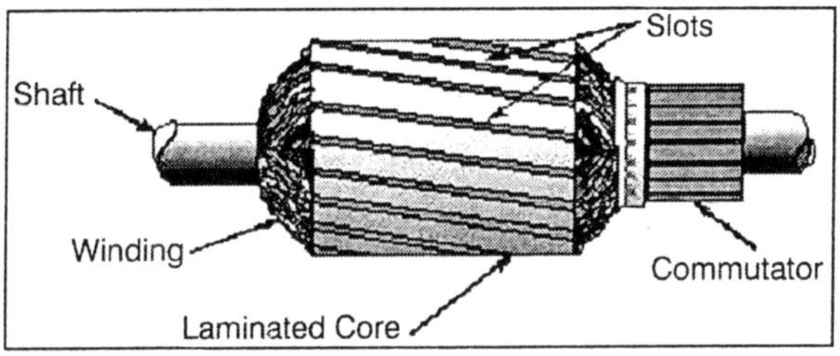

Fig. Drum-type Armature.

Drum-type armatures are wound with either of two types of windings – the lap winding or the wave winding.

The lap winding is used in dc generators designed for high-current applications. The windings are connected to provide several parallel paths for current in the armature. For this reason, lap-wound armatures used in dc generators require several pairs of poles and brushes.

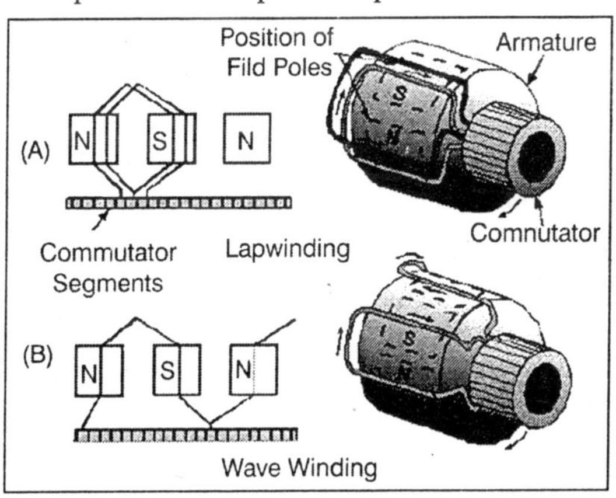

Fig. Types of Windings used on Drum-type Armatures.

A wave winding on a drum-type armature. This type of winding is used in dc generators employed in high-voltage applications. Notice that the two ends of each coil are connected to commutator segments separated by the distance between poles.

This configuration allows the series addition of the voltages in all the windings between brushes. This type of winding only requires one pair of brushes. In practice, a practical generator may have several pairs to improve commutation.

FIELD EXCITATION

When a dc voltage is applied to the field windings of a dc generator, current flows through the windings and sets up a steady magnetic field. This is called Field Excitation..

This excitation voltage can be produced by the generator itself or it can be supplied by an outside source, such as a battery. A generator that supplies its own field excitation is called a Self-Excited Generator. Self-excitation is possible only if the field pole pieces have retained a slight amount of permanent magnetism, called Resudual Magnetism. When the generator starts rotating, the weak residual magnetism causes a small voltage to be generated in the armature. This small voltage applied to the field coils causes a small field current.

Although small, this field current strengthens the magnetic field and allows the armature to generate a higher voltage. The higher voltage increases the field strength, and so on. This process continues until the output voltage reaches the rated output of the generator.

CLASSIFICATION OF GENERATORS

Self-excited generators are classed according to the type of field connection they use. There are three general types of field connections — *Series-Wound, Shunt-Wound* (parallel), and *Compound-Wound*. Compound-wound generators are further classified as cumulative-compound and differential-compound.

Series-Wound Generator: In the series-wound generator, the field windings are connected in series with the armature. Current that flows in the armature flows through the external circuit and through the field windings. The external circuit connected to the generator is called the load circuit.

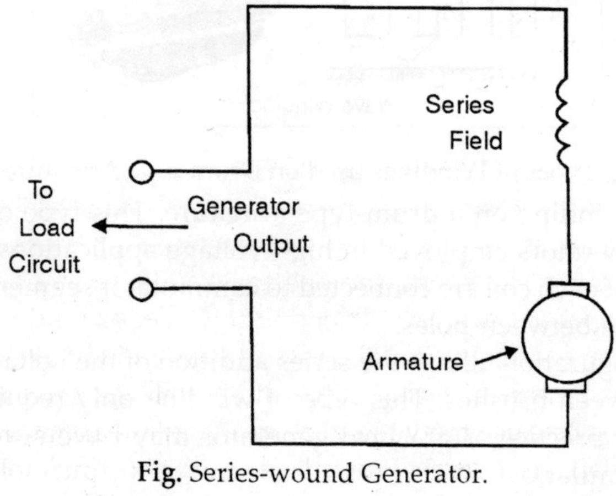

Fig. Series-wound Generator.

A series-wound generator uses very low resistance field coils, which consist of a few turns of large diameter wire. The voltage output increases as the load circuit starts drawing more current. Under low-load current conditions, the current that flows in the load and through the generator is small. Since small current means that a small magnetic field is set up by the field poles, only a small voltage is induced in the armature.

If the resistance of the load decreases, the load current increases. Under this condition, more current flows through the field. This increases the magnetic field and increases the output voltage. A series-wound dc generator has the characteristic that the output voltage varies with load current. This is undesirable in most applications. For this reason, this type of generator is rarely used in everyday practice. The series-wound generator has provided an easy method to introduce you to the subject of self- excited generators.

Shunt-Wound Generators

In a shunt-wound generator, the field coils consist of many turns of small wire. They are connected in parallel with the load. In other words, they are connected across the output voltage of the armature.

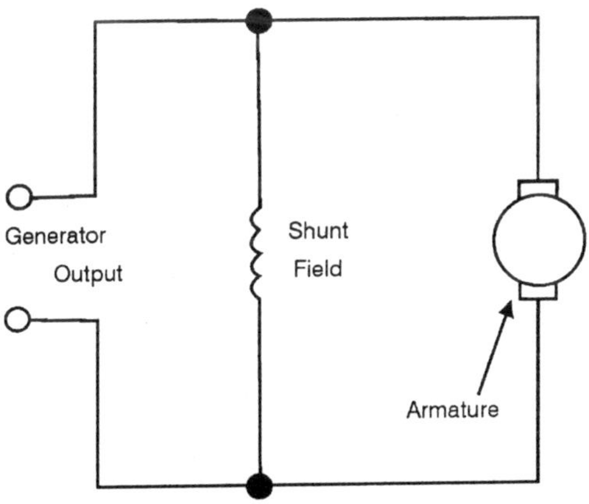

Fig. Shunt-wound Generator.

Current in the field windings of a shunt-wound generator is independent of the load current (currents in parallel branches are independent of each other). Since field current, and therefore field strength, is not affected by load current, the output voltage remains more nearly constant than does the output voltage of the series-wound generator.

In actual use, the output voltage in a dc shunt-wound generator varies inversely as load current varies. The output voltage decreases as load current increases because the voltage drop across the armature resistance increases (E = IR). In a series-wound generator, output voltage varies directly with load current. In the shunt-wound generator, output voltage varies inversely with load current. A combination of the two types can overcome the disadvantages of both. This combination of windings is called the compound-wound dc generator.

Compound-Wound Generators

Compound-wound generators have a series-field winding in addition to a shunt-field winding. The shunt and series windings are wound on the same pole pieces.

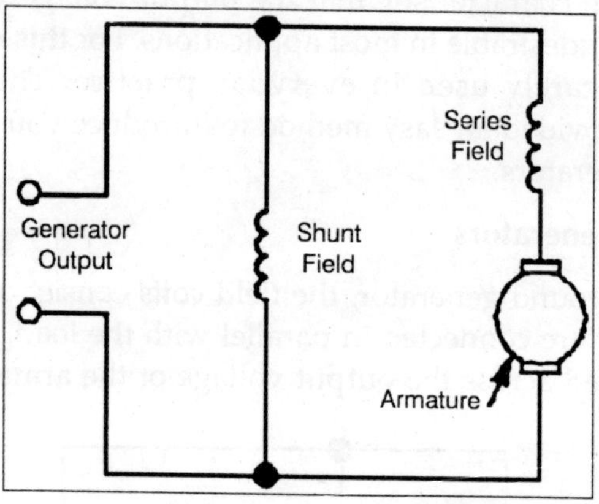

Fig. Compound-wound Generator.

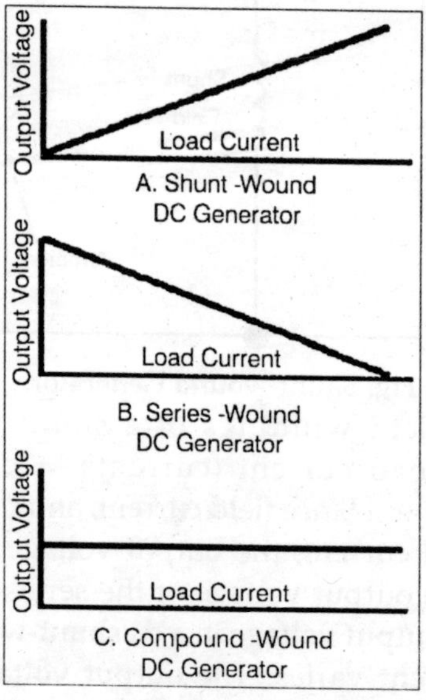

Fig. Voltage Output Characteristics of the Series-, Shunt-, and Compound-wound dc Generators.

In the compound-wound generator when load current increases, the armature voltage decreases just as in the shunt-wound generator. This causes the voltage applied to the shunt-field winding to decrease, which results in a decrease in the magnetic field.

This same increase in load current, since it flows through the series winding, causes an increase in the magnetic field produced by that winding. By proportioning the two fields so that the decrease in the shunt field is just compensated by the increase in the series field, the output voltage remains constant.

The voltage characteristics of the series-, shunt-, and compound-wound generators. The effects of the two fields (series and shunt), a compound-wound generator provides a constant output voltage under varying load conditions.

VOLTAGE REGULATION

The regulation of a generator refers to the VOLTAGE CHANGE that takes place when the load changes. It is usually expressed as the change in voltage from a no-load condition to a full-load condition, and is expressed as a percentage of full-load. It is expressed in the following formula:

$$\text{Per cent of Regulation} = \frac{(E_{nL} - E_{fL})}{E_{fL}} \times 100$$

where E_{nL} is the no-load terminal voltage and E_{fL} is the full-load terminal voltage of the generator. For example, to calculate the per cent of regulation of a generator with a no-load voltage of 462 volts and a full-load voltage of 440 volts

Given: No-load voltage 462 V

- Full-load voltage 440 V

Solution:

$$\text{Per cent of Regulation} = \frac{(E_{nL} - E_{fL})}{E_{fL}} \times 100$$

$$\text{Per cent of Regulation} = \frac{(462\text{V} - 440\text{V})}{440\text{V}} \times 100$$

$$\text{Per cent of Regulation} = \frac{22\text{V}}{440\text{V}} \times 100$$

$$\text{Per cent of Regulation} = .05 \times 100$$

$$\text{Regulation} = 5\%$$

Note: The lower the per cent of regulation, the better the generator. In the above example, the 5% regulation represented a 22-volt change from no load to full load. A 1% change would represent a change of 4.4 volts, which, of course, would be better.

GENERATOR CONSTRUCTION

Figure, views A through E, shows the component parts of dc generators. Figure shows the entire generator with the component parts installed. The cutaway drawing helps you to see the physical relationship of the components to each other.

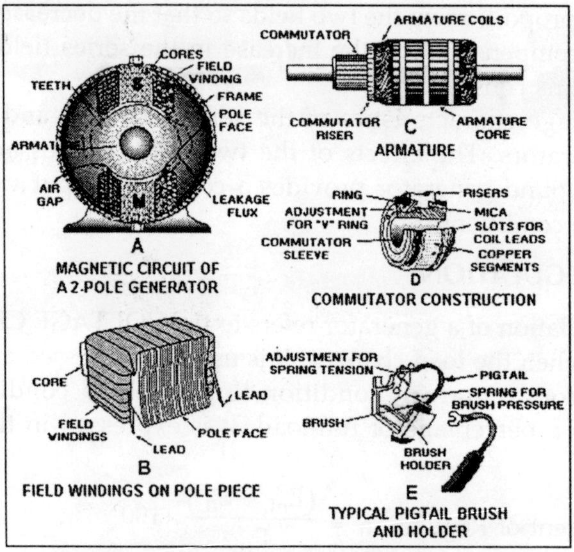

Fig. Components of a dc Generator.

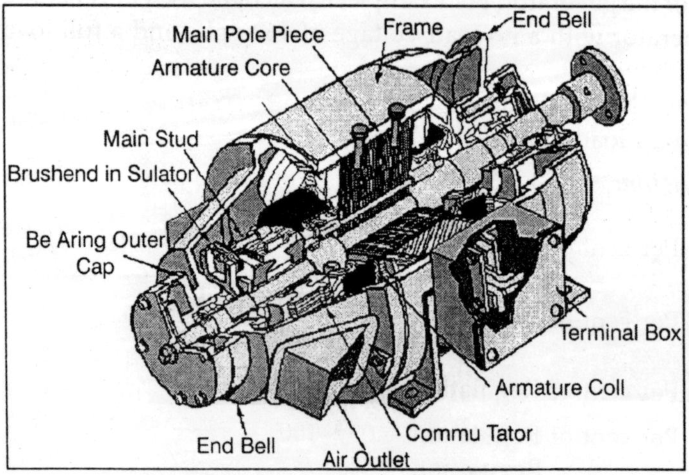

Fig. Construction of a dc Generator.

VOLTAGE CONTROL

Voltage control is either (1) manual or (2) automatic. In most cases the process involves changing the resistance of the field circuit. By changing the field circuit resistance, the field current is controlled. Controlling the field current permits control of the output voltage. The major difference between the various voltage control systems is merely the method by which the field circuit resistance and the current are controlled.

Voltage regulation should not be confused with voltage control. Voltage regulation is an internal action occurring within the generator whenever the load changes. Voltage control is an imposed action, usually through an external adjustment, for the purpose of increasing or decreasing terminal voltage.

Manual Voltage Control

The hand-operated field rheostat, is a typical example of manual voltage control. The field rheostat is connected in series with the shunt field circuit. This provides the simplest method of controlling the terminal voltage of a dc generator.

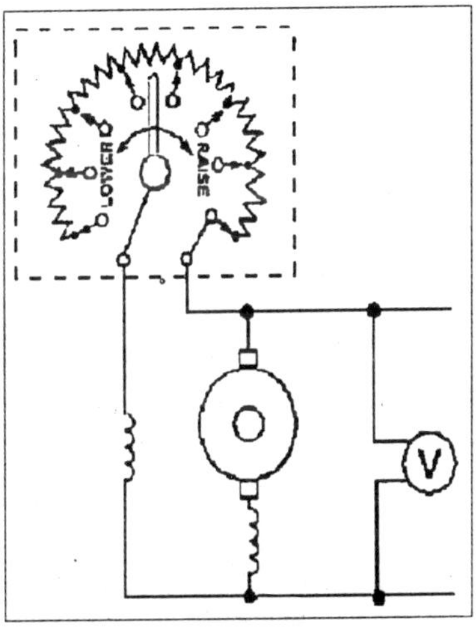

Fig. Hand-operated Field Rheostat.

This type of field rheostat contains tapped resistors with leads to a multiterminal switch. The arm of the switch may be rotated to make contact with the various resistor taps. This varies the amount of resistance in the field circuit. Rotating the arm in the direction of the LOWER arrow (counterclockwise) increases the resistance and lowers the output voltage. Rotating the arm in the direction of the RAISE arrow (clockwise) decreases the resistance and increases the output voltage.

Most field rheostats for generators use resistors of alloy wire. They have a high specific resistance and a low temperature coefficient. These alloys include copper, nickel, manganese, and chromium. They are marked under trade names such as Nichrome, Advance, Manganin, and so forth. Some very large generators use cast-iron grids in place of rheostats, and motor-operated switching mechanisms to provide voltage control.

Automatic Voltage Control

Automatic voltage control may be used where load current variations exceed the built-in ability of the generator to regulate itself. An automatic voltage control device "senses" changes in output voltage and causes a change in field resistance to keep output voltage constant. Whichever control method is used, the range over which voltage can be changed is a design characteristic of the generator. The voltage can be controlled only within the design limits.

PARALLEL OPERATION OF GENERATORS

When two or more generators are supplying a common load, they are said to be operating in parallel. The purpose of connecting generators in parallel is simply to provide more current than a single generator is capable of providing. The generators may be physically located quite a distance apart. However, they are connected to the common load through the power distribution system.

There are several reasons for operating generators in parallel. The number of generators used may be selected in accordance with the load demand. By operating each generator as nearly as possible to its rated capacity, maximum efficiency is achieved. A disabled or faulty generator may be taken off-line and replaced without interrupting normal operations.

AMPLIDYNES

Amplidynes are special-purpose dc generators. They supply large dc currents, precisely controlled, to the large dc motors used to drive heavy physical loads, such as gun turrets and missile launchers.

The amplidyne is really a motor and a generator. It consists of a constant-speed ac motor (the prime mover) mechanically coupled to a dc generator, which is wired to function as a high-gain amplifier (an amplifier is a device in which a small input voltage can control a large current source). For instance, in a normal dc generator, a small dc voltage applied to the field windings is able to control the output of the generator.

In a typical generator, a change in voltage from 0-volt dc to 3-volts dc applied to the field winding may cause the generator output to vary from 0-volt dc to 300-volts dc.

If the 3 volts applied to the field winding is considered an input, and the 300 volts taken from the brushes is an output, there is a gain of 100. Gain is expressed as the ratio of output to input:

$$\text{Gain} = \frac{\text{Output}}{\text{Input}}$$

In this case 300V ÷ 3V = 100. This means that the 3 volts output is 100 times larger than the input.

The following paragraphs explain how gain is achieved in a typical dc generator and how the modifications making the generator an amplidyne increase the gain to as high as 10,000. The schematic a separately excited dc generator. Because of the 10- volt controlling voltage, 10 amperes of current will flow through the 1-ohm field winding. This draws 100 watts of input power (P = IE).

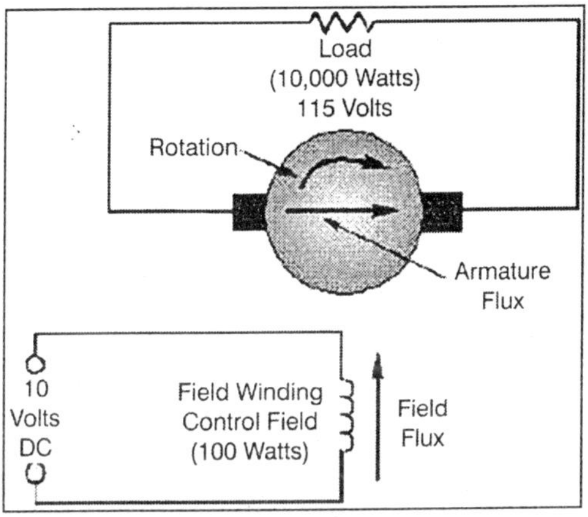

Fig. Ordinary dc Generator.

Assume that the characteristics of this generator enable it to produce approximately 87 amperes of armature current at 115 volts at the output terminals. This represents an output power of approximately 10,000 watts (P = IE). You can see that the power gain of this generator is 100. In effect, 100 watts controls 10,000 watts.

An amplidyne is a special type of dc generator. The following changes, for explanation purposes, will convert the typical dc generator above into an amplidyne. The first step is to short the brushes together. This removes nearly all of the resistance in the armature circuit. Because of the very low resistance in the armature circuit, a much lower control-field flux produces full-load armature current (full-load current in the armature is still about 87 amperes). The smaller control field now requires a control voltage of only 1 volt and an input power of 1 watt (1 volt across 1 ohm causes 1 ampere of current, which produces 1 watt of input power).

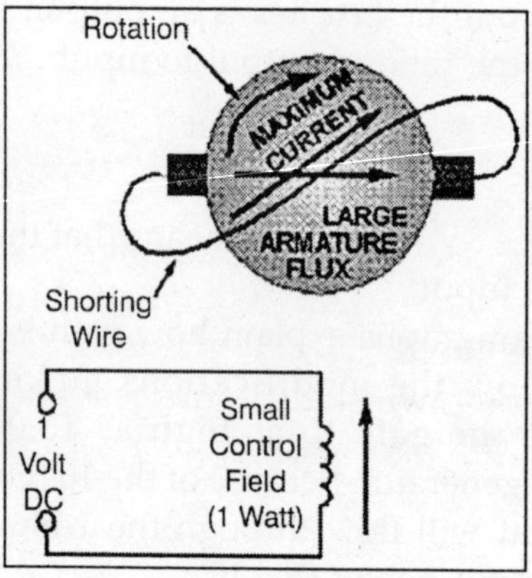

Fig. Brushes Shorted in a dc Generator.

The next step is to add another set of brushes. These now become the output brushes of the amplidyne. They are placed against the commutator in a position perpendicular to the original brushes. The shorted brushes are now called the "quadrature" brushes. This is because they are in quadrature (perpendicular) to the output brushes. The output brushes are in line with the armature flux.

Therefore, they pick off the voltage induced in the armature windings at this point. The voltage at the output will be the same as in the original generator, 115 volts in our example.

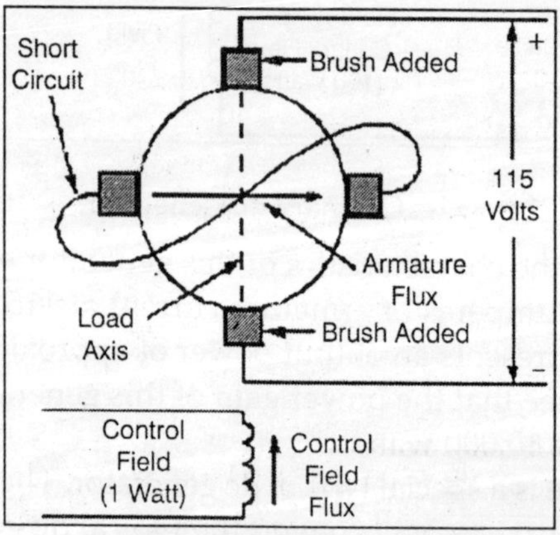

Fig. Amplidyne Load Brushes.

As you have seen, the original generator produced a 10,000-watt output with a 100-watt input. The amplidyne produces the same 10,000-watt output with only a 1-watt input. This represents a gain of 10,000. The gain of the original generator has been greatly increased.

An amplidyne is used to provide large dc currents. The primary use of an amplidyne is in the positioning of heavy loads through the use of synchro/servo systems. Synchro/servo systems will be studied in a later module.

Assume that a very large turning force is required to rotate a heavy object, such as an antenna, to a very precise position. A low-power, relatively weak voltage representing the amount of antenna rotation required can be used to control the field winding of an amplidyne.

Because of the amplidyne's ability to amplify, its output can be used to drive a powerful motor, which turns the heavy object (antenna). When the source of the input voltage senses the correct movement of the object, it drops the voltage to zero.

The field is no longer strong enough to allow an output voltage to be developed, so the motor ceases to drive the object (antenna).

The above is an oversimplification and is not meant to describe a functioning system. The intent is to show a typical sequence of events between the demand for movement and the movement itself. It is meant to strengthen the idea that with the amplidyne, something large and heavy can be controlled very precisely by something very small, almost insignificant.

SAFETY PRECAUTIONS

You must always observe safety precautions when working around electrical equipment to avoid injury to personnel and damage to equipment. Electrical equipment frequently has accessories that require separate sources of power. Lighting fixtures, heaters, externally powered temperature detectors, and alarm systems are examples of accessories whose terminals must be deenergized.

When working on dc generators, you must check to ensure that all such circuits have been de-energized and tagged before you attempt any maintenance or repair work. You must also use the greatest care when working on or near the output terminals of dc generators.

DIRECT CURRENT MOTORS

The dc motor is a mechanical workhorse, that can be used in many different ways. Many large pieces of equipment depend on a dc motor for their power to move. The speed and direction of rotation of a dc motor are easily controlled. This makes it especially useful for operating equipment, such as winches, cranes, and missile launchers, which must move in different directions and at varying speeds.

PRINCIPLES OF OPERATION

The operation of a dc motor is based on the following principle: A current-carrying conductor placed in a magnetic field, perpendicular to the lines of flux, tends to move in a direction perpendicular to the magnetic lines of flux. There is a definite relationship between the direction of the magnetic field, the direction of current in the conductor, and the direction in which the conductor tends to move. This relationship is best explained by using the *right-hand rule for motors*.

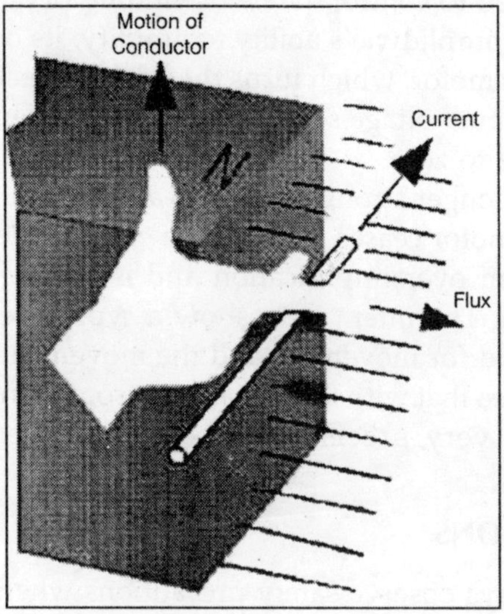

Fig. Right-hand Rule for Motors.

To find the direction of motion of a conductor, extend the thumb, forefinger, and middle finger of your right hand so they are at right angles to each other. If the forefinger is pointed in the direction of magnetic flux (north to south) and the middle finger is pointed in the direction of current flow in the conductor, the thumb will point in the direction the conductor will move.

Stated very simply, a dc motor rotates as a result of two magnetic fields interacting with each other. The armature of a dc motor acts like an electromagnet when current flows through its coils. Since the armature is located within the magnetic field of the field poles, these two magnetic fields interact.

Like magnetic poles repel each other, and unlike magnetic poles attract each other. As in the dc generator, the dc motor has field poles that are stationary and an armature that turns on bearings in the space between the field poles. The armature of a dc motor has windings on it just like the armature of a dc generator. These windings are also connected to commutator segments. A dc motor consists of the same components as a dc generator. In fact, most dc generators can be made to act as motors, and vice versa.

Look at the simple dc motor. It has two field poles, one a north pole and one a south pole. The magnetic lines of force extend across the opening between the poles from north to south.

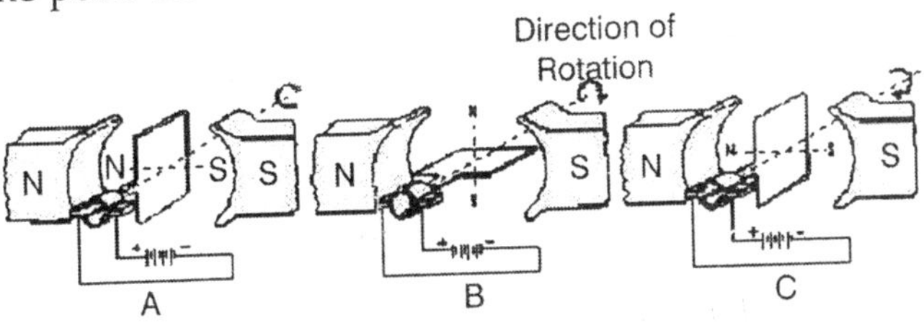

Fig. Dc motor Armature Rotation.

The armature in this simple dc motor is a single loop of wire. The loop of wire in the dc motor, however, has current flowing through it from an external source.

This current causes a magnetic field to be produced. This field is indicated by the dotted line through the loops. The loop (armature) field is both attracted and repelled by the field from the field poles. Since the current through the loop goes around in the direction of the arrows, the north pole of the armature is at the upper left, and the south pole of the armature is at the lower right. Of course, as the loop (armature) turns, these magnetic poles turn with it.

Now, the north armature pole is repelled from the north field pole and attracted to the right by the south field pole. Likewise, the south armature pole is repelled from the south field pole and is attracted to the left by the north field pole. This action causes the armature to turn in a clockwise direction.

After the loop has turned far enough so that its north pole is exactly opposite the south field pole, the brushes advance to the next segments. This changes the direction of current flow through the armature loop. Also, it changes the polarity of the armature field.

The magnetic fields again repel and attract each other, and the armature continues to turn.

In this simple motor, the momentum of the rotating armature carries the armature past the position where the unlike poles are exactly lined up. However, if these fields are exactly lined up when the armature current is turned on, there is no momentum to start the armature moving. In this case, the motor would not rotate.

It would be necessary to give a motor like this a spin to start it. This disadvantage does not exist when there are more turns on the armature, because there is more than one armature field. No two armature fields could be exactly aligned with the field from the field poles at the same time.

COUNTER EMF

While a dc motor is running, it acts somewhat like a dc generator. There is a magnetic field from the field poles, and a loop of wire is turning and cutting this magnetic field. For the moment, disregard the fact that there is current flowing through the loop of wire from the battery. As the loop sides cut the magnetic field, a voltage is induced in them, the same as it was in the loop sides of the dc generator. This induced voltage causes current to flow in the loop.

Now, consider the relative direction between this current and the current that causes the motor to run. First, check the direction the current flows as a result of the generator action taking place. Using the left hand, hold it so that the forefinger points in the direction of the magnetic field (north to south) and the thumb points in the direction that the black side of the armature moves (up).

Your middle finger then points out of the paper (toward you), showing the direction of current flow caused by the generator action in the black half of the armature. This is in the direction opposite to that of the battery current. Since this generator-action voltage is opposite that of the battery, it is called "counter emf." (The letters emf stand for electromotive force, which is another name for voltage.) The two currents are flowing in opposite directions.

This proves that the battery voltage and the counter emf are opposite in polarity. We disregarded armature current while explaining how counter emf was generated. Then, we showed that normal armature current flowed opposite to the current created by the counter emf. We talked about two opposite currents that flow at the same time. However, this is a bit oversimplified, as you may already suspect. Actually, only one current flows.

Because the counter emf can never become as large as the applied voltage, and because they are of opposite polarity as we have seen, the counter emf effectively cancels part of the armature voltage. The single current that flows is armature current, but it is greatly reduced because of the counter emf.

In a dc motor, there is always a counter emf developed. This counter emf cannot be equal to or greater than the applied battery voltage; if it were, the motor would not run.

The counter emf is always a little less. However, the counter emf opposes the applied voltage enough to keep the armature current from the battery to a fairly low value. If there were no such thing as counter emf, much more current would flow through the armature, and the motor would run much faster. However, there is no way to avoid the counter emf.

MOTOR LOADS

Motors are used to turn mechanical devices, such as water pumps, grinding wheels, fan blades, and circular saws. For example, when a motor is turning a water pump, the water pump is the load. The water pump is the mechanical device that the motor must move. This is the definition of a motor load. As with electrical loads, the mechanical load connected to a dc motor affects many electrical quantities.

Such things as the power drawn from the line, amount of current, speed, efficiency, etc., are all partially controlled by the size of the load. The physical and electrical characteristics of the motor must be matched to the requirements of the load if the work is to be done without the possibility of damage to either the load or the motor.

PRACTICAL DC MOTORS

As you have seen, dc motors are electrically identical to dc generators. In fact, the same dc machine may be driven mechanically to generate a voltage, or it may be driven electrically to move a mechanical load. While this is not normally done, it does point out the similarities between the two machines. These similarities will be used to introduce you to practical dc motors. You will immediately recognize series, shunt, and compound types of motors as being directly related to their generator counterparts.

SERIES DC MOTOR

In a series dc motor, the field is connected in series with the armature. The field is wound with a few turns of large wire, because it must carry full armature current.

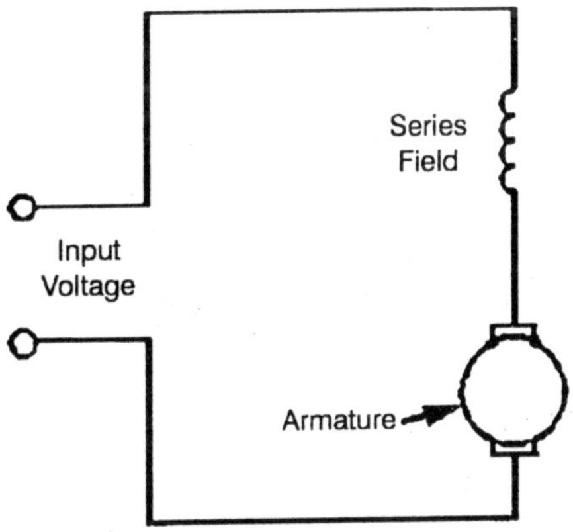

Fig. Series-wound dc Motor.

This type of motor develops a very large amount of turning force, called torque, from a standstill. Because of this characteristic, the series dc motor can be used to operate small electric appliances, portable electric tools, cranes, winches, hoists, and the like. Another characteristic is that the speed varies widely between no-load and full-load. Series motors cannot be used where a relatively constant speed is required under conditions of varying load.

A major disadvantage of the series motor is related to the speed characteristic mentioned in the last paragraph. The speed of a series motor with no load connected to it increases to the point where the motor may become damaged.

Usually, either the bearings are damaged or the windings fly out of the slots in the armature. There is a danger to both equipment and personnel. Some load must ALWAYS be connected to a series motor before you turn it on. This precaution is primarily for large motors. Small motors, such as those used in electric hand drills, have enough internal friction to load themselves.

SHUNT MOTOR

A shunt motor is connected in the same way as a shunt generator. The field windings are connected in parallel (shunt) with the armature windings.

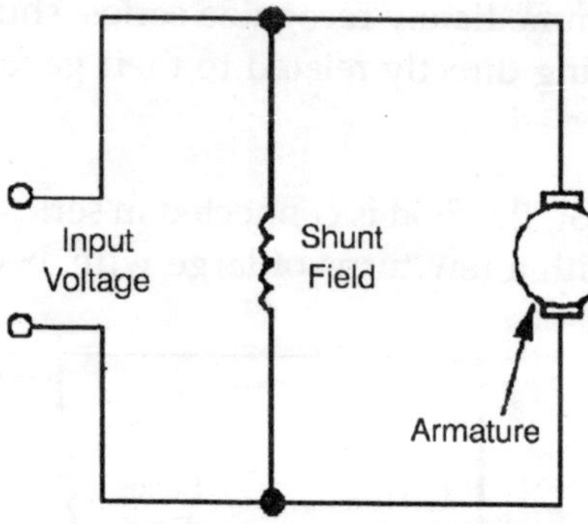

Fig. Shunt-wound dc Motor.

Once you adjust the speed of a dc shunt motor, the speed remains relatively constant even under changing load conditions. One reason for this is that the field flux remains constant. A constant voltage across the field makes the field independent of variations in the armature circuit.

If the load on the motor is increased, the motor tends to slow down. When this happens, the counter emf generated in the armature decreases. This causes a corresponding decrease in the opposition to battery current flow through the armature.

Armature current increases, causing the motor to speed up. The conditions that established the original speed are reestablished, and the original speed is maintained. Conversely, if the motor load is decreased, the motor tends to increase speed; counter emf increases, armature current decreases, and the speed decreases. In each case, all of this happens so rapidly that any actual change in speed is slight. There is instantaneous tendency to change rather than a large fluctuation in speed.

COMPOUND MOTOR

A compound motor has two field windings. One is a shunt field connected in parallel with the armature; the other is a series field that is connected in series with the armature.

The shunt field gives this type of motor the constant speed advantage of a regular shunt motor. The series field gives it the advantage of being able to develop a large torque when the motor is started under a heavy load. It should not be a surprise that the compound motor has both shunt- and series-motor characteristics.

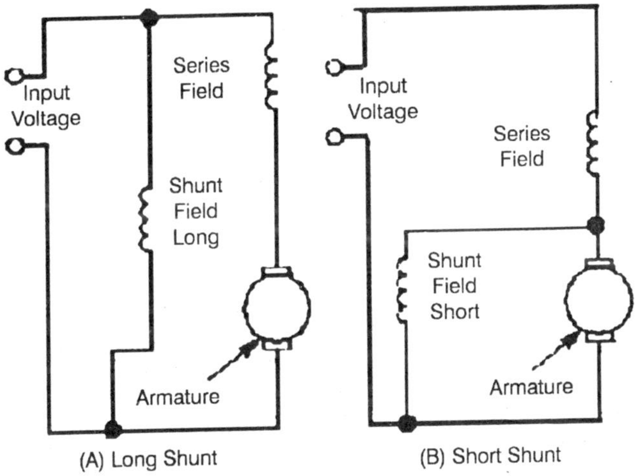

Fig. Compound-wound dc Motor.

When the shunt field is connected in parallel with the series field and armature, it is called a "long shunt", (view A). Otherwise, it is called a "short shunt", (view B).

TYPES OF ARMATURES

As with dc generators, dc motors can be constructed using one of two types of armatures. A brief review of the Gramme-ring and drum-wound armatures is necessary to emphasize the similarities between dc generators and dc motors.

GRAMME-RING ARMATURE

The Gramme-ring armature is constructed by winding an insulated wire around a soft-iron ring. Eight equally spaced connections are made to the winding.

Each of these is connected to a commutator segment. The brushes touch only the top and bottom segments. There are two parallel paths for current to follow — one up the left side and one up the right side. These paths are completed through the top brush back to the positive lead of the battery.

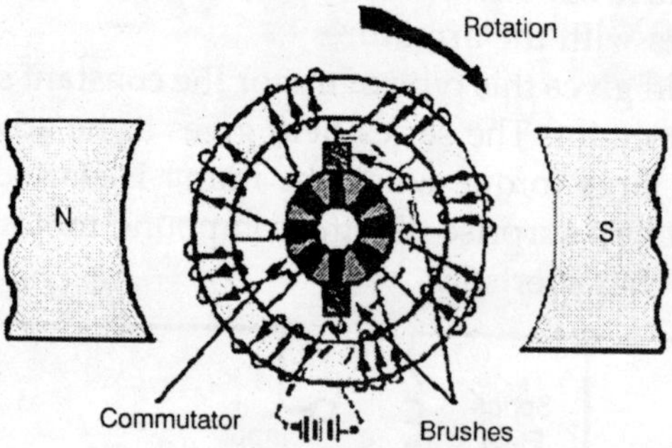

Fig. Gramme-ring Armature.

To check the direction of rotation of this armature, you should use the right-hand rule for motors. Hold your thumb, forefinger, and middle finger at right angles. Point your forefinger in the direction of field flux; in this case, from left to right.

Now turn your wrist so that your middle finger points in the direction that the current flows in the winding on the outside of the ring. Note that current flows into the page (away from you) in the left-hand windings and out of the page (toward you) in the right-hand windings. Your thumb now points in the direction that the winding will move.

The Gramme-ring armature is seldom used in modem dc motors. The windings on the inside of the ring are shielded from magnetic flux, which causes this type of armature to be inefficient.

DRUM-WOUND ARMATURE

The drum-wound armature is generally used in ac motors. It is identical to the drum winding. If the drum-wound armature were cut in half, an end view at the cut would resemble, (view A), (view B) is a side view of the armature and pole pieces.

Notice that the length of each conductor is positioned parallel to the faces of the pole pieces.

Therefore, each conductor of the armature can cut the maximum flux of the motor field. The inefficiency of the Gramme-ring armature is overcome by this positioning.

The direction of current flow is marked in each conductor in figure, (view A) as though the armature were turning in a magnetic field. The dots show that current is flowing toward you on the left side, and the crosses show that the current is flowing away from you on the right side. Strips of insulation are inserted in the slots to keep windings in place when the armature spins.

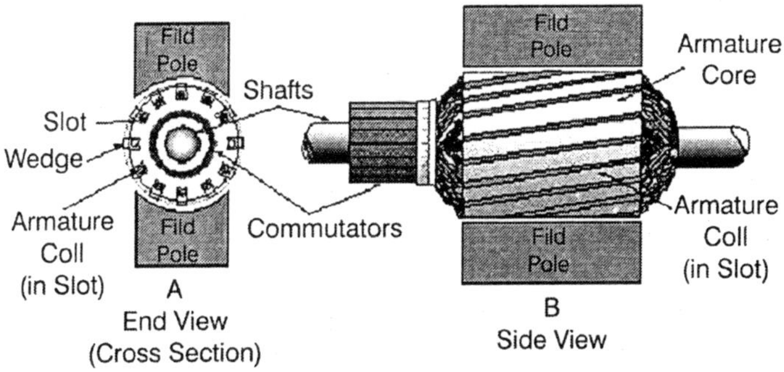

Fig. Drum-type Armature.

DIRECTION OF ROTATION

The direction of rotation of a dc motor depends on the direction of the magnetic field and the direction of current flow in the armature. If either the direction of the field or the direction of current flow through the armature is reversed, the rotation of the motor will reverse. However, if both of these factors are reversed at the same time, the motor will continue rotating in the same direction. In actual practice, the field excitation voltage is reversed in order to reverse motor direction.

Ordinarily, a motor is set up to do a particular job that requires a fixed direction of rotation.

However, there are times when it is necessary to change the direction of rotation, such as a drive motor for a gun turret or missile launcher. Each of these must be able to move in both directions. Remember, the connections of either the armature or the field must be reversed, but not both. In such applications, the proper connections are brought out to a reversing switch.

MOTOR SPEED

A motor whose speed can be controlled is called a variable-speed motor; dc motors are variable- speed motors. The speed of a dc motor is changed by changing the current in the field or by changing the current in the armature.

When the field current is decreased, the field flux is reduced, and the counter emf decreases.

This permits more armature current. Therefore, the motor speeds up. When the field current is increased, the field flux is increased. More counter emf is developed, which opposes the armature current. The armature current then decreases, and the motor slows down. When the voltage applied to the armature is decreased, the armature current is decreased, and the motor again slows down. When the armature voltage and current are both increased, the motor speeds up.

In a shunt motor, speed is usually controlled by a rheostat connected in series with the field windings.

When the resistance of the rheostat is increased, the current through the field winding is decreased. The decreased flux momentarily decreases the counter emf.

The motor then speeds up, and the increase in counter emf keeps the armature current the same. In a similar manner, a decrease in rheostat resistance increases the current flow through the field windings and causes the motor to slow down.

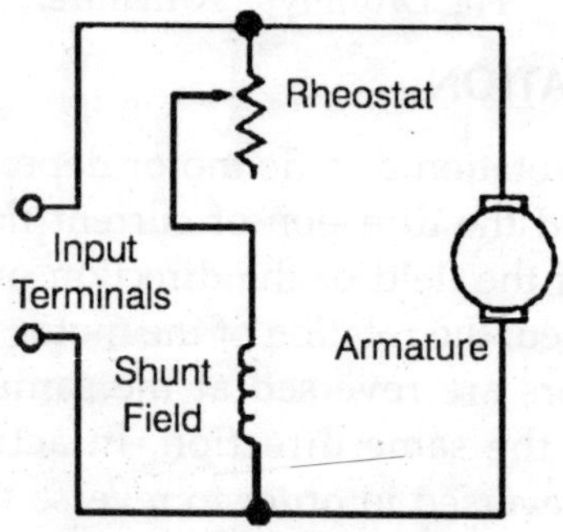

Fig. Controlling Motor Speed.

In a series motor, the rheostat speed control may be connected either in parallel or in series with the armature windings. In either case, moving the rheostat in a direction that lowers the voltage across the armature lowers the current through the armature and slows the motor. Moving the rheostat in a direction that increases the voltage and current through the armature increases motor speed.

ARMATURE REACTION

The reasons for armature reaction and the methods of compensating for its effects are basically the same for dc motors as for dc generators. Figure reiterates for you the distorting effect that the armature field has on the flux between the pole pieces.

Notice, however, that the effect has shifted the neutral plane backward, against the direction of rotation. This is different from a dc generator, where the neutral plane shifted forward in the direction of rotation. As before, the brushes must be shifted to the new neutral plane. The shift is counterclockwise. Again, the proper location is reached when there is no sparking from the brushes.

Compensating windings and interpoles, two more "old" subjects, cancel armature reaction in dc motors. Shifting brushes reduces sparking, but it also makes the field less effective. Canceling armature reaction eliminates the need to shift brushes in the first place. Compensating windings and interpoles are as important in motors as they are in generators.

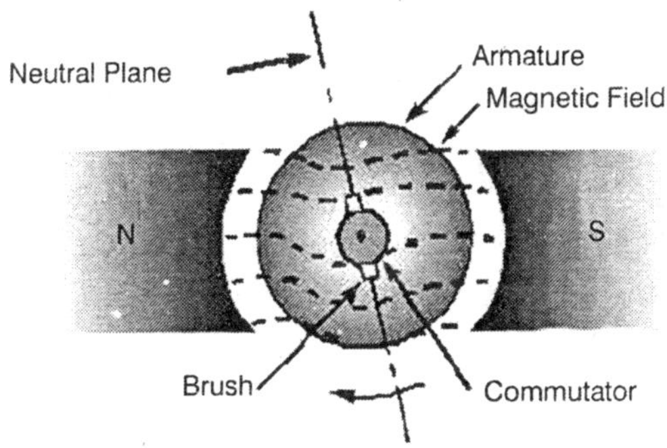

Fig. Armature Reaction.

Compensating windings are relatively expensive; therefore, most large dc motors depend on interpoles to correct armature reaction. Compensating windings are the same in motors as they are in generators. Interpoles, however, are slightly different. The difference is that in a generator the interpole has the same polarity as the main pole Ahead of it in the direction of rotation. In a motor the interpole has the same polarity as the main pole following it.

The interpole coil in a motor is connected to carry the armature current the same as in a generator.

As the load varies, the interpole flux varies, and commutation is automatically corrected as the load changes. It is not necessary to shift the brushes when there is an increase or decrease in load. The brushes are located on the no-load neutral plane. They remain in that position for all conditions of load.

The dc motor is reversed by reversing the direction of the current in the armature. When the armature current is reversed, the current through the interpole is also reversed. Therefore, the interpole still has the proper polarity to provide automatic commutation.

MANUAL AND AUTOMATIC STARTERS

Because the dc resistance of most motor armatures is low (0.05 to 0.5

ohm), and because the counter emf does not exist until the armature begins to turn, it is necessary to use an external starting resistance in series with the armature of a dc motor to keep the initial armature current to a safe value. As the armature begins to turn, counter emf increases; and, since the counter emf opposes the applied voltage, the armature current is reduced.

The external resistance in series with the armature is decreased or eliminated as the motor comes up to normal speed and full voltage is applied across the armature.

Controlling the starting resistance in a dc motor is accomplished either manually, by an operator, or by any of several automatic devices. The automatic devices are usually just switches controlled by motor speed sensors. Automatic starters are not covered in detail in this module.

ALTERNATING CURRENT GENERATORS

Most of the electrical power used aboard Navy ships and aircraft as well as in civilian applications is ac. As a result, the ac generator is the most important means of producing electrical power. Ac generators, generally called alternators, vary greatly in size depending upon the load to which they supply power.

For example, the alternators in use at hydroelectric plants, such as Hoover Dam, are tremendous in size, generating thousands of kilowatts at very high voltage levels. Another example is the alternator in a typical automobile, which is very small by comparison. It weighs only a few pounds and produces between 100 and 200 watts of power, usually at a potential of 12 volts.

Basic AC Generators Regardless of size, all electrical generators, whether dc or ac, depend upon the principle of magnetic induction. An emf is induced in a coil as a result of:

- A coil cutting through a magnetic field, or
- A magnetic field cutting through a coil. As long as there is relative motion between a conductor and a magnetic field, a voltage will be induced in the conductor.

That part of a generator that produces the magnetic field is called the field.

That part in which the voltage is induced is called the armature. For relative motion to take place between the conductor and the magnetic field, all generators must have two mechanical parts — a rotor and a stator. The ROTor is the part that ROTates; the STATor is the part that remains STATionary.

In a dc generator, the armature is always the rotor. In alternators, the armature may be either the rotor or stator.

ROTATING-ARMATURE ALTERNATORS

The rotating-armature alternator is similar in construction to the dc generator in that the armature rotates in a stationary magnetic field, view A. In the dc generator, the emf generated in the armature windings is converted from ac to dc by means of the commutator.

In the alternator, the generated ac is brought to the load unchanged by means of slip rings. The rotating armature is found only in alternators of low power rating and generally is not used to supply electric power in large quantities.

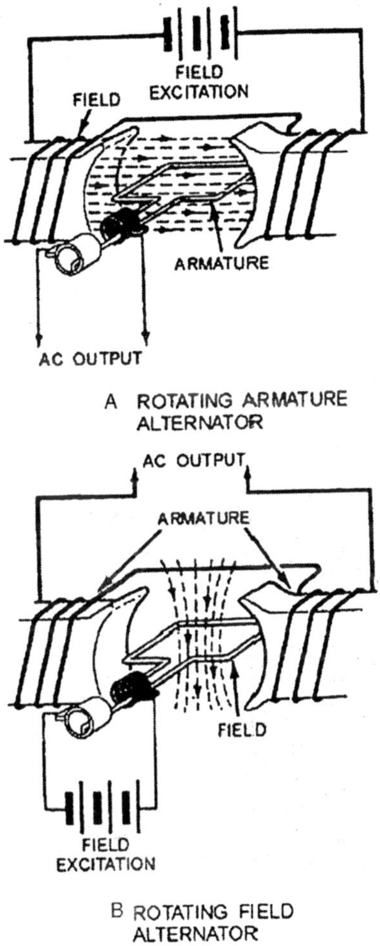

Fig. Types of ac Generators.

ROTATING-FIELD ALTERNATORS

The rotating-field alternator has a stationary armature winding and a rotating-field winding, view B The advantage of having a stationary armature winding is that the generated voltage can be connected directly to the load.

A rotating armature requires slip rings and brushes to conduct the current from the armature to the load. The armature, brushes, and slip rings are difficult to insulate, and arc-overs and short circuits can result at high voltages.

For this reason, high-voltage alternators are usually of the rotating-field type. Since the voltage applied to the rotating field is low voltage dc, the problem of high voltage arc-over at the slip rings does not exist.

The stationary armature, or stator, of this type of alternator holds the windings that are cut by the rotating magnetic field.

The voltage generated in the armature as a result of this cutting action is the ac power that will be applied to the load.

The stators of all rotating-field alternators are about the same. The stator consists of a laminated iron core with the armature windings embedded in this core. The core is secured to the stator frame.

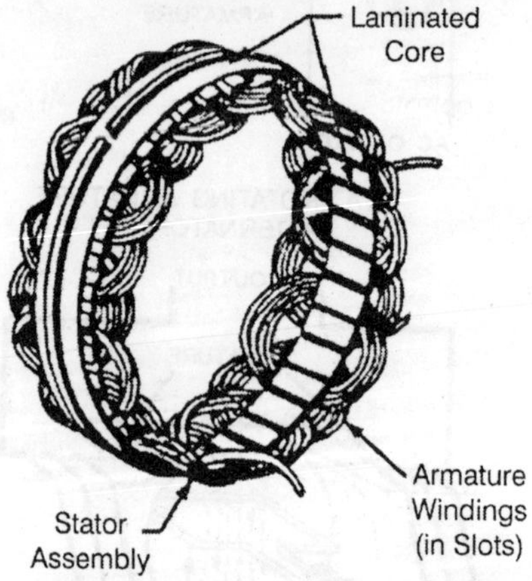

Fig. Stationary Armature Windings.

PRACTICAL ALTERNATORS

The alternators described so far are elementary in nature; they are seldom used except as examples to aid in understanding practical alternators.

The principles of the elementary alternator to the alternators actually in use in the civilian community, as well as aboard Navy ships and aircraft. The following paragraphs will introduce such concepts as prime movers, field excitation, armature characteristics and limitations, single-phase and polyphase alternators, controls, regulation, and parallel operation.

FUNCTIONS OF ALTERNATOR COMPONENTS

A typical rotating-field ac generator consists of an alternator and a smaller dc generator built into a single unit. The output of the alternator section supplies alternating voltage to the load. The only purpose for the dc exciter generator is to supply the direct current required to maintain the alternator field. This dc generator is referred to as the exciter. A typical alternator is shown in figure, view A; figure, view B, is a simplified schematic of the generator.

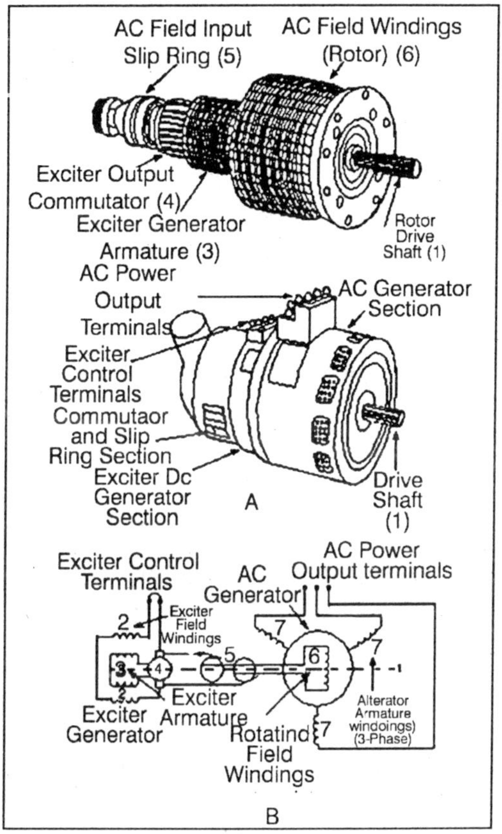

Fig. Ac Generator Pictorial and Schematic Drawings.

- The exciter is a dc, shunt-wound, self-excited generator. The exciter shunt field.
- Creates an area of intense magnetic flux between its poles. When the exciter armature.
- Is rotated in the exciter-field flux, voltage is induced in the exciter armature windings. The output from the exciter commutator
- Is connected through brushes and slip rings.

- To the alternator field. Since this is direct current already converted by the exciter commutator, the current always flows in one direction through the alternator field.
- Thus, a fixed-polarity magnetic field is maintained at all times in the alternator field windings. When the alternator field is rotated, its magnetic flux is passed through and across the alternator armature windings.

The armature is wound for a three-phase output. Remember, a voltage is induced in a conductor if it is stationary and a magnetic field is passed across the conductor, the same as if the field is stationary and the conductor is moved. The alternating voltage in the ac generator armature windings is connected through fixed terminals to the ac load.

PRIME MOVERS

All generators, large and small, ac and dc, require a source of mechanical power to turn their rotors. This source of mechanical energy is called a prime mover. Prime movers are divided into two classes for generators-high-speed and low-speed. Steam and gas turbines are high-speed prime movers, while internal-combustion engines, water, and electric motors are considered low-speed prime movers.

The type of prime mover plays an important part in the design of alternators since the speed at which the rotor is turned determines certain characteristics of alternator construction and operation.

ALTERNATOR ROTORS

There are two types of rotors used in rotating-field alternators. They are called the turbine-driven and salient-pole rotors.

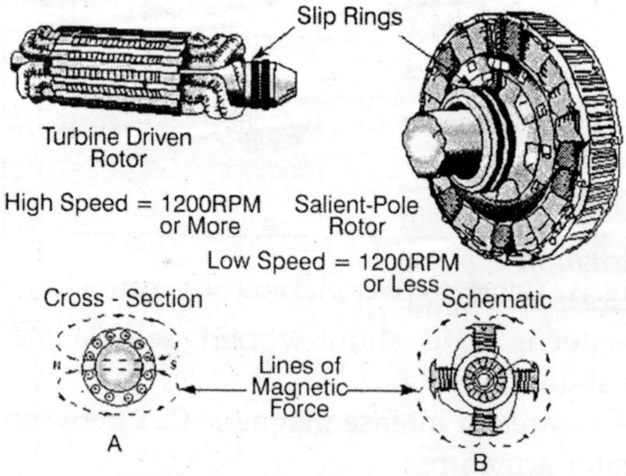

Fig. Types of Rotors used in Alternators.

As you may have guessed, the turbine-driven rotor, view A, is used when the prime mover is a high-speed turbine. The windings in the turbine-driven rotor are arranged to form two or four distinct poles. The windings are firmly embedded in slots to withstand the tremendous centrifugal forces encountered at high speeds. The salient-pole rotor is used in low-speed alternators. The salient-pole rotor often consists of several separately wound pole pieces, bolted to the frame of the rotor.

If you could compare the physical size of the two types of rotors with the same electrical characteristics, you would see that the salient-pole rotor would have a greater diameter. At the same number of revolutions per minute, it has a greater centrifugal force than does the turbine-driven rotor. To reduce this force to a safe level so that the windings will not be thrown out of the machine, the salient pole is used only in low-speed designs.

ALTERNATOR CHARACTERISTICS AND LIMITATIONS

Alternators are rated according to the voltage they are designed to produce and the maximum current they are capable of providing. The maximum current that can be supplied by an alternator depends upon the maximum heating loss that can be sustained in the armature. This heating loss (which is an I^2R power loss) acts to heat the conductors, and if excessive, destroys the insulation.

Thus, alternators are rated in terms of this current and in terms of the voltage output — the alternator rating in small units is in volt- amperes; in large units it is kilovolt-amperes. When an alternator leaves the factory, it is already destined to do a very specific job. The speed at which it is designed to rotate, the voltage it will produce, the current limits, and other operating characteristics are built in. This information is usually stamped on a nameplate on the case so that the user will know the limitations.

SINGLE-PHASE ALTERNATORS

A generator that produces a single, continuously alternating voltage is known as a *single-phase* alternator. The stator (armature) windings are connected in series. The individual voltages, therefore, add to produce a single-phase ac voltage. A basic alternator with its single-phase output voltage.

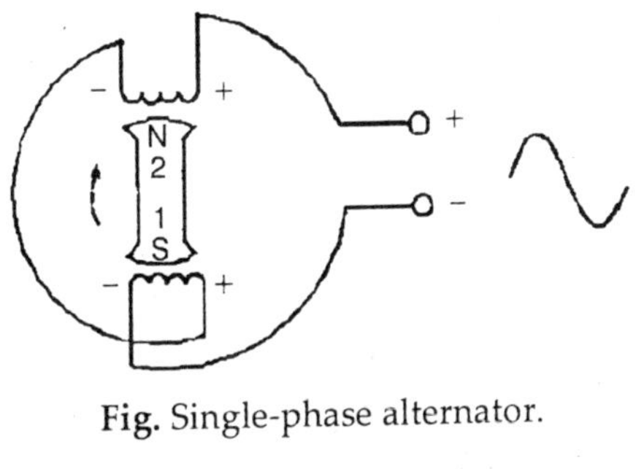

Fig. Single-phase alternator.

The definition of phase as you learned it in studying ac circuits may not help too much right here. Remember, "out of phase" meant "out of time." Now, it may be easier to think of the word phase as meaning voltage as in single voltage. The need for a modified definition of phase in this usage will be easier to see as we go along.

Single-phase alternators are found in many applications. They are most often used when the loads being driven are relatively light. The reason for this will be more apparent as we get into multiphase alternators (also called polyphase).

Power that is used in homes, shops, and ships to operate portable tools and small appliances is single-phase power. Single-phase power alternators always generate single-phase power. However, all single-phase power does not come from single-phase alternators. This will sound more reasonable to you as we get into the next subjects.

TWO-PHASE ALTERNATORS

Two phase implies two voltages if we apply our new definition of phase. And, it's that simple. A two-phase alternator is designed to produce two completely separate voltages. Each voltage, by itself, may be considered as a single-phase voltage. Each is generated completely independent of the other. Certain advantages are gained. These and the mechanics of generation will be covered in the following paragraphs.

Generation of Two-Phase Power

A simplified two-pole, two-phase alternator. Note that the windings of the two phases are physically at right angles (90°) to each other. You would expect the outputs of each phase to be 90° apart, which they are. The graph shows the two phases to be 90° apart, with A leading B. Note that by using our original definition of phase (from previous modules), we could say that A and B are 90° out of phase. There will always be 90° between the phases of a two-phase alternator. This is by design.

Now, let's go back and see the similarities and differences between our original (single-phase) alternators and this new one (two-phase). Note that the principles applied are not new.

The stator consists of two single-phase windings completely separated from each other. Each winding is made up of two windings that are connected in series so that their voltages add. The rotor is identical to that used in the single-phase alternator. In the left-hand schematic, the rotor poles are opposite all the windings of phase A.

Therefore, the voltage induced in phase A is maximum, and the voltage induced in phase B is zero. As the rotor continues rotating counterclockwise, it moves away from the A windings and approaches the B windings. As a result, the voltage induced in phase A decreases from its maximum value, and the voltage induced in phase B increases from zero.

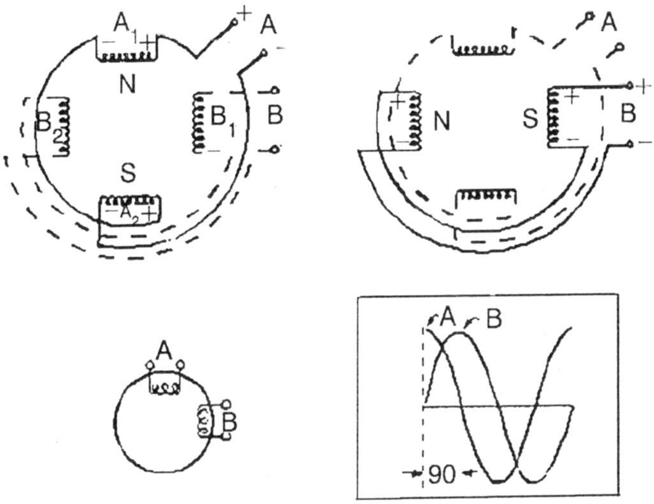

Fig. Two-phase Alternator.

In the right-hand schematic, the rotor poles are opposite the windings of phase B. Now the voltage induced in phase B is maximum, whereas the voltage induced in phase A has dropped to zero. Notice that a 90-degree rotation of the rotor corresponds to one-quarter of a cycle, or 90 electrical degrees. The waveform picture shows the voltages induced in phase A and B for one cycle.

The two voltages are 90° out of phase. Notice that the two outputs, A and B, are independent of each other. Each output is a single-phase voltage, just as if the other did not exist.

The obvious advantage, so far, is that we have two separate output voltages. There is some saving in having one set of bearings, one rotor, one housing, and so on, to do the work of two. There is the disadvantage of having twice as many stator coils, which require a larger and more complex stator.

The large schematic shows four separate wires brought out from the A and B stator windings. Notice, however, that the dotted wire now connects one end of B1 to one end of A2. The effect of making this connection is to provide a new output voltage. This sine- wave voltage, C in the picture, is larger than either A or B. It is the result of adding the instantaneous values of phase A and phase B. For this reason it appears exactly half way between A and B. Therefore, C must lag A by 45° and lead B by 45°.

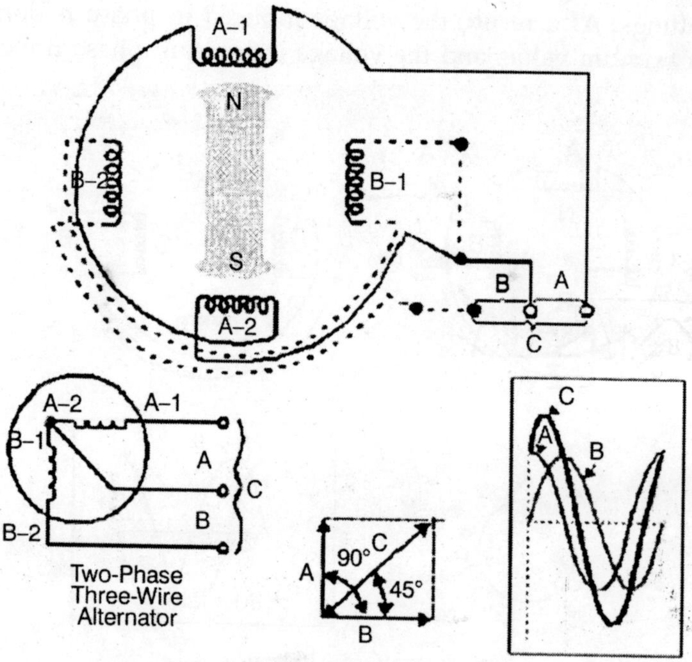

Fig. Connections of a Two-phase, Three-wire Alternator Output.

Only three connections have been brought out from the stator. Electrically, this is the same as the large diagram above it. Instead of being connected at the output terminals, the B1-A2 connection was made internally when the stator was wired.

A two- phase alternator connected in this manner is called a two-phase, three-wire alternator.

The three-wire connection makes possible three different load connections: A and B (across each phase), and C (across both phases). The output at C is always 1.414 times the voltage of either phase. These multiple outputs are additional advantages of the two-phase alternator over the single-phase type.

Now, you can understand why single-phase power doesn't always come from single-phase alternators. It can be generated by two-phase alternators as well as other multiphase (polyphase) alternators. However, the operation of polyphase alternators is more easily explained using two phases than three phases.

THREE-PHASE ALTERNATOR

The three-phase alternator, as the name implies, has three single-phase windings spaced such that the voltage induced in any one phase is displaced by 120° from the other two. A schematic diagram of a three-phase stator showing all the coils becomes complex, and it is difficult to see what is actually happening. The windings of each phase lumped together as one winding. The rotor is omitted for simplicity. The voltage waveforms generated across each phase are drawn on a graph, phase-displaced 120° from each other. The three-phase alternator as shown in this schematic is made up of three single-phase alternators whose generated voltages are out of phase by 120°. The three phases are independent of each other.

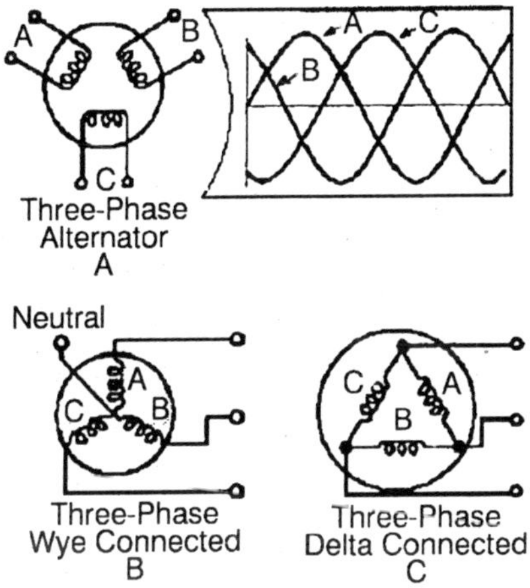

Fig. Three-phase Alternator Connections.

Rather than having six leads coming out of the three-phase alternator, the same leads from each phase may be connected together to form a wye (Y) connection. It is called a wye connection because, without the neutral, the windings appear as the letter Y, in this case sideways or upside down.

The neutral connection is brought out to a terminal when a single-phase load must be supplied. Single-phase voltage is available from neutral to A, neutral to B, and neutral to C.

In a three-phase, Y-connected alternator, the total voltage, or line voltage, across any two of the three line leads is the vector sum of the individual phase voltages.

Each line voltage is 1.73 times one of the phase voltages. Because the windings form only one path for current flow between phases, the line and phase currents are the same (equal).

A three-phase stator can also be connected so that the phases are connected end-to-end; it is now delta connected. (Delta because it looks like the Greek letter delta, D.)

In the delta connection, line voltages are equal to phase voltages, but each line current is equal to 1.73 times the phase current. Both the wye and the delta connections are used in alternators. The majority of all alternators in use in the Navy today are three-phase machines. They are much more efficient than either two-phase or single-phase alternators.

Three-Phase Connections

The stator coils of three-phase alternators may be joined together in either wye or delta connections. With these connections only three wires come out of the alternator. This allows convenient connection to three-phase motors or power distribution transformers. It is necessary to use three-phase transformers or their electrical equivalent with this type of system.

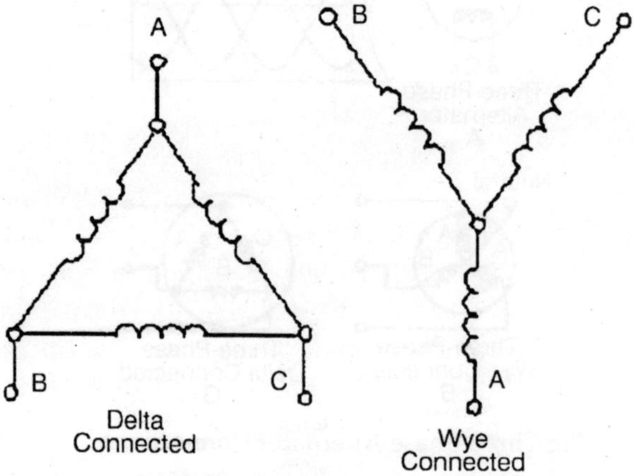

Fig. Three-phase Alternator or Transformer Connections.

A three-phase transformer may be made up of three, single-phase transformers connected in delta, wye, or a combination of both. If both the primary and secondary are connected in wye, the transformer is called a wye-wye. If both windings are connected in delta, the transformer is called a delta-delta.

Single-phase transformers connected delta-delta for operation in a three-phase system.

You will note that the transformer windings are not angled to illustrate the typical delta (D) as has been done with alternator windings. Physically, each transformer in the diagram stands alone. There is no angular relationship between the windings of the individual transformers. However, if you follow the connections, you will see that they form an electrical delta.

The primary windings, for example, are connected to each other to form a closed loop. Each of these junctions is fed with a phase voltage from a three-phase alternator. The alternator may be connected either delta or wye depending on load and voltage requirements, and the design of the system.

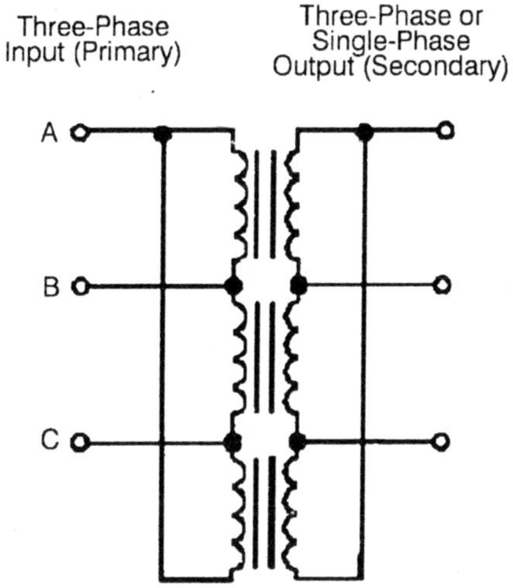

Fig. Three Single-phase Transformers Connected Delta-delta.

Three single-phase transformers connected wye-wye. Again, note that the transformer windings are not angled. Electrically, a Y is formed by the connections. The lower connections of each winding are shorted together. These form the common point of the wye. The opposite end of each winding is isolated. These ends form the arms of the wye.

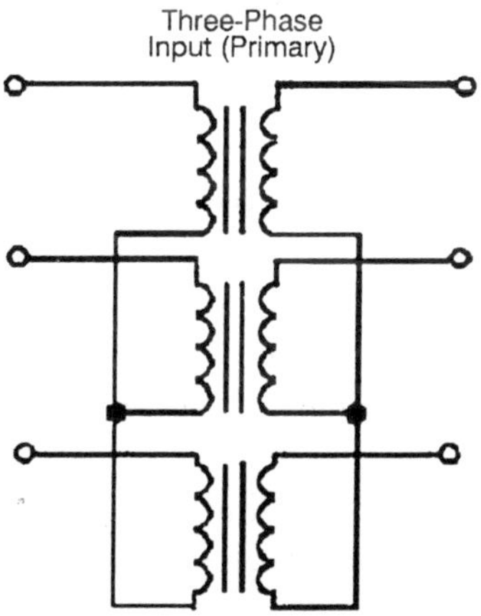

Fig. Three Single-phase Transformers Connected Wye-wye.

The ac power on most ships is distributed by a three-phase, three-wire, 450-volt system. The single- phase transformers step the voltage down to 117 volts. These transformers are connected delta-delta. With a delta-delta configuration, the load may be a three-phase device connected to all phases; or, it may be a single-phase device connected to only one phase.

At this point, it is important to remember that such a distribution system includes everything between the alternator and the load. Because of the many choices that three-phase systems provide, care must be taken to ensure that any change of connections does not provide the load with the wrong voltage or the wrong phase.

FREQUENCY

The output frequency of alternator voltage depends upon the speed of rotation of the rotor and the number of poles. The faster the speed, the higher the frequency. The lower the speed, the lower the frequency. The more poles there are on the rotor, the higher the frequency is for a given speed. When a rotor has rotated through an angle such that two adjacent rotor poles (a north and a south pole) have passed one winding, the voltage induced in that winding will have varied through one complete cycle.

For a given frequency, the more pairs of poles there are, the lower the speed of rotation. A two-pole generator must rotate at four times the speed of an eight-pole generator to produce the same frequency of generated voltage.

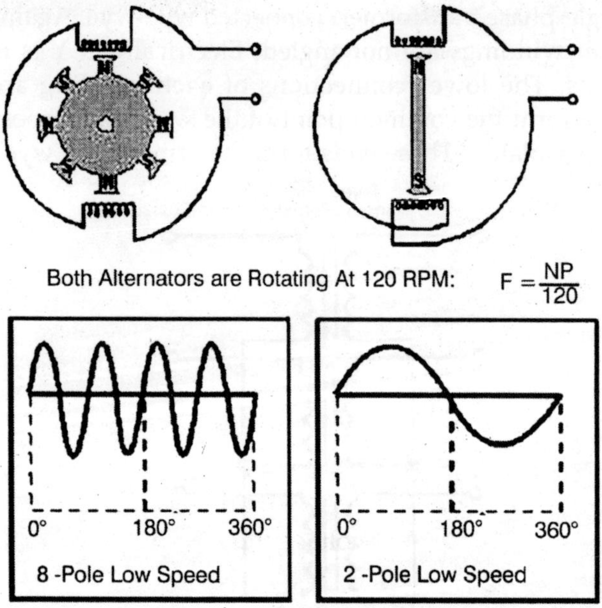

Fig. Frequency Regulation.

+The frequency of any ac generator in hertz (Hz), which is the number of cycles per second, is related to the number of poles and the speed of rotation, as expressed by the equation $F = \frac{NP}{120}$ where P is the number of poles, N is the speed of rotation in revolutions per minute (rpm), and 120 is a constant to allow for the conversion of minutes to seconds and from poles to pairs of poles. For example, a 2-pole, 3600-rpm alternator has a frequency of 60 Hz; determined as follows:

$$\frac{2 \times 3600}{120} = 60\text{Hz}$$

A 4-pole, 1800-rpm generator also has a frequency of 60 Hz. A 6-pole, 500-rpm generator has a frequency of $\frac{6 \times 500}{120} = 25\text{Hz}$ A 12-pole, 4000-rpm generator has a frequency of $\frac{12 \times 4000}{120} = 400\text{Hz}$.

VOLTAGE REGULATION

As we have seen before, when the load on a generator is changed, the terminal voltage varies. The amount of variation depends on the design of the generator.

The voltage regulation of an alternator is the change of voltage from full load to no load, expressed as a percentage of full-load volts, when the speed and dc field current are held constant.

$$\frac{E_{nL} - E_{fL}}{E_{fL}} \times 100 = \text{Per cent of Regulation}$$

Assume the no-load voltage of an alternator is 250 volts and the full-load voltage is 220 volts. The per cent of regulation is

$$\frac{250 - 220}{220} \times 100 = 13.6\%.$$

Remember, the lower the per cent of regulation, the better it is in most applications.

PRINCIPLES OF AC VOLTAGE CONTROL

In an alternator, an alternating voltage is induced in the armature windings when magnetic fields of alternating polarity are passed across these windings. The amount of voltage induced in the windings depends mainly on three things:

- The number of conductors in series per winding,
- The speed (alternator rpm) at which the magnetic field cuts the winding, and
- The strength of the magnetic field.

Any of these three factors could be used to control the amount of voltage induced in the alternator windings. The number of windings, of course, is fixed when the alternator is manufactured. Also, if the output frequency is required to be of a constant value, then the speed of the rotating field must be held constant. This prevents the use of the alternator rpm as a means of controlling the voltage output.

Thus, the only practical method for obtaining voltage control is to control the strength of the rotating magnetic field. The strength of this electromagnetic field may be varied by changing the amount of current flowing through the field coil. This is accomplished by varying the amount of voltage applied across the field cod.

PARALLEL OPERATION OF ALTERNATORS

Alternators are connected in parallel to

- Increase the output capacity of a system beyond that of a single unit,
- Serve as additional reserve power for expected demands, or
- Permit shutting down one machine and cutting in a standby machine without interrupting power distribution.

When alternators are of sufficient size, and are operating at different frequencies and terminal voltages, severe damage may result if they are suddenly connected to each other through a common bus. To avoid this, the machines must be synchronized as closely as possible before connecting them together.

This may be accomplished by connecting one generator to the bus (referred to as bus generator), and then synchronizing the other (incoming generator) to it before closing the incoming generator's main power contactor. The generators are synchronized when the following conditions are set:

- Equal terminal voltages. This is obtained by adjustment of the incoming generator's field strength.
- Equal frequency. This is obtained by adjustment of the incoming generator's prime-mover speed.
- Phase voltages in proper phase relation. At this point, it is enough for you to know that the above must be accomplished to prevent damage to the machines.

ALTERNATING CURRENT MOTORS

Most of the power-generating systems, ashore and afloat, produce ac. For this reason a majority of the motors used throughout the Navy are designed to operate on ac. There are other advantages in the use of ac motors besides the wide availability of ac power. In general, ac motors cost less than dc motors.

Some types of ac motors do not use brushes and commutators. This eliminates many problems of maintenance and wear. It also eliminates the problem of dangerous sparking. An ac motor is particularly well suited for constant-speed applications. This is because its speed is determined by the frequency of the ac voltage applied to the motor terminals. The dc motor is better suited than an ac motor for some uses, such as those that require variable- speeds. An ac motor can also be made with variable speed characteristics but only within certain limits. Industry builds ac motors in different sizes, shapes, and ratings for many different types of jobs.

These motors are designed for use with either polyphase or single-phase power systems. It is not possible here to cover all aspects of the subject of ac motors.

AC motors will be divided into (1) series, (2) synchronous, and (3) induction motors. Single-phase and polyphase motors, may be considered as polyphase motors, of constant speed, whose rotors are energized with dc voltage. Induction motors, single-phase or polyphase, whose rotors are energized by induction, are the most commonly used ac motor. The series ac motor, in a sense, is a familiar type of motor. It is very similar to the dc motor and will serve as a bridge between the old and the new.

SERIES AC MOTOR

A series ac motor is the same electrically as a dc series motor. The left- hand rule for the polarity of coils.

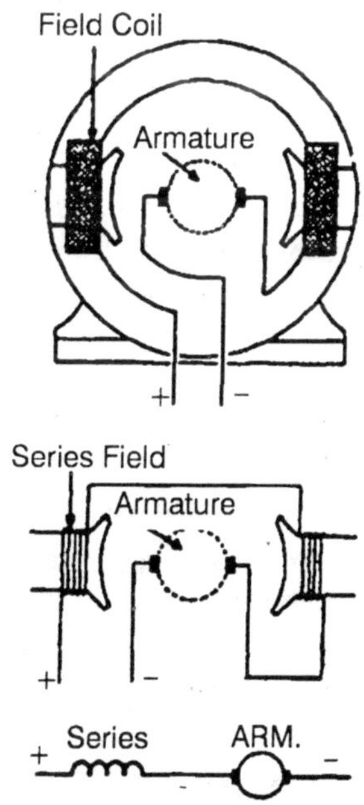

Fig. Series ac motor.

You can see that the instantaneous magnetic polarities of the armature and field oppose each other, and motor action results. Now, reverse the current by reversing the polarity of the input. Note that the field magnetic polarity still opposes the armature magnetic polarity. This is because the reversal effects both the armature and the field. The ac input causes these reversals to take place continuously.

The construction of the ac series motor differs slightly from the dc series motor. Special metals, laminations, and windings are used. They reduce losses caused by eddy currents, hysteresis, and high reactance. Dc power can be used to drive an ac series motor efficiently, but the opposite is not true.

The characteristics of a series ac motor are similar to those of a series dc motor. It is a varying-speed machine. It has low speeds for large loads and high speeds for light loads.

The starting torque is very high. Series motors are used for driving fans, electric drills, and other small appliances. Since the series ac motor has the same general characteristics as the series dc motor, a series motor has been designed that can operate both on ac and dc. This ac/dc motor is called a universal motor. It finds wide use in small electric appliances. Universal motors operate at lower efficiency than either the ac or dc series motor. They are built in small sizes only. Universal motors do not operate on polyphase ac power.

ROTATING MAGNETIC FIELDS

The principle of rotating magnetic fields is the key to the operation of most ac motors.

Both synchronous and induction types of motors rely on rotating magnetic fields in their stators to cause their rotors to turn.

The idea is simple. A magnetic field in a stator can be made to rotate electrically, around and around. Another magnetic field in the rotor can be made to chase it by being attracted and repelled by the stator field. Because the rotor is free to turn, it follows the rotating magnetic field in the stator. Let's see how it is done.

Rotating magnetic fields may be set up in two-phase or three-phase machines. To establish a rotating magnetic field in a motor stator, the number of pole pairs must be the same as (or a multiple of) the number of phases in the applied voltage. The poles must then be displaced from each other by an angle equal to the phase angle between the individual phases of the applied voltage.

TWO-PHASE ROTATING MAGNETIC FIELD

A rotating magnetic field is probably most easily seen in a two-phase stator. The stator of a two- phase induction motor is made up of two windings (or a multiple of two). They are placed at right angles to each other around the stator. The simplified drawing in figure illustrates a two-phase stator.

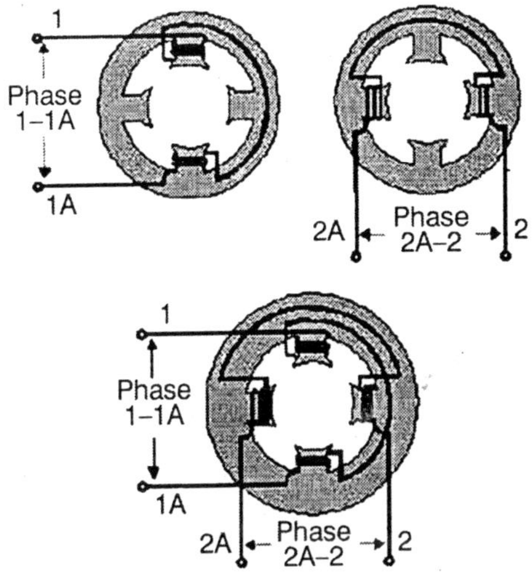

Fig. Two-phase Motor Stator.

If the voltages applied to phases 1-1A and 2-2A are 90° out of phase, the currents that flow in the phases are displaced from each other by 90°.

Since the magnetic fields generated in the coils are in phase with their respective currents, the magnetic fields are also 90° out of phase with each other. These two out-of-phase magnetic fields, whose coil axes are at right angles to each other, add together at every instant during their cycle. They produce a resultant field that rotates one revolution for each cycle of ac.

To analyse the rotating magnetic field in a two-phase stator. The arrow represents the rotor. For each point set up on the voltage chart, consider that current flows in a direction that will cause the magnetic polarity indicated at each pole piece.

Note that from one point to the next, the polarities are rotating from one pole to the next in a clockwise manner. One complete cycle of input voltage produces a 360-degree rotation of the pole polarities. Let's see how this result is obtained.

The waveforms are of the two input phases, displaced 90° because of the way they were generated in a two-phase alternator. The waveforms are numbered to match their associated phase.

The windings for the poles 1-1A and 2-2A. At position 1, the current flow and magnetic field in winding 1-1A is at maximum (because the phase voltage is maximum). The current flow and magnetic field in winding 2-2A is zero (because the phase voltage is zero).

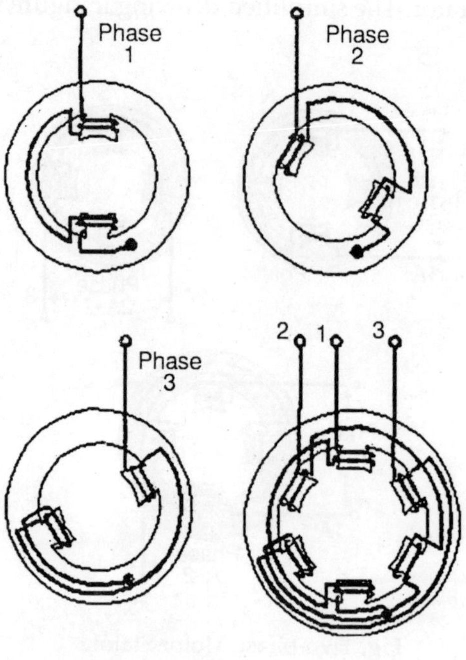

Fig. Two-phase Rotating Field.

The resultant magnetic field is therefore in the direction of the 1-1A axis. At the 45-degree point (position 2), the resultant magnetic field lies midway between windings 1-1A and 2-2A. The coil currents and magnetic fields are equal in strength. At 90° (position 3), the magnetic field in winding 1-1A is zero.

The magnetic field in winding 2-2A is at maximum. Now the resultant magnetic field lies along the axis of the 2-2A winding.

The resultant magnetic field has rotated clockwise through 90° to get from position 1 to position 3. When the two-phase voltages have completed one full cycle (position 9), the resultant magnetic field has rotated through 360°. Thus, by placing two windings at right angles to each other and exciting these windings with voltages 90° out of phase, a rotating magnetic field results.

THREE-PHASE ROTATING FIELDS

The three-phase induction motor also operates on the principle of a rotating magnetic field. The stator windings can be connected to a three-phase ac input and have a resultant magnetic field that rotates. A-C show the individual windings for each phase.

View D, shows how the three phases are tied together in a Y-connected stator. The dot in each diagram indicates the common point of the Y-connection. You can see that the individual phase windings are equally spaced around the stator. This places the windings 120° apart.

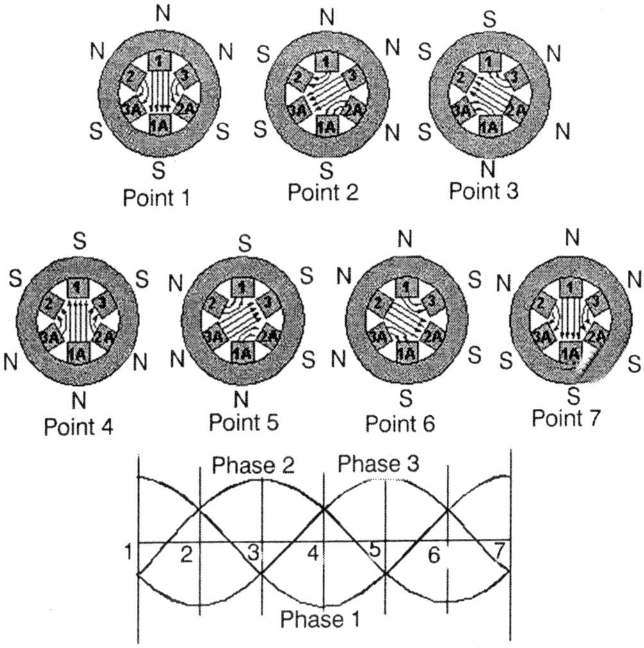

Fig. Three-phase, Y-connected Stator.

The three-phase input voltage to the stator of figure is shown in the graph of figure. Use the left-hand rule for determining the electromagnetic polarity of the poles at any given instant. In applying the rule to the coils, consider that current flows toward the terminal numbers for positive voltages, and away from the terminal numbers for negative voltages.

ROTOR BEHAVIOUR IN A ROTATING FIELD

For purposes of explaining rotor movement, let's assume that we can place a bar magnet in the centre of the stator. We'll mount this magnet so that it is free to rotate in this area. Let's also assume that the bar magnet is aligned so that at point 1 its south pole is opposite the large N of the stator field. You can see that this alignment is natural. Unlike poles attract, and the two fields are aligned so that they are attracting. Now, go from point 1 through point 7.

As before, the stator field rotates clockwise. The bar magnet, free to move, will follow the stator field, because the attraction between the two fields continues to exist.

A shaft running through the pivot point of the bar magnet would rotate at the same speed as the rotating field. This speed is known as synchronous speed. The shaft represents the shaft of an operating motor to which the load is attached. Remember, this explanation is an oversimplification.

It is meant to show how a rotating field can cause mechanical rotation of a shaft. Such an arrangement would work, but it is not used. There are limitations to a permanent magnet rotor. Practical motors use other methods, as we shall see in the next paragraphs.

SYNCHRONOUS MOTORS

The construction of the synchronous motors is essentially the same as the construction of the salient- pole alternator. In fact, such an alternator may be run as an ac motor.

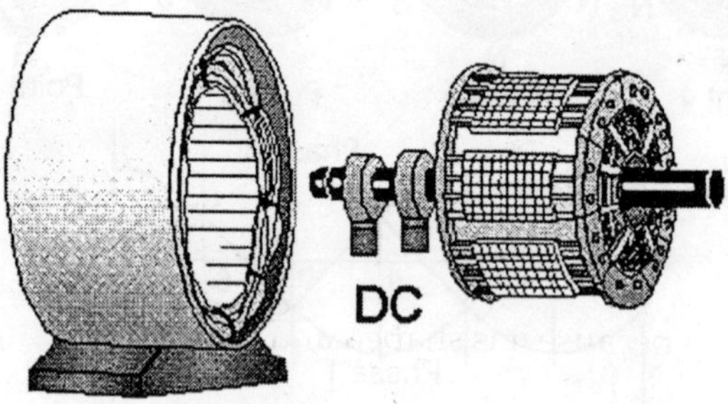

Fig. Revolving-field Synchronous Motor.

Synchronous motors have the characteristic of constant speed between no load and full load. They are capable of correcting the low power factor of an inductive load when they are operated under certain conditions.

They are often used to drive dc generators. Synchronous motors are designed in sizes up to thousands of horsepower. They may be designed as either single-phase or multiphase machines. The following is based on a three-phase design.

To understand how the synchronous motor works, assume that the application of three-phase ac power to the stator causes a rotating magnetic field to be set up around the rotor. The rotor is energized with dc (it acts like a bar magnet).

The strong rotating magnetic field attracts the strong rotor field activated by the dc. This results in a strong turning force on the rotor shaft. The rotor is therefore able to turn a load as it rotates in step with the rotating magnetic field.

It works this way once it's started. However, one of the disadvantages of a synchronous motor is that it cannot be started from a standstill by applying three-phase ac power to the stator. When ac is applied to the stator, a high-speed rotating magnetic field appears immediately. This rotating field rushes past the rotor poles so quickly that the rotor does not have a chance to get started. In effect, the rotor is repelled first in one direction and then the other. A synchronous motor in its purest form has no starting torque. It has torque only when it is running at synchronous speed.

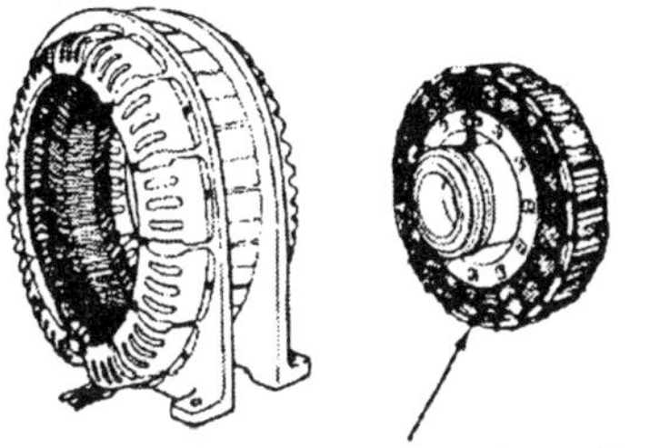

Fig. Self-starting Synchronous ac Motor.

A squirrel-cage type of winding is added to the rotor of a synchronous motor to cause it to start. The squirrel cage is shown as the outer part of the rotor in figure.

It is so named because it is shaped and looks something like a turnable squirrel cage. Simply, the windings are heavy copper bars shorted together by copper rings. A low voltage is induced in these shorted windings by the rotating three-phase stator field. Because of the short circuit, a relatively large current flows in the squirrel cage.

This causes a magnetic field that interacts with the rotating field of the stator. Because of the interaction, the rotor begins to turn, following the stator field; the motor starts. We will run into squirrel cages again in other applications, where they will be covered in more detail.

To start a practical synchronous motor, the stator is energized, but the dc supply to the rotor field is not energized. The squirrel-cage windings bring the rotor to near synchronous speed. At that point, the dc field is energized. This locks the rotor in step with the rotating stator field. Full torque is developed, and the load is driven. A mechanical switching device that operates on centrifugal force is often used to apply dc to the rotor as synchronous speed is reached.

The practical synchronous motor has the disadvantage of requiring a dc exciter voltage for the rotor. This voltage may be obtained either externally or internally, depending on the design of the motor.

INDUCTION MOTORS

The induction motor is the most commonly used type of ac motor. Its simple, rugged construction costs relatively little to manufacture. The induction motor has a rotor that is not connected to an external source of voltage. The induction motor derives its name from the fact that ac voltages are induced in the rotor circuit by the rotating magnetic field of the stator. In many ways, induction in this motor is similar to the induction between the primary and secondary windings of a transformer. Large motors and permanently mounted motors that drive loads at fairly constant speed are often induction motors.

Examples are found in washing machines, refrigerator compressors, bench grinders, and table saws. The stator construction of the three-phase induction motor and the three-phase synchronous motor are almost identical. However, their rotors are completely different. The induction rotor is made of a laminated cylinder with slots in its surface. The windings in these slots are one of two types.

The most common is the squirrel-cage winding. This entire winding is made up of heavy copper bars connected together at each end by a metal ring made of copper or brass. No insulation is required between the core and the bars. This is because of the very low voltages generated in the rotor bars. The other type of winding contains actual coils placed in the rotor slots. The rotor is then called a wound rotor.

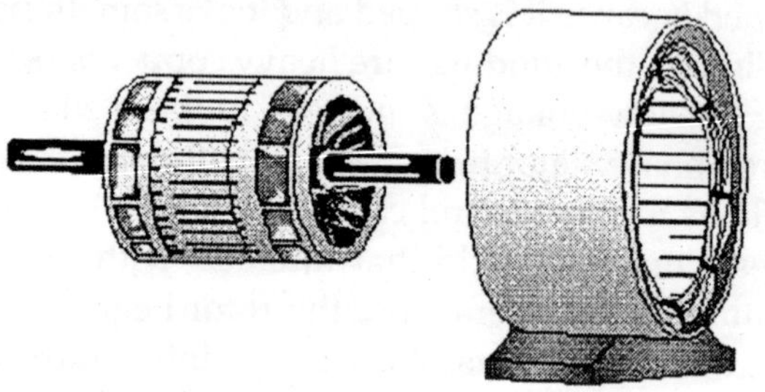

Fig. Induction Motor.

Types of ac induction motor rotors. Regardless of the type of rotor used, the basic principle is the same. The rotating magnetic field generated in the stator induces a magnetic field in the rotor. The two fields interact and cause the rotor to turn. To obtain maximum interaction between the fields, the air gap between the rotor and stator is very small.

As you know from Lenz's law, any induced emf tries to oppose the changing field that induces it. In the case of an induction motor, the changing field is the motion of the resultant stator field.

A force is exerted on the rotor by the induced emf and the resultant magnetic field. This force tends to cancel the relative motion between the rotor and the stator field. The rotor, as a result, moves in the same direction as the rotating stator field. It is impossible for the rotor of an induction motor to turn at the same speed as the rotating magnetic field.

If the speeds were the same, there would be no relative motion between the stator and rotor fields; without relative motion there would be no induced voltage in the rotor. In order for relative motion to exist between the two, the rotor must rotate at a speed slower than that of the rotating magnetic field.

The. difference between the speed of the rotating stator field and the rotor speed is called slip. The smaller the slip, the closer the rotor speed approaches the stator field speed.

The speed of the rotor depends upon the torque requirements of the load. The bigger the load, the stronger the turning force needed to rotate the rotor. The turning force can increase only if the rotor- induced emf increases. This emf can increase only if the magnetic field cuts through the rotor at a faster rate.

To increase the relative speed between the field and rotor, the rotor must slow down. Therefore, for heavier loads the induction motor turns slower than for lighter loads. You can see from the previous statement that slip is directly proportional to the load on the motor. Actually only a slight change in speed is necessary to produce the usual current changes required for normal changes in load. This is because the rotor windings have such a low resistance. As a result, induction motors are called constant-speed motors.

SINGLE-PHASE INDUCTION MOTORS

There are probably more single-phase ac induction motors in use today than the total of all the other types put together. It is logical that the least expensive, lowest maintenance type of ac motor should be used most often. The single-phase ac induction motor fits that description. Unlike polyphase induction motors, the stator field in the single-phase motor does not rotate. Instead it simply alternates polarity between poles as the ac voltage changes polarity.

Voltage is induced in the rotor as a result of magnetic induction, and a magnetic field is produced around the rotor. This field will always be in opposition to the stator field (Lenz's law applies). The interaction between the rotor and stator fields will not produce rotation, however. The interaction is shown by the double-ended arrow view A. Because this force is across the rotor and through the pole pieces, there is no rotary motion, just a push and/or pull along this line.

Now, if the rotor is rotated by some outside force (a twist of your hand, or something), the push-pull along the line, view A, is disturbed. Look at the fields, view B. At this instant the south pole on the rotor is being attracted by the left-hand pole. The north rotor pole is being attracted to the right-hand pole. All of this is a result of the rotor being rotated 90°

by the outside force. The pull that now exists between the two fields becomes a rotary force, turning the rotor toward magnetic correspondence with the stator.

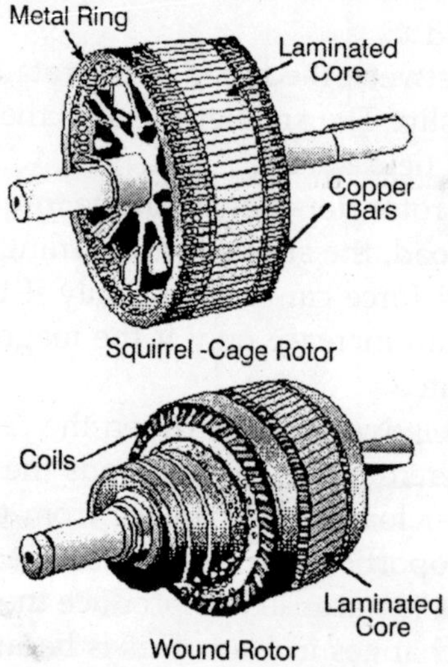

Fig. Rotor Currents in a Single-phase ac Induction Motor.

Because the two fields continuously alternate, they will never actually line up, and the rotor will continue to turn once started. It remains for us to learn practical methods of getting the rotor to start.

There are several types of single-phase induction motors in use today. Basically they are identical except for the means of starting. The split-phase and shaded-pole motors; so named because of the methods employed to get them started. Once they are up to operating speed, all single-phase induction motors operate the same.

Split-Phase Induction

Motors One type of induction motor, which incorporates a starting device, is called a split-phase induction motor. Split-phase motors are designed to use inductance, capacitance, or resistance to develop a starting torque. The principles are those that you learned in your study of alternating current.

CAPACITOR-START

The first type of split-phase induction motor that will be covered is the capacitor-start type. A simplified schematic of a typical capacitor-start motor. The stator consists of the main winding and a starting winding (auxiliary). The starting winding is connected in parallel with the main winding and is placed physically at right angles to it.

A 90-degree electrical phase difference between the two windings is obtained by connecting the auxiliary winding in series with a capacitor and starting switch. When the motor is first energized, the starting switch is closed.

This places the capacitor in series with the auxiliary winding. The capacitor is of such value that the auxiliary circuit is effectively a resistive-capacitive circuit (referred to as capacitive reactance and expressed as X_C). In this circuit the current leads the line voltage by about 45° (because X C about equals R).

The main winding has enough resistance-inductance (referred to as inductive reactance and expressed as X_L) to cause the current to lag the line voltage by about 45° (because X L about equals R). The currents in each winding are therefore 90° out of phase - so are the magnetic fields that are generated. The effect is that the two windings act like a two-phase stator and produce the rotating field required to start the motor.

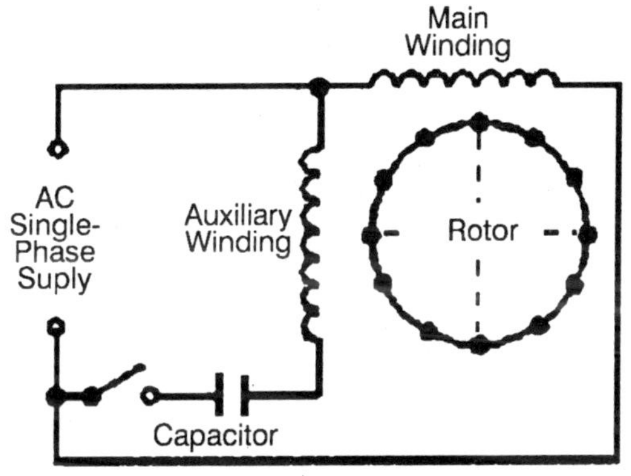

Fig. Capacitor-start, ac Induction Motor.

When nearly full speed is obtained, a centrifugal device (the starting switch) cuts out the starting winding.

The motor then runs as a plain single-phase induction motor. Since the auxiliary winding is only a light winding, the motor does not develop sufficient torque to start heavy loads. Split-phase motors, therefore, come only in small sizes.

RESISTANCE-START

Another type of split-phase induction motor is the resistance-start motor. This motor also has a starting winding in addition to the main winding. It is switched in and out of the circuit just as it was in the capacitor-start motor.

The starting winding is positioned at right angles to the main winding. The electrical phase shift between the currents in the two windings is obtained by making the impedance of the windings unequal. The main winding has a high inductance and a low resistance. The current, therefore, lags the voltage by a large angle. The starting winding is designed to have a fairly low inductance and a high resistance. Here the current lags the voltage by a smaller angle. For example, suppose the current in the main winding lags the voltage by 70°. The current in the auxiliary winding lags the voltage by 40°. The currents are, therefore, out of phase by 30°. The magnetic fields are out of phase by the same amount.

Although the ideal angular phase difference is 90° for maximum starting torque, the 30-degree phase difference still generates a rotating field. This supplies enough torque to start the motor. When the motor comes up to speed, a speed-controlled switch disconnects the starting winding from the line, and the motor continues to run as an induction motor. The starting torque is not as great as it is in the capacitor-start.

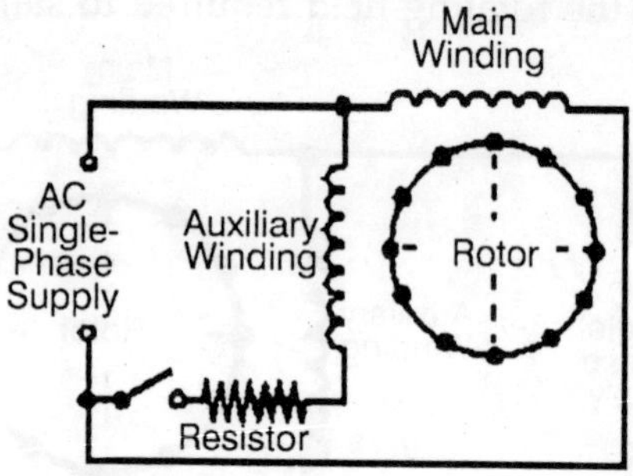

Fig. Resistance-start ac Induction Motor.

Shaded-Pole Induction Motors

The shaded-pole induction motor is another single-phase motor. It uses a unique method to start the rotor turning. The effect of a moving magnetic field is produced by constructing the stator in a special way. This motor has projecting pole pieces just like some dc motors. In addition, portions of the pole piece surfaces are surrounded by a copper strap called a shading coil. The strap causes the field to move back and forth across the face of the pole piece. Note the numbered sequence and points on the magnetization curve in the figure. As the alternating stator field starts increasing from zero:

- The lines of force expand across the face of the pole piece and cut through the strap. A voltage is induced in the strap. The current that results generates a field that opposes the cutting action (and decreases the strength) of the main field.

This produces the following actions: As the field increases from zero to a maximum at 90°, a large portion of the magnetic lines of force are concentrated in the unshaded portion of the pole:

- At 90° the field reaches its maximum value. Since the lines of force have stopped expanding, no emf is induced in the strap, and no opposing magnetic field is generated. As a result, the main field is uniformly distributed across the pole
- From 90° to 180°, the main field starts decreasing or collapsing inward. The field generated in the strap opposes the collapsing field. The effect is to concentrate the lines of force in the shaded portion of the pole face.
- You can see that from 0° to 180°, the main field has shifted across the pole face from the unshaded to the shaded portion. From 180° to 360°, the main field goes through the same change as it did from 0° to 180°; however, it is now in the opposite direction.
- The direction of the field does not affect the way the shaded pole works. The motion of the field is the same during the second half-cycle as it was during the first half of the cycle.

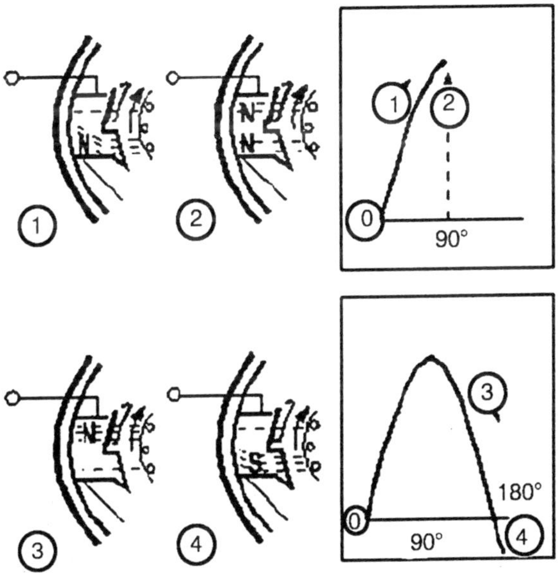

Fig. Shaded Poles as used in Shaded-pole ac Induction Motors.

The motion of the field back and forth between shaded and unshaded portions produces a weak torque to start the motor.

Because of the weak starting torque, shaded-pole motors are built only in small sizes. They drive such devices as fans, clocks, blowers, and electric razors.

Speed of Single-Phase Induction Motors

The speed of induction motors is dependent on motor design. The synchronous speed (the speed at which the stator field rotates) is determined by the frequency of the input ac power and the number of poles in the stator.

The greater the number of poles, the slower the synchronous speed. The higher the frequency of applied voltage, the higher the synchronous speed. Remember, however, that neither frequency nor number of poles are variables. They are both fixed by the manufacturer.

The relationship between poles, frequency, and synchronous speed is as follows:

$$n\ (\text{rpm}) = \frac{120f}{p}$$

where n is the synchronous speed in rpm, f is the frequency of applied voltage in hertz, and p is the number of poles in the stator.

Let's use an example of a 4-pole motor, built to operate on 60 hertz. The synchronous speed is determined as follows:

$$n = \frac{120f}{p}$$

$$n = \frac{120 \times 60}{4}$$

$$n = 1800\ \text{rpm}$$

Common synchronous speeds for 60-hertz motors are 3600, 1800, 1200, and 900 rpm, depending on the number of poles in the original design.

As we have seen before, the rotor is never able to reach synchronous speed. If it did, there would be no voltage induced in the rotor. No torque would be developed.

The motor would not operate. The difference between rotor speed and synchronous speed is called slip. The difference between these two speeds is not great. For example, a rotor speed of 3400 to 3500 rpm can be expected from a synchronous speed of 3600 rpm.

Index